# Global Energy
*Perspectives*

# Global Energy
*Perspectives*

*Edited by*

Nebojša Nakićenović, Arnulf Grübler, and Alan McDonald

IIASA
International Institute for
Applied Systems Analysis

World Energy Council
CONSEIL MONDIAL DE L'ENERGIE

CAMBRIDGE
UNIVERSITY PRESS

PUBLISHED BY THE PRESS SYNDICATE OF THE UNIVERSITY OF CAMBRIDGE
The Pitt Building, Trumpington Street, Cambridge CB2 1RP, United Kingdom

CAMBRIDGE UNIVERSITY PRESS
The Edinburgh Building, Cambridge CB2 2RU, UK http://www.cup.cam.ac.uk
40 West 20th Street, New York, NY 10011-4211, USA http://www.cup.org
10 Stamford Road, Oakleigh, Melbourne 3166, Australia

First published 1998

Printed in the United Kingdom at the University Press, Cambridge

*A catalogue record for this book is available from the British Library*

ISBN 0 521 64200 0 hardback
ISBN 0 521 64569 7 paperback

|                          |                        |
|-------------------------:|------------------------|
| **Editors:**             | Nebojša Nakićenović    |
|                          | Arnulf Grübler         |
|                          | Alan McDonald          |
| **Lead Authors:**        | Arnulf Grübler         |
|                          | Michael Jefferson      |
|                          | Nadejda Makarova       |
|                          | Alan McDonald          |
|                          | Sabine Messner         |
|                          | Nebojša Nakićenović    |
|                          | Hans-Holger Rogner     |
|                          | Leo Schrattenholzer    |
|                          | Manfred Strubegger     |
| **Contributing Authors:**| Markus Amann           |
|                          | Günther Fischer        |
|                          | Andrii Gritsevskyi     |
|                          | Hisham Khatib          |
| **Study Chair:**         | Nebojša Nakićenović    |
| **Steering Group:**      | Martin Bekkeheien      |
|                          | Klaus Brendow          |
|                          | Peter E. de Jánosi     |
|                          | Ian D. Lindsay         |
|                          | John S. Foster         |
|                          | Jean-Romain Frisch     |
|                          | Michael Jefferson      |
|                          | Gordon J. MacDonald    |
|                          | Gerhard Ott            |
|                          | Keiichi Yokobori       |
| **Regional Coordinators:**| José Luiz Alquéres    |
|                          | Richard E. Balzhiser   |
|                          | Klaus Brendow          |
|                          | Claude Destival        |
|                          | Yoichi Kaya            |
|                          | Hisham Khatib          |
|                          | Steve J. Lennon        |
|                          | Rajendra K. Pachauri   |
|                          | Yuan Shuxun            |

# Contents

# Preface

This book presents the results of a five-year study conducted jointly by the International Institute for Applied Systems Analysis (IIASA) and the World Energy Council (WEC). IIASA and WEC both have a 20-year history in analyzing energy futures. Starting with the 1981 study *Energy in a Finite World* (Häfele, 1981), IIASA has gained distinction for its research on the interactions between energy systems and the environment. WEC is well-known for its influential series of global energy perspectives culminating in the 1993 Commission Report *Energy for Tomorrow's World* (WEC, 1993).

The starting point for the study was the three alternative cases and one variant of future economic and energy development presented in *Energy for Tomorrow's World*. There, the principal analysis covered the period through 2020 with some extensions to 2100. This study describes three cases of alternative energy futures that diverge into a total of six scenarios and their implications for 11 world regions. The objective of this study was to examine more thoroughly the period beyond 2020, where the real potential for change lies. The goals were to integrate the near-term strategies through 2020 with the long-term opportunities to 2050 and beyond; to analyze alternative futures with a unified methodological framework using formal models and databases to ensure consistency and reproducibility; to incorporate a dynamic treatment of technological change; to harmonize regional aspirations with global possibilities; and to take account of new data and changes since 1993.

The study was conducted in two phases. The first phase, from 1993 to 1995, developed the six scenarios for the 11 world regions and analyzed their implications. It involved a core modeling team and approximately 50 additional experts, amounting to 10 person-years of effort. The results were presented in the 1995 IIASA–WEC report *Global Energy Perspectives to 2050 and Beyond*. The second phase, from 1995 to 1998, was devoted to an extensive review of the study assumptions, results, and implications for the 11 world regions. Teams of regional experts and reviewers, totaling over 100 individuals, were convened to provide more thorough regional assessments and alternative perspectives. Subsequently, the findings of this review process were incorporated into the scenarios and are reflected in their implications. The more specific regional perspectives are presented in Chapter 7. The results are thus a unique view of the future that offers both long-term global perspectives and more detailed regional realities and assessments. More detailed information about the data and quantitative findings of the study are available on the Internet (see Appendix E).

We believe that the issues addressed by this study are vital. An understanding of long-term energy perspectives, and how near-term decisions can expand or narrow the range of future opportunities, is essential if we are to build a future that is more prosperous and more equitable. We know of no other study that integrates long-term perspectives, near-term energy strategies, their implications, detailed modeling, and a broad range of regional assessments into one unified framework. IIASA and WEC are to be commended for their foresight and dedication in sponsoring such studies. No analysis can ever turn an uncertain future into a sure thing. However, we believe the study has identified patterns that are robust across a purposefully broad range of scenarios and regional perceptions.

This study has benefited from the experience and earlier work of both organizations on global energy perspectives. Some of this earlier work was only possible with external support for which we are grateful. We deeply appreciate the contributions of all authors to the analysis that made the study possible and for their contributions to numerous drafts of the book. It is a pleasure to also thank others who made the study possible: the members of the steering group who guided the study and the regional experts and reviewers, listed in Appendix D, for providing the regional perceptions and assessments that helped shape the final study findings. We would also like to acknowledge other colleagues at IIASA and WEC who provided assistance and support during the long process of conducting the analysis and drafting the book. In addition, we thank our colleagues in IIASA's Publications Department for the preparation of the manuscript.

Finally a personal note. This book is the product of a closely cooperating, interdisciplinary, and multinational team. Consequently, the authorship of various chapters is sometimes diffuse and individual authors cannot be held responsible for all the views and the opinions expressed in the study. And while there are many who can share in the credit for the study's successes and insights, the responsibility for any shortcomings or errors in the analysis as a whole and its presentation in this book are solely mine.

*Nebojša Nakićenović*
Project Leader
Environmentally Compatible Energy Strategies
International Institute for Applied Systems Analysis (IIASA)
A-2361 Laxenburg, Austria
naki@iiasa.ac.at

# Summary

It is easier to anticipate the forms in which energy will be demanded by consumers in the future than to estimate the absolute level of energy demand, or which energy sources will supply that demand. With increasing per capita incomes around the world, people will demand higher levels of more efficient, cleaner, and environmentally less obtrusive energy services. Thus, we believe we see reasonably clearly the direction in which energy consumers are headed. However, the question of what kind of companies will supply energy services, and how, is wide open.

That is the central message of the three cases, subdivided into six scenarios, presented in this book. They cover a wide range of global energy developments – from a tremendous expansion of coal production to strict limits, from a phaseout of nuclear energy to a substantial increase, from carbon emissions in 2100 that are only one-third of today's levels to increases by more than a factor of three. Yet, for all the variation explored, all alternatives managed to match the expected demand pull for more flexible, more convenient, and cleaner forms of energy. The odds are thus good that consumers will indeed get what they want – flexibility, convenience, and cleanliness. Who their suppliers will be, and which energy sources will be tapped, depends on economic development in the world, on progress in science and technology, and on policies and institutions. And it depends very much on which suppliers make the near-term decisions that prove most effective in pairing up their services with evolving consumer preferences.

The six scenarios build both on the analysis presented in the IIASA–WEC study report *Global Energy Perspectives to 2050 and Beyond* published in 1995 and on the WEC Commission Report *Energy for Tomorrow's World* published in 1993. These studies concluded that, at least through 2020, the world will have to rely largely upon fossil fuels, with relatively few opportunities for alternatives. The results presented here reinforce this earlier conclusion – all six scenarios look much the same through 2020, and all rely heavily on fossil fuels. But after 2020, the six scenarios start to diverge.

Part of that divergence will depend on policy choices and development strategies. For example, two scenarios that assume dynamic, progressive international cooperation focused on environmental protection and international equity lead to less fossil fuel use than other scenarios. Most of the post–2020 divergence will depend on technological developments, industrial strategies, and consumer choice. Which energy sources will best match consumers' preference for flexible, more convenient, and cleaner energy services? Which companies will have made the investments in research, development, and demonstration (RD&D) that will give them a technological edge? Which will have refocused their businesses from providing just tons of coal or kilowatt-hours of electricity, to providing convenient and clean energy services to consumers?

The answers to those questions will be determined between now and 2020. Because of the long lifetimes of power plants, refineries, and other energy investments, there is not a sufficient turnover of such facilities to reveal large differences in our scenarios prior to 2020, but the seeds of the post–2020 world will have been sown by then. The choice of the world's post–2020 energy systems may be wide open now. It will be a lot narrower by 2020.

## World energy needs will increase

World population is expected to double by the middle of the 21st century, and economic development needs to continue, particularly in the South. According to the scenarios of this study, this results in a 3- to 5-fold increase in world economic output by 2050 and a 10- to 15-fold increase by 2100. By 2100, per capita income in most of the currently developing countries will have reached and surpassed the levels of today's developed countries. Disparities are likely to persist, and despite rapid economic development, adequate energy services may not be available to everyone, even in 100 years. Nonetheless the distinction between "developed" and "developing" countries in today's sense will no longer be appropriate. Primary and final energy use will grow much less than the demand for energy services due to improvements in energy intensities. We expect a 1.5- to 3-fold increase in primary energy requirements by 2050, and a 2- to 5-fold increase by 2100.

## Energy intensities will improve significantly

As individual technologies improve, conversion processes and end-use devices progress along their learning curves, and as inefficient technologies are retired in favor of more efficient ones, the amount of primary energy needed per unit of economic output – the energy intensity – decreases. Other things being equal, the faster the economic growth, the shorter the obsolescence time, the higher the turnover of capital, and the greater the energy intensity improvements. In the six scenarios of this study, improvements in individual technologies are varied across a range derived from historical trends. Combined with the economic growth patterns of the different scenarios, the average global reductions in energy intensity range between

0.8% per year and 1.4% per year. These bracket the historical rate of approximately 1% per year and cumulatively lead to substantial energy intensity decreases. Some regions improve faster, especially where current intensities are high and economic growth and capital turnover are rapid.

### *Resource availability will not be a major global constraint*

The resource scarcity perceived in the 1970s did not occur as originally assumed. With technological and economic development, estimates of the ultimately available energy resource base will continue to increase. A variety of assumptions about the timing and extent of new discoveries of fossil energy reserves and resources (conventional and unconventional), and about improvements in the economics of their recoverability, are reflected in the range of scenarios reported here. All, however, indicate that economic development over the next century will not be constrained by geological resources. Regional shortages and price increases can occur, due to the unequal distribution of fossil resources, but globally they are not a constraint. Environmental concerns, financing, and technological needs appear more likely sources of future limits. The short-term volatility of international politics, speculation, and business cycles will periodically upset the long-term expansion of resources.

### *Quality of energy services and forms will increasingly shape future energy systems*

The energy system is service driven, from consumer to producer, while energy flows are resource and conversion-process driven, from producer to consumer. The scenarios demonstrate the need to consider energy end use and energy supply simultaneously, both from an analytical and a policy perspective. In addition to prices and quantities, energy *quality* matters increasingly. Quality considerations include convenience, flexibility, efficiency, and environmental cleanliness. The energy system is end-use driven. Under market liberalization, special competitive advantages will arise for those companies prepared to deliver a full range of energy services beyond just fuels and electricity.

### *Energy end-use patterns will converge, even as energy supply structures diverge*

The six scenarios indicate that the historical drive toward ever more convenient, flexible, and clean fuels reaching the consumer can continue for a wide range of possible future energy supply structures. In all scenarios there is a shift toward electricity and toward higher-quality fuels, such as natural gas, oil products, methanol, and, eventually, hydrogen. In contrast to these converging trends, primary energy supply structures diverge, particularly after 2020. Fossil sources continue to provide most of the world's energy well into the next century but to a varying extent across the scenarios. There is a shift away from noncommercial and mostly unsustainable uses of biomass, and direct uses of coal virtually disappear. Sustainable uses of renewables including modern biomass come to hold a prominent place in all scenarios. They reach the consumer as electricity, liquids, or gases, rather than as solids.

### Technological change will be critical for future energy systems

Technological change drives productivity growth and economic development. Across all scenarios the role of technological progress is critical. But progress has a price. RD&D of new energy technologies and the accumulation of experience in niche markets require upfront expenditures of money and effort. These are increasingly viewed as too high a price to pay in liberalized markets where the maximization of short-term shareholder value generally takes precedence over longer-term socioeconomic development and environmental protection. Yet, it is the RD&D investments of the next few decades that will shape the technology options available after 2020. A robust hedging strategy focuses on generic technologies at the interface between energy supply and end use, including gas turbines, fuel cells, and photovoltaics. These could become as important as today's gasoline engines, electric motors, and microchips.

### Rates of change in global energy systems will remain slow

Capital turnover rates in end-use applications are comparatively short – one to two decades. Therefore, pervasive changes can be implemented rather quickly, and missed opportunities may be revisited. Conversely, the lifetimes of energy supply technologies, and particularly of infrastructures, are five decades or longer. Thus, at most one or two replacements can occur during the time horizon of this study. Betting on the wrong horse will have serious, possibly irreversible consequences. The RD&D and investment decisions made now and in the immediate future will determine which long-term options become available after 2020 and which are foreclosed. Initiating long-term changes requires action sooner rather than later.

### Interconnectivity will enhance cooperation, systems flexibility, and resilience

Despite energy globalization, *market exclusion* remains a serious challenge. To date, some two billion people do not have access to modern energy services due to poverty and a lack of energy infrastructures. Many regions are overly dependent on a single, locally available resource, such as traditional fuelwood or coal, and have limited access to the clean flexible energy forms required for economic and social development. Policies to deregulate markets and get "prices right" ignore the poor.

Even the best functioning energy markets will not reach those who cannot pay. To include today's poor in energy markets, poverty must be eradicated, and that requires policy action that goes beyond energy policies alone. What energy policies can accomplish is the improvement of old infrastructures – the backbone of the energy system – and the development of new ones. New infrastructures are needed in Eurasia, in particular, to match the large available resources of oil and gas in the Caspian region and Siberia with the newly emerging centers of energy consumption in Asia. Extended interconnections are also needed in Latin America and Africa. Governments and industry need an expanded spirit of regional cooperation and a shared commitment to infrastructure investments now if benefits are to accrue in the long term.

### *Capital requirements will present major challenges for all energy strategies*

For all scenarios the capital requirements of the energy sector are large, but not infeasible. Although investment requirements expand more slowly than overall economic growth, the energy sector will have to raise an increasing fraction of its capital from the private sector. It will face stiffer competition and return-on-investment criteria than it has in the past. Moreover, the greatest investment needs are in the now developing world, where current trends in the availability of both international development capital and private investment capital are not auspicious. How available capital is best mobilized remains a critical issue.

### *Regional differences will persist in global energy systems*

Regional energy supply trends diverge across the scenarios even more than those at the global level due to differences in resource availability, trade possibilities, and national and regional development strategies. Regional development aspirations often exceed even the wide range of possibilities outlined by the six global scenarios analyzed here. Yet, for all their diversity, regional perspectives confirm the essential global conclusion: while a range of possible energy sources can fuel the future, there is a persistent trend toward cleaner, more convenient energy forms reaching the consumer. In all regions, success will depend on improved efficiency, continued technological progress, a favorable investment climate, free trade, and enhanced regional and international cooperation, particularly in shared energy infrastructures.

### *Local environmental impacts will take precedence over global change*

The natural capacity of the environment to absorb higher levels of pollution, particularly in densely populated metropolitan areas, will become the limiting factor for the unconstrained use of fossil fuels. Local environmental problems are of greater concern to local decision makers than global problems and therefore will have a greater near-term impact on policy. In the developing world indoor air pollution is an urgent environmental problem. A shift away from cooking with wood in open fireplaces will reduce indoor pollution levels currently estimated to be 20 times higher than in industrialized countries. A second urgent problem is the high concentration of particulate matter and sulfur dioxide in many urban areas. Regional air pollution could also prove problematic, especially in the rapidly growing, densely populated coal-intensive economies of Asia. Without abatement measures, sulfur emissions could cause serious public health problems and subject key agricultural crops to acid deposition 10 times sustainable levels.

## *Decarbonization will improve the environment at local, regional, and global levels*

The continuing shift to higher-quality fuels means a continuing reduction in the carbon content of fuels, that is, decarbonization of the energy system. Decarbonization means lower adverse environmental impacts per unit of energy consumed, independent of any active policies specifically designed to protect the environment. And at the global level, it translates directly into lower carbon dioxide emissions. But decarbonization is not enough – additional active policies will be required. In some cases energy and environmental policies are mutually reinforcing. Policies to reduce global carbon dioxide emissions, for example, also reduce acidification risks. In others, energy and environmental policies work at cross purposes. Restrictions on nuclear power, for example, mean a possibly greater dependence on fossil fuels, and vice versa. In all cases, however, more rapid technological improvement means quicker progress toward cleaner fuels, and cleaner fuels mean a cleaner environment.

# Acronyms

| | |
|---|---|
| Adv. Coal | – advanced coal |
| AFR | – Sub-Saharan Africa |
| bbl | – barrels (oil equivalent, 1 toe = 7 bbl) |
| BIGSTIG | – biomass integrated-gasifier steam-injected gas turbine |
| BLS | – IIASA's Basic Linked System |
| BOO | – build-own-operate |
| BOT | – build-own-transfer |
| CC | – combined cycle |
| CFCs | – chlorofluorocarbons |
| CH$_3$OH | – methanol |
| CH$_4$ | – methane |
| CHP | – combined heat and power generation |
| CIS | – Commonwealth of Independent States |
| CO | – carbon monoxide |
| CO$_2$ | – carbon dioxide |
| CO2DB | – IIASA's Carbon Dioxide Mitigation Technology Database |
| CPA | – Centrally planned Asia and China |
| CPE | commercial primary energy |
| DCs | – developing countries |
| DSM | – demand-side management |
| EEU | – Central and Eastern Europe |
| EIA | – Energy Information Administration |
| EU | – European Union |
| FAO | – UN Food and Agriculture Organization |
| FBRs | fast breeder reactors |
| FCCC | – Framework Convention on Climate Change |
| FSU | Newly independent states of the former Soviet Union |
| GATT | – General Agreement on Tariffs and Trade |
| GDP | – gross domestic product |
| GDP$_{mer}$ | – GDP calculated on the basis of market exchange rates |
| GDP$_{ppp}$ | – GDP calculated on the basis of purchasing power parities |
| GNP | – gross national product |
| GtC | – gigatons carbon |
| Gtoe | – giga [billion ($10^9$)] tons oil equivalent |
| GW | – gigawatts |
| GW$_e$ | – gigawatts electric |
| GWP | – gross world product |
| H$_2$ | – hydrogen |
| ha | – hectares |
| HTFC | – high temperature fuel cell |
| IEA | – International Energy Agency |
| IGCC | – integrated gasifier combined cycle |

| | |
|---|---|
| IIASA | – International Institute for Applied Systems Analysis |
| IND | – industrialized countries |
| IPCC | – Intergovernmental Panel on Climate Change |
| IRP | – integrated resource planning |
| kgoe | – kilograms oil equivalent |
| $kW_e$ | – kilowatts electric |
| kWh | – kilowatt hours |
| LAM | – Latin America and the Caribbean |
| LESS | – Low $CO_2$-Emitting Energy Supply Systems |
| LNG | – liquefied natural gas |
| LPG | – liquefied petroleum gas |
| mbd | – million barrels per day |
| MEA | – Middle East and North Africa |
| mer | – market exchange rate |
| MtC | – megatons carbon |
| Mtoe | – megatons oil equivalent |
| MtS | – megatons sulfur |
| $MW_e$ | – megawatts electric |
| $N_2O$ | – nitrous oxide |
| NAFTA | – North American Free Trade Agreement |
| NAM | – North America |
| NGL | – natural gas liquids |
| $NO_x$ | – nitrogen oxide |
| OECD | – Organisation for Economic Co-operation and Development |
| OLADE | – Latin American Energy Organization |
| OPEC | – Organization of the Petroleum Exporting Countries |
| PAO | – Pacific OECD |
| PAS | – Other Pacific Asia |
| PE | – primary energy |
| $PM_{10}$ | – particulate matter less than 10 microns in diameter |
| ppl | – power plants |
| ppmv | – parts per million by volume |
| PSA | – production sharing agreement |
| PV | – photovoltaics |
| RAINS | – IIASA's Regional Acidification INformation and Simulation model |
| RD&D | – research, development, and demonstration |
| REFs | – reforming economies (EEU plus FSU) |
| SAS | – South Asia |
| SG | – Scenario Generator |
| $SO_2$ | – sulfur dioxide |
| $SO_x$ | – sulfur oxide |
| tC | – tons carbon |
| tce | – tons coal equivalent |
| toe | – tons oil equivalent |
| TPE | – total primary energy |
| tU | – tons uranium |
| $TW_e$ | – terawatts electric |
| TWh | – terawatt hours |
| UN | – United Nations |
| UNCED | – United Nations Conference on Environment and Development |
| UNDP | – United Nations Development Programme |
| VOC | – volatile organic compound |
| WEC | – World Energy Council |
| WEU | – Western Europe |
| WHO | – World Health Organization |

# Chapter 1

# Introduction

This study examines long-term energy perspectives, their constraints, and opportunities by formulating scenarios. A scenario is a narrative, in this case with illustrations and quantitative characteristics, describing one possible way the future might unfold. As discussed in *Box 2.1*, it is not a prediction or a forecast. Nor is it just any narrative. It has to be an internally consistent and reproducible narrative, which means checking to make sure all the numbers add up. This is done largely with formal models. The formal models used for this study are described briefly in Appendix A on methodology. In the analysis and modeling work, the countries of the world were grouped into 11 different regions. The groupings are shown in *Figure 1.1* and discussed in *Box 1.1*. A detailed listing is given in Appendix B.

To draw robust conclusions, we need to formulate a range of scenarios covering very different possible futures. Each must be internally consistent, as checked by the formal models, and each must be reproducible and plausible to someone exercising open-minded, but realistic, judgment. If there are common tendencies across very different scenarios, those provide one useful set of analytic conclusions. And if there are discernible connections between the long-term divergence among scenarios and differences in their near-term policy or investment actions, those form another useful set of conclusions. In this study the scenarios provided conclusions of both sorts.

Chapter 2 lays out the scenarios and, in particular, how they were put together to purposely cover a broad range of possible futures. There are six scenarios grouped into three cases, Cases A, B, and C. The definition of cases came before the definition of scenarios, and the three cases are based on the three cases presented in *Energy for Tomorrow's World* (WEC, 1993) and *Global Energy Perspectives to 2050 and Beyond* (IIASA–WEC, 1995). Each is summarized in Chapter 2. Case A is basically a high-growth future in terms of income, energy, and technology. Case B has more modest but perhaps more realistic growth. Case C assumes

**Box 1.1: World regions**

The analysis and modeling work underlying the scenarios developed within this study was done for 11 world regions (for their definition see map below and Appendix B), which follow the regional detail of IIASA–WEC, 1995.

In this report some results are presented at the level of three "macro-regions" and others for all 11 study regions. The macroregions are the former centrally planned economies now in the process of economic reform (REFs, in shades of blue in *Figure 1.1*), the developing countries (DCs, in shades of green), and a group that approximates the Organisation for Economic Co-operation and Development (OECD, in shades of red and orange). Developing countries are sometimes referred to in the text as the "South" to distinguish them from the industrialized regions of the "North" (i.e., OECD and REFs).

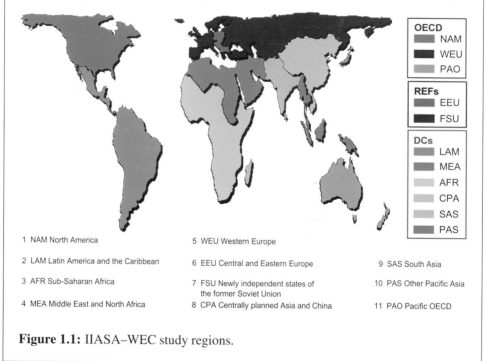

| | |
|---|---|
| 1 NAM North America | 5 WEU Western Europe |
| 2 LAM Latin America and the Caribbean | 6 EEU Central and Eastern Europe | 9 SAS South Asia |
| 3 AFR Sub-Saharan Africa | 7 FSU Newly independent states of the former Soviet Union | 10 PAS Other Pacific Asia |
| 4 MEA Middle East and North Africa | 8 CPA Centrally planned Asia and China | 11 PAO Pacific OECD |

**OECD**
NAM
WEU
PAO

**REFs**
EEU
FSU

**DCs**
LAM
MEA
AFR
CPA
SAS
PAS

**Figure 1.1:** IIASA–WEC study regions.

dynamic progressive international cooperation focused on environmental protection and international equity. Energy use in Case C is the lowest of the three cases, but economic and technological growth are higher than in Case B. As the study proceeded, it became apparent that more variety was needed than was represented by the three basic cases. Case A was therefore subdivided into three alternative scenarios (A1, A2, and A3), and Case C into two alternatives (C1 and C2). The six scenarios are all introduced in Chapter 2 and are based on the analysis presented in *Global Energy Perspectives to 2050 and Beyond* (IIASA–WEC, 1995).

All three cases – all six scenarios – start from the base year 1990 and the historical developments leading up to that year. Chapter 3 reviews the empirical basis for the study in terms of historical changes and tendencies. It emphasizes current disparities around the world in income, economic structure, energy use, and energy intensity, and compares them with variations over time in individual countries. The chapter also presents the basic interpretation of past data that underlies our extensions of historical tendencies and structural transformations into the future in the different cases. Finally, it summarizes key developments that have taken place since the base year, 1990.

Chapter 4 looks at the determinants of future energy use and describes how they vary across the six scenarios. The presentation is divided into six headings: population growth, economic growth, energy intensity improvements, technological change, the resource base, and environment.

Chapter 5 presents the essential quantitative characteristics of all six scenarios using many illustrations. It describes how they differ in their primary energy structures, their resource use, their electricity sectors, and their final energy use.

Chapter 6 turns to important implications on a number of fronts: investments and financing, international trade, costs, technology, the constraints and opportunities facing different energy industries (coal, oil, gas, renewables, and nuclear), and environmental impacts at the local, regional, and global levels.

Chapter 7 presents scenario developments and findings for each of the 11 study regions and compares them with the global picture provided by Chapter 5. Chapter 7 also presents a summary of regional expert reviews – a unique feature of this study. Detailed regional reviews provide a "bottom-up" interpretation of the "top-down" global scenario analysis. This sets the stage for evaluating and comparing regional aspirations with global possibilities and provides a final test of the consistency of the global and regional findings. The chapter also summarizes the range of energy-related concerns across regions as identified in a detailed poll of World Energy Council (WEC) regional experts.

In Chapter 8, we come to conclusions – of both types mentioned at the beginning of this chapter. First, there are some pervasive developments across all scenarios and in most regions. We believe strategies that take such developments carefully into account will benefit no matter how the future turns out. Second, there are also apparent connections between near-term strategies and the different futures of the scenarios. Particular near-term strategies improve the odds of particular long-term futures. Chapter 8 lays out these conclusions and relates them to near-term decisions. It identifies future opportunities and how best to exploit them. Finally, the chapter gives a different, visual interpretation of the scenarios.

# Chapter 2

# An Overview of Scenarios

This study presents three cases of future developments subdivided into six alternative scenarios (see *Box 2.1* for a discussion of scenarios). The principal focus is on the period between 2020 and 2050, but some results are also presented out to 2100. In brief, Case A presents a future of impressive technological improvements and consequent high economic growth. Case B describes a future with less ambitious, though perhaps more realistic, technological improvements, and consequently more intermediate economic growth. Case C presents a "rich and green" future. It includes both substantial technological progress and unprecedented international cooperation centered explicitly on environmental protection and international equity. Key characteristics of the three cases are given in *Table 2.1*. The following paragraphs provide details on what the scenarios have in common and where they differ.

## 2.1. Commonalities

All three cases provide for substantial social and economic development, particularly in the developing world. They provide for improved energy efficiencies and environmental compatibility, and thus for associated growth in both the quantity and quality of energy services. Across all three cases, the structure of final energy develops in the same way and energy intensities improve steadily. To facilitate comparisons among the three cases, all share the same central demographic baseline assumption in which global population grows to 10 billion ($10^9$) by 2050 and to nearly 12 billion by 2100. All have been checked for internal consistency with the aid of formal models and through a lengthy regional expert review process.

**Table 2.1**: Summary of the three cases in 2050 and 2100 compared with 1990.

| | Case | | |
|---|---|---|---|
| | A | B | C |
| | High growth | Middle course | Ecologically driven |
| Population, billion | | | |
| 1990 | 5.3 | 5.3 | 5.3 |
| 2050 | 10.1 | 10.1 | 10.1 |
| 2100 | 11.7 | 11.7 | 11.7 |
| GWP, trillion US(1990)$ | | | |
| 1990 | 20 | 20 | 20 |
| 2050 | 100 | 75 | 75 |
| 2100 | 300 | 200 | 220 |
| Global primary energy intensity improvement, percent per year | Medium | Low | High |
| 1990 to 2050 | −0.9 | −0.8 | −1.4 |
| 1990 to 2100 | −1.0 | −0.8 | −1.4 |
| Primary energy demand, Gtoe | | | |
| 1990 | 9 | 9 | 9 |
| 2050 | 25 | 20 | 14 |
| 2100 | 45 | 35 | 21 |
| Resource availability | | | |
| Fossil | High | Medium | Low |
| Non-fossil | High | Medium | High |
| Technology costs | | | |
| Fossil | Low | Medium | High |
| Non-fossil | Low | Medium | Low |
| Technology dynamics | | | |
| Fossil | High | Medium | Medium |
| Non-fossil | High | Medium | High |
| Environmental taxes | No | No | Yes |
| $CO_2$ emission constraint | No | No | Yes |
| Net carbon emissions, GtC | | | |
| 1990 | 6 | 6 | 6 |
| 2050 | 9–15 | 10 | 5 |
| 2100 | 6–20 | 11 | 2 |
| Number of scenarios | 3 | 1 | 2 |

Abbreviations: GWP = gross world product; Gtoe = gigatons oil equivalent; $CO_2$ = carbon dioxide; GtC = gigatons of carbon.

---

**Box 2.1: Scenarios: Alternative views of the future**

In designing scenarios we devise images of the future, or better of alternative futures. Scenarios are neither predictions nor forecasts. Rather, each scenario is one alternative image of how the future could unfold. Each is based on an internally consistent and reproducible set of assumptions about key relation-ships and driving forces of change that is derived from our understanding of both history and our current situation. Often scenarios are formulated with the help of formal models. Most scenarios, including those presented in this book, make one particular strong assumption about the future: the absence of major discontinuities and catastrophes. These are not only inherently difficult to anticipate, but also offer little policy guidance on managing an orderly tran-sition from today's energy system, which relies largely on fossil fuels, toward a more sustainable system with more equitable access to energy services.

Scenarios can be both *descriptive* and *normative*. Descriptive scenarios outline possible developments in the absence of significant changes in poli-cies, economics, or technology. Normative (or *prescriptive*) scenarios attempt to incorporate the consequences of specific modifications in current policies, institutions, and technologies. In the study reported here, Case B is a prime illustration of a descriptive scenario, while Case C is the prototype of a nor-mative scenario. Case A contains both descriptive and normative elements.

---

## 2.2. Differences

### 2.2.1. Case A: "High Growth"

Case A presents a future designed around ambitiously high rates of economic growth and technological progress. It incorporates the conviction that there are essentially no limits to human technological ingenuity. Case A presumes favor-able geopolitics and free markets. Economic growth runs about 2% per year in the OECD countries and is twice as high in the developing countries. High growth facilitates a more rapid turnover of capital stock and changes in economic struc-ture, both of which spur efficiency improvements and technological progress. If Case A is extended all the way to 2100 – again with the heroic assumption of 100 years of favorable geopolitics – global average per capita income surpasses even the highest levels observed today, making the distinction between "developed" and "developing" regions in today's sense no longer appropriate.

Case A includes three scenarios that address key developments in energy sup-ply. They vary principally in the future they envision for coal, on one side, and nuclear and renewables, on the other. In Scenario A1, there is high future availability of oil and gas resources. Dominance of oil and gas is perpetuated to the end of the 21st century. At the other end of the spectrum, Scenario A2 assumes oil and gas resources to be scarce, resulting in a massive return to coal. Finally, in Scenario A3 rapid technological change in nuclear and renewable energy technologies results in a phaseout of fossil fuels for economic reasons rather than due to resource scarcity.

Scenario A1 describes a case with no remarkable developments favoring either coal or nuclear. As a result, technological change focuses on tapping the vast potential of conventional and unconventional oil and gas occurrences. The result challenges conventional wisdom about the possible exhaustion of fossil resources. Indeed, they appear sufficient to allow a smooth, comfortable transition over the next century to alternative supply sources based on nuclear and new renewables matched with high-quality energy carriers in the form of electricity, liquids, gas, and later hydrogen. There is little need to resort to the "backstop" fossil fuel *par excellence* – coal. It continuously loses market share.

The world of Scenario A2 is one in which the greenhouse warming debate is resolved in favor of coal. The benefits of climate change and $CO_2$ fertilization prove to exceed the disadvantages. This leaves little policy incentive to phase out fossil fuels early, particularly in areas endowed with large, cheap coal resources. Sulfur and nitrogen emissions are mitigated through control technologies, and coal's vast resources make it the fossil fuel of choice. Coal also provides the "backstop" for dwindling resources of conventional oil and gas, which are assumed to be limited to currently known reserves and resources. With the depletion of open-cast coal deposits, significant technological challenges remain concerning ever deeper coal mines and the elaborate chemistry of converting coal to synliquids. *In situ* gasification and remote-controlled, robotic mining operations characterize a capital-intensive "coal economy."

We have labeled the future of Scenario A3 "bio-nuc." Large-scale renewables and a new generation of nuclear power lead a technology-driven transition to a post-fossil-fuel age. The market penetration dynamics of the transition match those characterizing the historical replacement of traditional 19th century energy forms by fossil fuels. Natural gas is the transitional fossil fuel of choice, and no challenges are imposed on the availability of economically competitive oil resources. Pressure to dig deeply into nonconventional fossils is limited. By 2100, there is an almost equal reliance on nuclear energy, natural gas, biomass, and a fourth category composed mostly of solar energy, but also containing wind and other "new" renewables.

### 2.2.2.   Case B: "Middle Course"

Case B has a single scenario. It incorporates more modest estimates of economic growth and technological development, and the demise of trade barriers and expansion of new arrangements facilitating international exchange. Compared with the Case A scenarios (and the Case C scenarios), it is more "pragmatic," which is its main appeal. Case B manages to fulfill the development aspirations of the South, but less uniformly and at a slower pace than in the other cases. For regions such as Africa, progress is painfully slow. Overall, economic growth rates to 2020 are more modest than in many studies with shorter time horizons, for example, Case B of *Energy for Tomorrow's World* (WEC, 1993). This change reflects recent setbacks and slow economic restructuring in Eastern Europe, the former Soviet Union, and many developing countries. These short-term setbacks are, however, ultimately counterbalanced by higher growth rates in the long run.

Case B's more modest energy demand and slower technology improvements result in the greatest reliance on fossil fuels of any of the scenarios, except the coal-intensive Scenario A2. Through the medium term, the structure of Case B's energy supply and use remains much closer to the current situation than those of Cases A and C. Beyond 2020, however, the depletion of fossil resources without counterbalancing technological progress forces more dramatic changes in energy supply structures. Nonetheless, a transition away from fossil fuel use is feasible and manageable. Contrary to perceptions of imminent resource shortages, oil and gas maintain a significant share in the global primary energy mix up to about 2070 by moving into costlier categories of conventional and unconventional resources. Constraints prove to be based less on geology and more on financial and environmental considerations. As a scenario that might be characterized as "muddling through," Case B inevitably contains features that are less attractive, though perhaps more realistic, than corresponding characteristics in the other cases.

### 2.2.3.   Case C: "Ecologically Driven"

Case C is the most challenging. It is optimistic about technology and geopolitics, but unlike Case A, it assumes unprecedented progressive international cooperation focused explicitly on environmental protection and international equity. The future described by Case C includes a broad portfolio of environmental control technologies and policies, including incentives to encourage energy producers and consumers to utilize energy more efficiently and carefully, "green" taxes, international environmental and economic agreements, and technology transfer. It reflects substantial resource transfers from industrialized to developing countries, spurring growth in the South. These resource transfers reflect stringent international environmental taxes or incentives, which recycle funds from the OECD to developing countries. While economic output is less than in Case A, Case C still describes a positive-sum game relative to Case B, with a total GWP greater than in Case B and a significant reduction in present economic disparities.

Case C incorporates policies to reduce carbon emissions in 2100 to 2 GtC per year, one-third of today's level. One option is a carbon tax that gradually increases well above US$100 per ton of carbon (tC) in 2100 to a value comparable with average current gasoline taxes in Western Europe. The 2 GtC emission ceiling can also be imposed by command-and-control regulations, such as inflexible emission limits and mandated technologies, but model checks confirm that these create inefficiencies that show up in the form of higher and irregular costs (shadow prices) for emission reductions. Other policy alternatives also exist. In general, we believe that the more policies focus on carrots, rather than sticks – that is, the more they create incentives to change behavior and use more energy-efficient equipment, and the less they focus on penalties such as fines and taxes – the more likely consumers and industry are to respond positively and quickly.

In Case C, nuclear energy is at a crossroads. Two scenarios are included. They both meet the $CO_2$ emissions ceiling in 2100, but they describe two very different paths that nuclear power might take. In one path a new generation of nuclear reactors is developed (Scenario C2) that is inherently safe and small scale – 100 to 300 megawatts electric ($MW_e$) installed capacity – and that finds widespread social acceptability, particularly in areas of scarce land resources and high population densities that limit the potential supply from renewables. In the other (Scenario C1), nuclear power proves a transient technology that is eventually phased out entirely by the end of the 21st century.

# Chapter 3

# Global Energy Needs:
# Past and Present

Energy needs – in the past, for the present, and in the future – are driven by three principal factors: population growth, economic development, and technological progress. Each of these is addressed in Chapter 4. These driving forces develop in different, but internally consistent, ways in each of the six possible futures laid out in this report – the three Case A scenarios, the single Case B scenario, and the two Case C scenarios. For each case, it is important to understand how the three driving forces fit together and how they tie into historical developments leading up to today. We start, in this section, with some essential features from history.

## 3.1.    Two Grand Transitions

Prior to the Industrial Revolution, the energy system relied on harnessing natural energy flows and animal and human power to provide required energy services in the form of heat, light, and work. Power densities and availability were constrained by site-specific factors, with mechanical energy sources limited to draft animals, water, and windmills. The only form of energy conversion was from chemical energy to heat and light – through burning fuelwood, for example, or tallow candles. Energy consumption typically did not exceed 0.5 tons oil equivalent (toe) per capita per year (Smil, 1994).

Two "grand transitions" have since shaped structural changes in the energy system at all levels. The first was initiated with a radical technological end-use innovation: the steam engine powered by coal. The steam cycle represented the first conversion of fossil energy sources into work; it allowed the provision of energy services to be site independent, as coal could be transported and stored as needed;

and it permitted power densities previously only possible in exceptional locations of abundant hydropower. Stationary steam engines were first introduced for lifting water from coal mines, thereby facilitating increased coal production. Later, they provided stationary power for what was to become an entirely new form of organizing production: the factory system. Mobile steam engines, on locomotives and steam ships, enabled the first transport revolution, as railway networks were extended to even the most remote locations and ships were converted from sail to steam. Characteristic energy consumption levels during the "steam age," approximately the mid-19th century in England, were about 2 toe per capita per year. By the turn of the 20th century, coal had replaced traditional non-fossil energy sources and supplied virtually all the primary energy needs of industrialized countries.

The second grand transition was the greatly increased diversification of both energy end-use technologies and energy supply sources. Perhaps the most important single innovation was the introduction of electricity as the first energy carrier that could be easily converted to light, heat, or work at the point of end use. A second key innovation was the internal combustion engine, which revolutionized individual and collective mobility through the use of cars, buses, and aircraft. Like the transition triggered by the steam engine, this "diversification transition" was led by technological innovations in energy end use, such as the electric light bulb, the electric motor, the internal combustion engine, and aircraft. However, changes in energy supply have been equally far-reaching. In particular, oil emerged from its place as an expensive curiosity at the end of the 19th century to occupy the dominant global position, where it has remained for the past 30 years.

*Figure 3.1* (see also *Figure 5.4*) illustrates these two grand transitions by showing the changing shares of different primary energy sources in the global energy supply, including the long transition away from traditional renewable energy forms toward fossil fuels; the emergence and culmination of coal, which supplied close to two-thirds of global energy needs by the eve of World War I; the introduction of oil and, later, natural gas, first as a by-product of oil production and then as an energy carrier in its own right; the peak in the market share of oil in the 1970s; and finally a reduction in the dynamics of change in the primary energy supply structure during the past two decades. This reduced dynamism may be partly due to the increased regulatory interest received by the energy sector in recent decades, partly due to oil's attractiveness for the transportation sector, where demand has risen steadily, and partly due to the delayed switch in power generation away from coal and oil to natural gas.

The two grand transitions have made possible far-reaching structural changes in employment, the spatial division of labor, and international trade. These are associated with modernizing traditional economic and social structures, and include the following, in particular:

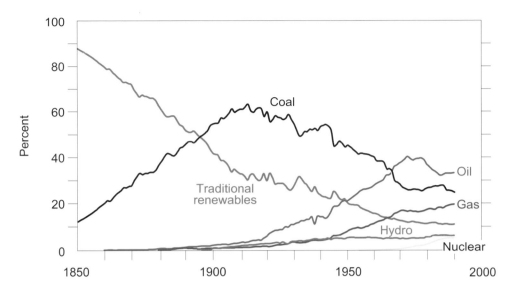

**Figure 3.1**: World primary energy shares from 1850 to 1990, in percent. Source: Nakićenović, 1984; updated using BP, 1995 and earlier volumes.

- *Industrialization:* Employment and value generation move progressively away from agriculture toward industry and manufacturing in particular. Subsequently there are structural shifts away from "smoke-stack" industries toward services and industries characterized by the increasing importance of information generation and handling.

- *Urbanization:* Spatially, urbanization implies a drastic relocation from rural to urban residence, employment, and economic activities. Socially, urbanization entails deep changes in the social fabric and the emergence of new values and lifestyles. Economically, urbanization is driven by the large and diverse economic *opportunities* that the agglomeration of many enterprises and consumers entails. As a result, the largest portion of a country's gross domestic product (GDP) is generated in urban areas and agglomerations. As an extreme case, it is estimated that 80% of the GDP of Indonesia is generated in the national capital. For energy, urbanization imposes strict quality requirements for the energy carriers used, including higher power density and cleanliness.

Important changes in energy demand and supply associated with these two grand transitions include the following:

- *Commercial energy:* There has been a transition from noncommercial to commercial energy forms that reflects the structural economic shift from agriculture to industry, the related monetarization of the economy, and increasing urbanization.

- *Increasing energy "quality"*: There has also been a transition from direct use of solid energy forms, such as traditional biomass and coal, to liquids and grid-dependent energy forms that are more flexible, more convenient, and cleaner. This is a function of three major underlying trends in the "quality ladder" of different energy currencies:

   - Industrial processes and technologies are becoming ever more complex, with increasing requirements for ease of handling, storability, continuity, and flexible availability of energy supplies.

   - Requirements for convenience increase with rising levels of affluence, so that in addition to price there is an increasing "convenience premium" determining fuel choices in residential and commercial end uses.

   - Both of these trends lead to higher "form value" (quality) of the energy currencies that are at the interface between energy supply and demand, favoring flexible, clean energy forms such as electricity, gas, and ultimately hydrogen. Put succinctly: one cannot operate a computer directly with fuelwood or coal.

- *Decreasing energy intensity*: Although per capita energy needs have increased with economic development, the specific energy needs per unit of economic activity have decreased. This ratio is referred to as energy intensity.

The trends triggered by the two grand transitions of the Industrial Revolution, and how we are likely to see them extended in the future, lie at the heart of the six scenarios explored in this report. We will return to them in detail. First, however, we should recognize that, because the Industrial Revolution started in Europe and spread at different speeds to other parts of the world, there are now substantial disparities around the globe. These disparities among regions will translate into important differences in their future patterns of development. Thus the next section examines current disparities.

## 3.2. Disparities in Income and Energy Consumption

Levels of economic development, standards of living, and access to energy services are distributed distinctly unevenly around the world. Disparities are evident even at high levels of regional aggregation, as shown in *Figure 3.2* (based on the 11 regions

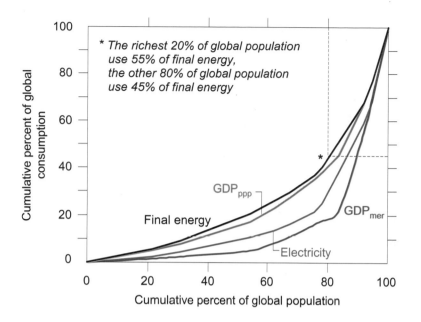

**Figure 3.2**: Disparities in economic activity and energy consumption in 1990. Cumulative percentage of global economic product, $GDP_{mer}$ and $GDP_{ppp}$, and consumption of final energy and electricity by percentage of cumulative global population.

used in this study), and become accentuated as we consider more disaggregated regions, individual countries, and eventually different social strata within countries.

Comparisons based on GDP show the richest 20% of the world's population producing and consuming 80% of the value of all goods and services globally. The poorest 20% dispose of only 1% of global GDP. In 1990, GDP per capita varied by a factor of 70, from US$22,800 per capita to US$330 per capita, between PAO and SAS, respectively. Disparities among individual countries and among different social strata are even more pronounced. The poorest 20% in Bangladesh, for example, have a per capita GDP of less than US$90. That is a factor of 700 lower than the US$60,000 annual income of the top 20% in Switzerland.

Such disparities are somewhat reduced when comparisons are based on GDP calculated on the basis of purchasing power parities ($GDP_{ppp}$) and not on GDP calculated at market exchange rates ($GDP_{mer}$). $GDP_{ppp}$ corrects for divergences between formal exchange rates and the relative purchasing power of different currencies (see *Box 3.1*). Nonetheless, disparities remain significant. The richest 20% consume 60% of global $GDP_{ppp}$, while the poorest 20% consume only 5%. Per capita $GDP_{ppp}$ differs by a factor of 17, for example, between SAS and NAM. The relative per capita income ranking of regions remains quite stable, however,

---

**Box 3.1: Purchasing power parities**

A nation's GDP is defined as the money equivalent of all products and services generated, sold, and bought in that nation's economy in a given year. The gross national product (GNP) is equal to the GDP *plus* the net balance of international payments to and from the country.

A major difficulty in comparing GDPs across countries is the need to translate everything into a common currency. Most often this is done using market exchange rates, with the US dollar as the common currency. Problems arise for several reasons. First, not all economies have a free market for foreign currency exchange. Second, the use of market exchange rates implicitly assumes that domestic prices are comparable with international prices. This is not the case for most developing countries, where prices for food and basic services, for example, are substantially below international levels. Third, many transactions are not accounted for in the formal economy, especially in less developed economies.

To get around some of these problems, the concept of purchasing power parities has been developed (see UNDP, 1993). A country's $GDP_{ppp}$ is simply its GDP in US dollars, corrected for the differences between domestic and international prices. In this report we use the notation $GDP_{mer}$ and $GDP_{ppp}$ to distinguish between calculations based on market exchange rates and those based on purchasing power parities, respectively. Where GDP appears without a subscript, it refers to $GDP_{mer}$.

International comparisons using $GDP_{ppp}$ present a picture different from those based on $GDP_{mer}$. For example, 1990 $GDP_{mer}$ per capita is higher in EEU than in CPA by about a factor of six (US\$2,400 versus US\$380). The ratio drops to about three using 1990 $GDP_{ppp}$ per capita (US\$5,800 versus US\$2,000). Taken together, the $GDP_{mer}$ of the OECD countries equals 80% of the world total. OECD $GDP_{ppp}$, however, is 55% of the global total. In this report, Case A's ambitious economic growth rates result in CPA's $GDP_{ppp}$ surpassing the current $GDP_{ppp}$ for NAM by 2010. In terms of $GDP_{ppp}$, CPA becomes the largest regional economy of the world around 2040. If progress is measured by $GDP_{mer}$, however, this catch-up process is shifted toward the end of the 21st century.

---

regardless of whether the comparison is based on per capita $GDP_{mer}$ or per capita $GDP_{ppp}$.

Disparities in energy availability mirror the economic disparities among regions. The richest 20% of the world's population use 55% of final and primary energy, while the poorest 20% use only 5%. (For a definition of final energy, see *Box 5.1*.) In 1990, per capita use of final energy varied by a factor of 18 between SAS and NAM – from 0.3 toe per capita to 5.3 toe per capita, respectively (*Figure 3.3*). Of all energy carriers, the disparities are largest for electricity. The richest 20% use 75% of all electricity, while the poorest 20% use less than 3% (see

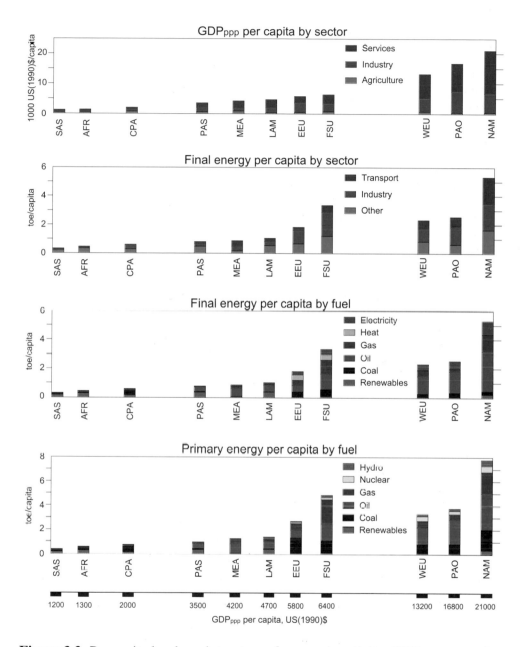

**Figure 3.3**: Per capita levels and structure of economic activity, GDP$_{ppp}$ per capita in US(1990)$1,000, and primary and final energy use, in toe per capita, for 11 world regions in 1990. Regions are grouped from left to right by their per capita GDP$_{ppp}$. Data sources: Hall and Rosillo-Calle, 1991; IEA, 1993, 1994; World Bank, 1993, 1994; UN, 1993b, 1993c.

*Figures 3.2* and *3.3*), reflecting their much more limited access to commercial energy in general, and to electricity in particular.

These large disparities, combined with future population growth, are important drivers of future energy demand and supply. However, it would be both simplistic and misleading to expect that present disparities will disappear rapidly and that global consumption levels will approach the upper bounds of today's consumption in industrialized countries. Such rapid convergence models have been the source of serious overestimates in some past energy demand scenarios, notwithstanding the ethical appeal of global consumption levels rapidly "catching up" to those prevailing in the industrialized countries.

## 3.3.    Economic and Energy Structures

Regional disparities in per capita GDP (GDP$_{mer}$ and GDP$_{ppp}$ alike) are correlated with differences in economic structure (*Figure 3.3*). Lower per capita GDP is associated with a high share of agriculture. Industry's share of both GDP and employment increases with increasing incomes until high levels of per capita GDP are reached. Then the share of industry begins to decline and that of services increases, illustrating the emergence of "post-industrial" or "service" economies.

A similar structural change can be seen in final energy use. At low per capita GDP, residential energy uses dominate – based largely on traditional energy carriers. With increasing incomes, the shares of industry and of commercial energy carriers, particularly liquids, increase. At high per capita GDP, final energy demand is dominated by the transportation sector and, again, the residential sector, this time based on commercial rather than noncommercial fuels. The energy portfolio is dominated by high-quality, grid-dependent energy forms like electricity, district heat, and gas, while the share of solid fuels (i.e., traditional biomass and coal) declines.

Similar structural differences can be seen in primary energy supplies. The correspondence is not perfect, however, because of differences in regional resource availability and the energy sector's ability to draw on a variety of primary energy sources to meet final energy demands.

## 3.4.    Energy Intensities

Energy intensities, measured as energy requirements per unit of economic activity, also show great differences across regions. They can be expressed in four different ways, depending on whether the numerator includes or excludes noncommercial energy, and depending on whether the denominator is GDP$_{mer}$ or GDP$_{ppp}$. All four are shown in *Figure 3.4* for both primary and final energy use.

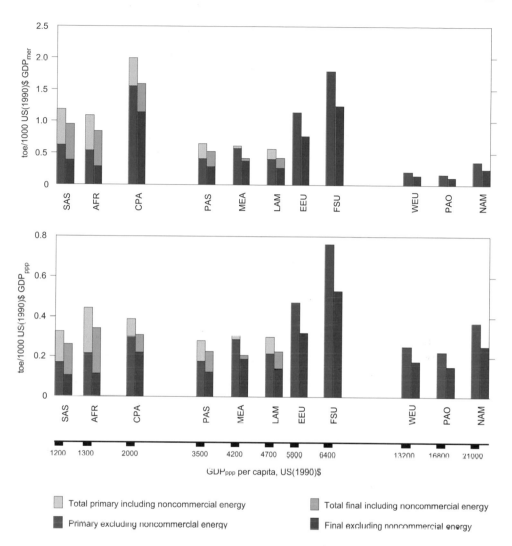

**Figure 3.4**: Primary and final energy intensities, including and excluding noncommercial energy, in energy per GDP$_{mer}$ (*top*) and per GDP$_{ppp}$ (*bottom*). Regions are grouped from left to right according to their per capita GDP$_{ppp}$ in 1990.

The influence of different accounting conventions is apparent in *Figure 3.4*. For example, the primary energy intensities of CPA and PAO differ by a factor of 16 when GDP$_{mer}$ is used and noncommercial energy is included. The difference decreases to a factor of 2.7 when GDP$_{ppp}$ is used. The variations are less striking, though still significant, when comparing energy intensities including noncommercial energy with those leaving it out. For example, in the extreme cases of SAS and AFR, including traditional biomass use doubles primary energy intensity. Overall, we believe that energy intensities using $GDP_{ppp}$ more accurately reflect relative differences in the efficiency of energy use than do those using $GDP_{mer}$.

An important characteristic of *Figure 3.4* is that, with the exception of the former centrally planned regions, energy intensities are lower at higher per capita GDP. This correlation is the result, first, of more efficient energy conversion and end-use technologies that become affordable at higher income levels, and, second, of the higher quality of energy carriers made available by the energy sector in high-income economies. Consider the case of cooking: while the thermal efficiency of a traditional fuelwood cooking stove is less than 10%, an electric oven can achieve efficiencies in excess of 70%. However, the electric oven requires both the availability of capital (as it is at least five times more capital intensive than the traditional cooking stove) and high incomes (given the high price of electricity).

The disparities across regions highlighted in this section largely mirror the changes that present-day industrialized countries have witnessed during their development over the past 150 years. In particular, the trend across regions toward lower energy intensities at higher incomes shown in *Figure 3.4* is consistent with long-term historical trends within high-income countries. Energy intensities in today's developing economies are quite similar to levels that characterized many of the OECD countries when they had similarly low levels of per capita income and were developing economies themselves. This consistent trend is incorporated in all three of the cases described in the following sections.

## 3.5.    Recent Developments: 1990 to 1998

To ease comparability with *Energy for Tomorrow's World* (WEC, 1993) and *Global Energy Perspectives to 2050 and Beyond* (IIASA–WEC, 1995), we have retained 1990 as the base year for this study. For completeness, this section summarizes actual short-term developments during the past few years, relying both on statistics through 1997 (BP, 1995, 1997; World Bank, 1995, 1997a) and on short-term forecasts through 2006.

### 3.5.1.    Economic growth

Between 1991 and 1995, the world economy continued to expand, although at a sluggish rate of about 2% per year. GWP grew by 2.5% in 1995 and was estimated to grow by almost 2.9% in 1996. The World Bank projects an even more robust average growth rate of 3.4% from 1997 to 2006, as the rapidly growing economies of large developing and "transition" countries tend to push GWP upward (World Bank, 1997a, 1997b).

Regional experience has been varied (see Chapter 7). Growth in the OECD regions (NAM, WEU, and PAO) between 1991 and 1995 averaged 1.8% per year, with a tendency toward higher growth in more recent years: 2.0% for 1995, an estimated 2.3% for 1996, and a projected 2.7% for 1997 through 2006 (OECD, 1996; World Bank, 1997a). CPA's GDP grew at about 10% per year, while GDP in the FSU region declined by nearly 10% per year between 1991 and 1995 (World Bank, 1997a). Indeed, since the early 1990s the reforming economies have suffered an economic depression unprecedented in recent history, which has been mirrored by drastic reductions in energy demand (see Bashmakov, 1990; Commission for the Energy Strategy of Russia, 1995; Dienes *et al.*, 1994; EIA, 1994; Government of the Russian Federation, 1995). In the developing regions, average annual GDP growth has been close to 5%. There is substantial variation among countries; Asia has seen the most impressive growth, while growth in Africa and the Middle East has been sluggish.

Although barriers to economic cooperation still exist, geopolitical tensions have diminished somewhat since 1990. The 1994 agreement on trade liberalization and the formation of the World Trade Organization (GATT, 1994) have advanced prospects for global economic integration, and world trade is expected to be the "engine of growth" (World Bank, 1995) of the next decade. At the regional level, economic integration has also advanced, as exemplified by the North American Free Trade Agreement (NAFTA) and the expansion of the European Union (EU). New member countries are expected to join the EU in coming years, and a common European currency will be introduced in early 1999. There has also been a significant increase in the number of international environmental agreements, most recently the Kyoto Protocol to the United Nations Framework Convention on Climate Change (see WEC, 1995a, 1995b; UN/FCCC, 1997; Bolin, 1998).

### 3.5.2. Energy developments

Global primary energy demand has grown more slowly than GDP. Primary energy growth averaged slightly above 1% per year between 1990 and 1995, about half the rate of growth of global GDP. Thus, global energy intensities continue to improve, except in some transitional economies.

Nearly all the additional primary energy demand growth has been for commercial energy forms, though there was variation across energy sources. Petroleum continues to be the world's primary energy source, followed by coal and natural gas. Global demand for coal stabilized between 1990 and 1995. Gas, renewables, and nuclear energy, on the other hand, expanded their shares between 1991 and 1996 (BP, 1997; EIA, 1998a, 1998b).

Regional differences have been enormous. In the OECD region, primary energy demand grew at nearly the same rate as the economy; in other words, energy intensity improvements came close to a standstill. In reforming economies, primary energy demand fell by approximately one-quarter. But the even greater drop in economic output meant that energy intensities increased significantly. Since 1990, CPA's total primary energy use increased by about 4% per year and close to 5% per year for commercial energy. Compared with a GDP growth rate of 10% per year, this translates into energy intensity improvements of 6% per year (5% per year for commercial energy). In developing countries outside CPA, total primary energy growth was some 3% per year (over 4% per year for commercial energy), compared with a GDP growth rate of 4% per year. Thus, aggregate energy intensity declined by about 1% per year, whereas commercial energy use grew at roughly the same rate as the economy (see also Chapter 7). The overall pattern is therefore one of decreasing energy intensities where economic growth is strong, stagnating energy intensities where growth is sluggish, and increasing energy intensities where growth is negative. This matches the patterns discussed in both this section and the next and is incorporated in all three cases in this study.

In the mid-1990s many OECD countries and to a lesser extent LAM began the process of deregulating their electricity markets to introduce greater competition and reduce costs. This process involves changes in the structure, ownership, and regulation of the power sector. For example, production, transmission, and local distribution have been unbundled into separate accounting units, and third parties have been granted access to transportation and distribution grids (OECD, 1996). Privatization of large, state-owned energy companies is a key element of deregulation in several countries. Although it is too soon to know precisely what the impacts of this will be on future energy consumption patterns, restructuring may well lead to diversification of energy supplies. Restructuring has also raised environmental concerns in some countries, as increased use of comparatively inexpensive coal could result in greater emissions of $CO_2$, sulfur oxides ($SO_x$), and nitrogen oxides ($NO_x$).

Taken together, recent short-term developments in the economy and energy most closely resemble the middle-course Case B.

# Chapter 4

# Determinants of Future Energy Systems

The historical developments and current disparities summarized in Chapter 3 provide the starting point for all three cases described in this book. In this chapter we summarize the key driving forces that influence how the three cases evolve from their common starting point. We divide these driving forces into six categories – population growth, economic growth, energy intensity improvements, technological change, the resource base, and environment.

## 4.1. Population Growth

In all three sets of scenarios – Cases A, B, and C – a common central (i.e., medium) projection of the world's population is assumed. We chose to use the same pattern and growth of population in all cases so that the differences that emerge among our cases are more easily connected to differences in their energy systems and their driving forces. Here, we describe the central population projection we chose and how it compares with available alternatives.

We used the single projection published by the World Bank in 1992 (Bos *et al.*, 1992), which is the same as that used in *Global Energy Perspectives to 2050 and Beyond* (IIASA–WEC, 1995). First we compare it with alternative projections and recent developments. We then assess the implications of the uncertainty in the World Bank and other population projections for the six scenarios and overall study results presented later.

The World Bank projection published in 1992 is consistent with other central projections developed at the time. Most important for the choice made in this study is the fact that the World Bank has been the only source that provides long-term, country-by-country projections. These we judged essential for this study, not least because of the high degree of regional detail required for analyzing the 11 world regions adopted in this study. In more recent studies, the central projections for the year 2100 have declined somewhat but are essentially the same as the World Bank projection for the next 50 years. These more recent studies include the United Nations (UN, 1998), the US Census Bureau (McDevitt, 1996), and the International Institute for Applied Systems Analysis (IIASA; Lutz, 1994, 1996). This is also consistent with the World Bank's own subsequent updates (Bos and Vu, 1994) – now no longer published.

Overall, the good news in the 1992 World Bank and other global projections is that population growth is slowing down. The next doubling of the world's population will take longer than the last one. In absolute numbers, of course, the next doubling will be the biggest ever. In the 1992 World Bank projection, global population doubles from 5.3 billion people in 1990 to 10.5 billion in 2060, a doubling time of 70 years. For comparison, the last doubling took approximately 40 years. Beyond 2050, population growth slows significantly and global population stabilizes at around 12 billion. In 2100, the value is 11.7 billion. Virtually all of this population growth is in the South (*Figure 4.1*).

The inevitable demographic momentum implies that uncertainties in demographic projections translate into noticeable differences in global population numbers only in the very long term (i.e., after 2050). For instance, the more recent assessments by the UN (1998) project global population by 2050 to increase to 9.4 billion and those by IIASA to 9.9 billion people (Lutz, 1996), compared with the 10 billion people of the 1992 World Bank projection used in this study. After 2050, the more recent demographic projections begin to diverge from the older ones. For instance, the latest UN (1998) medium-low and medium-high projections indicate a range of between 7.2 and 14.6 billion by 2100, with the medium scenario at 10.4 billion. As is the case with most projections, the UN does not attach probabilities to its population variants. The one study that does attach probabilities to alternative projections is IIASA's 1996 revision of estimates it originally published in 1994 (Lutz *et al.*, 1997). For the 1996 revision IIASA canvassed a group of demographic experts to assign subjective probabilities to possible future trends in fertility, mortality, and immigration in different regions of the world. These were then combined, taking into account possible correlations among the three factors, to yield probabilistic projections at both the regional and global levels. The 1992 World Bank projection used in this study corresponds approximately to IIASA's 70th percentile for 2100 (see *Figure 4.2*). That is, based on the judgment of the demographic experts it assembled, IIASA estimates that there is a 70% chance that the world's population growth will be below the 1992 World Bank projections used in this study. Conversely, there is a 30% chance that actual population growth will be higher than the 1992 World Bank projection.

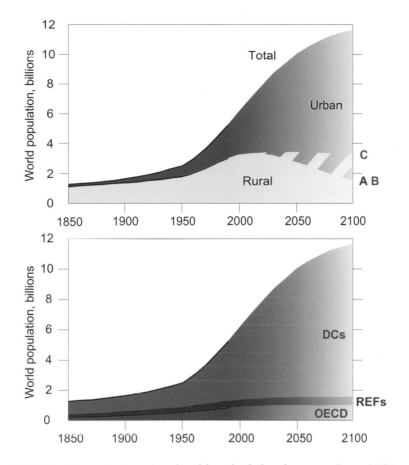

**Figure 4.1**: World population showing historical development from 1850 to 1990 and World Bank projection to 2100 (Bos *et al.*, 1992), (*top*) rural–urban and (*bottom*) by macroregion, in billion people. Urbanization trends are based on UN (1994) and Berry (1990).

One reason the 1992 World Bank projection used in this study is slightly higher than IIASA's 1996 and UN's 1998 central projections is that fertility has declined faster during the 1990s than originally anticipated (Cleland, 1996). More recent projections incorporate this faster fertility decline in their calculations and thus project lower future populations than do earlier studies. Thus IIASA's own projections dropped from a 1994 central estimate of 12.6 billion people in 2100 to a revised central estimate of 10.4 billion. The 1998 UN long-range population projection of 10.4 billion by 2100 is also lower than their 1992 estimate of 11.2 billion.

Despite these various adjustments and updates among demographers, we have retained the 1992 World Bank projections for four reasons. First, we have no objection to being a bit conservative. If actual population growth comes in slightly below what is assumed in this study, so much the better. Lower than anticipated population growth would result in somewhat lower energy growth. That lower population growth globally results in lower energy growth is not as uncontentious a statement as it might seem at first blush. Among the study's reviewers are those who argue that too rapid population growth may well slow economic growth. The corollary is that slower population growth could speed economic growth and thus growth in energy demand. We agree that a reduction in population growth does not automatically mean a *proportional* reduction in energy growth, but it does mean some reduction in energy demand. (Conversely, higher population growth would not necessarily lead to a proportional growth in energy demand, either.)

A second reason we retained the original 1992 World Bank population projections is that several of the study's regional review groups considered these too *low* and several other considered them too *high* for their regions, the latter being consistent with the demographic updates described above. The regions that thought the projections are too low included AFR, MEA, and NAM, and regions that thought them too high included CPA, EEU, and SAS. The regional demographic patterns and the implications of regional reviews are discussed in greater detail in Chapter 7. In the face of these competing forces tugging in opposite directions, we initially considered developing variations of the study's six scenarios incorporating alternative high and low population projections. But this quickly led to the third reason to retain the 1992 World Bank population projection.

The third reason is that varying population blurs the study's conclusions about *energy*. Moreover, if population is varied independently of energy developments, adding high and low population variants turns six scenarios into 18. If the basic conclusions of the study are unlikely to change (a point we return to below), the 12 new scenarios would add little new information, only many more pages of text and likely confusion. On the other hand, if population variations are *not* independent of energy developments (e.g., if Case C is believed to be more consistent with low population variants than Case B), then the addition of population variants overlays the energy picture with demographic debates. The relationship between population and economic growth continues to be controversial (NRC, 1986; MacKellar *et al.*, 1998; Gaffin, 1998). Perspectives range between those who argue that population growth hinders economic development ("more mouths to feed") and those who consider population growth an important driver of economic growth ("more hands to work").

The fourth and final reason we retained the 1992 World Bank population projection is that sensitivity analyses indicate little reason to expect that the study's major conclusions would change significantly if population were varied within a reasonable range of uncertainty as indicated by IIASA's 1996 probabilistic projections. (At present no one can really venture into the full implications of

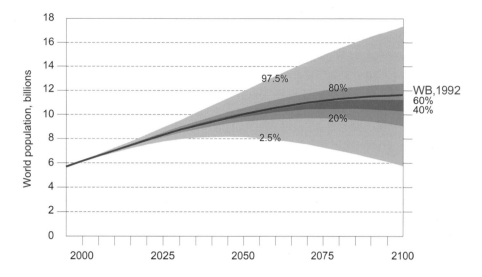

**Figure 4.2**: Probabilistic population projections for the world compared with the 1992 World Bank projection used in this study (shown in red). Probabilities of alternative population projections are given in percent. Source: adapted from Lutz, 1996.

extreme demographic outcomes ranging between collapse, for example, a decrease from 10 to 5 billion between 2050 and 2100, and a veritable explosion, such as an increase to 20 billion by 2100.[1]) Indeed the range of uncertainty is relatively narrow. As shown in *Figure 4.2*, the range between the 80th percentile and 20th percentile in 2100 runs only from 12.6 to 9.0 billion people, which is less than 8% higher and 23% lower than the 1992 World Bank projection used here. Sensitivity analyses performed with coupled demographic and energy models indicate that changing the population assumptions in the scenarios, in either direction, would lead to less than proportional changes in energy demand. For example, if we were to use IIASA's 80th percentile population projection of 12.6 billion people in 2100 in Case B, energy demand would likely be no more than 5% higher – that is, it would remain within the inevitable uncertainty range of any individual scenario, and more important it would be well below the range spanned by alternative developments, such as those described in Case A. Similarly, adopting IIASA's 20th percentile population projection of 9 billion by 2100 would result in a decrease of energy demand on the order of 10% in Case B. The weak inverse relationship between population growth and per capita economic growth incorporated in our scenarios would lead to higher per capita incomes (and energy demand); faster per

---

[1]Population was varied in the 1992 IPCC emissions scenarios across the full range of projections as a kind of sensitivity analysis, see Pepper *et al.*, 1992.

capita economic growth would in turn lead to quicker improvements in energy intensities in a less populated but more affluent world (which in turn would tend to lower energy demand). Thus, little additional insight is provided beyond that from Case C, which focuses on international development, equity, and environmental protection and thus on accelerated energy efficiency improvements and resulting low energy demand.

The above illustrations of increases and decreases in energy demand that would result from a reasonable range of alternative demographic projections are not enough to alter the basic patterns of change in the six scenarios, only to accelerate them slightly. In particular, the implied changes are not enough to open significantly different perspectives from those spanned between Case C (low demand) and Case A (high demand). They also would not cause any of the scenarios to encounter new and different resource or capital constraints. We thus believe that the results that are robust vis à vis the uncertainty range spanned by our three cases and six scenarios remain equally so in the face of demographic scenario variations.

Based therefore on the single 1992 World Bank population projection, the six scenarios examined in this study reflect only a weak relationship between population growth and economic growth. Two hypotheses are incorporated. First, the takeoff period of maximum growth of GDP happens sooner in those developing countries already more advanced in their demographic transition. Thus, the economic takeoff occurs sooner in China, where recent GDP growth rates have been approximately 10% per year and fertility rates are already low. In regions like SAS, where modest declines in population growth rates are just beginning and rapid declines come only well into the next century, the economic takeoff occurs later. Second, for industrialized countries, economic growth rates are lower in regions with possible declining populations, such as Japan after 2050, than in regions with stable or slightly increasing populations.

For future energy development, the two most important features of world population growth are consistent across all global projections available in the literature: World Bank, IIASA, UN, and US Census Bureau. The first is urbanization; the second is the concentration of future growth in the developing countries.

Increasing urbanization (*Figure 4.1*, top) is a pervasive trend in all countries. More than 80% of the population of industrialized countries live in urban environments, and many developing countries show similar high urbanization rates. According to the UN, 2.2 of 5.3 billion people lived in urban agglomerations in 1990. Over the next 35 years the urban population is projected to increase to 5.2 billion, an amount equal to the total global population in 1990. That increase of 3 billion accounts for nearly all the projected population growth of 3.2 billion over the next 35 years (UN, 1994). Thus, almost all additional global population growth will be urban. According to the UN (1994), 60% of the world's population will live in urban areas by 2025, and, if historical tendencies continue, three-quarters of the global population (approximately 8 billion people) will live in urban agglomerations by 2050. An increasing fraction will live in "megacities" with over 10 million inhabitants. It is estimated that shortly after the year 2000, eight cities will have more than 15 million inhabitants, only two of which, Tokyo and New York, are in highly industrialized countries. The remaining six (Beijing, Bombay, Calcutta, Mexico City, São Paulo, and Shanghai) are in the now developing world.

Providing adequate and clean energy services for a world whose population lives predominantly in urban areas will be a daunting task. Per capita energy use in urban areas is much higher than national averages, largely as a result of higher urban incomes (Sathaye and Meyers, 1990). Energy transport and conversion infrastructures that match urban energy demand densities will need to be put in place, and energy efficiencies will need to improve to at least partly offset demand growth and urban environmental impacts, especially in developing countries. Case C, with its greater emphasis on decentralized, small-scale, renewable sources of energy, provides a basis for slower rates of urbanization and for satisfying rural energy requirements at lower costs than in Cases A and B, but the broad trend is in the same direction in all cases.

The second critical feature of global population growth is by now well known and needs little elaboration – future growth will be concentrated in the developing countries (see *Figure 4.1*). By 2100, the population of the USA, Canada, and the whole of Europe combined drops to less than 10% of the world total, as indeed suggested by all central scenarios of the World Bank, IIASA, and the UN. One possible consequence is a major shift in the world's geopolitical balance in favor of the developing world, which may strengthen its ability to obtain and retain both internationally traded energy forms and access to technology.

## 4.2. Economic Growth

Economic development and growth are fundamental prerequisites for achieving an increase in living standards with the kind of vigorous global population growth described in the previous section. Although the definitive (and quantitative) long-term history of world economic output has yet to be written, it is fair to say that current disparities between North and South and between East and West are rooted largely in differential growth rates of economic output over extended historical periods (see *Table 4.1*). Small differences in these growth rates, when compounded over a generation or more, have much greater consequences for standards of living than even the most dramatic short-term business fluctuations.

Both classical and modern economic growth theory offer some insights into the mechanisms of successful economic development and the remaining disparities

**Table 4.1**: Economic growth rates, historical and 1990 to 2050, in percent per year.

| | Historical | | | Case | | | | | |
| | GDP$_{mer}$ | | GDP$_{ppp}$ | 1990 to 2020 | | | 2020 to 2050 | | |
| Region | Since 1850 | Since 1950 | since 1950 | A | B | C | A | B | C |
|---|---|---|---|---|---|---|---|---|---|
| NAM | 3.5 | 3.3 | 2.1 | 2.3 | 2.0 | 1.7 | 1.6 | 1.3 | 1.1 |
| WEU | 2.4 | 3.7 | 2.2 | 2.2 | 1.9 | 1.7 | 1.7 | 1.3 | 1.1 |
| PAO | 3.9 | 6.2 | 3.6 | 1.9 | 1.5 | 1.4 | 1.3 | 0.9 | 0.8 |
| EEU | 2.1 | 3.9 | 2.4 | 2.3 | 0.9 | 1.3 | 4.6 | 3.6 | 3.2 |
| FSU | 3.5 | 5.2 | 3.5 | 1.2 | 0.7 | 1.1 | 5.4 | 3.8 | 3.3 |
| CPA | 2.9 | 6.1 | 4.3 | 7.2 | 5.0 | 6.7 | 4.4 | 4.0 | 4.0 |
| SAS | 2.0 | 4.5 | 3.1 | 3.9 | 3.5 | 3.7 | 4.6 | 3.5 | 4.3 |
| PAS | n.a. | 9.8 | 6.8 | 5.7 | 4.4 | 5.3 | 3.3 | 3.1 | 3.1 |
| MEA | n.a. | 4.6 | 3.1 | 3.6 | 3.3 | 3.3 | 3.9 | 3.0 | 3.0 |
| AFR | n.a. | 2.7 | 2.0 | 3.3 | 3.0 | 3.1 | 4.7 | 3.5 | 3.9 |
| LAM | 3.7 | 4.2 | 2.9 | 3.1 | 3.0 | 3.1 | 3.2 | 2.8 | 2.8 |
| World | n.a | 4.0 | 2.8 | 2.7 | 2.2 | 2.2 | 2.6 | 2.0 | 2.1 |

n.a. = not available.
Sources for historical data: Maddison, 1989; UN, 1993a, 1993c.

among countries.[2] In addition to the traditional "tangible" factors driving economic development, such as technology, capital, natural resources, and trade possibilities, "intangible" factors such as the quality of human capital and institutions, limited income disparities, and social cohesion increasingly explain the historically uneven paths of economic development. The bottom line is that growth possibilities do not fall like "manna from heaven" but are largely determined by endogenous factors including knowledge (research, development, and demonstration, RD&D), skilled human capital (education), and a favorable social and institutional climate (e.g., social protection of the work force and a fair distribution of productivity gains).

Combining the insights of classical and modern economic growth theory, the fundamental forces driving developing economies to catch up to developed economies are differences in productivity and in knowledge and technologies applied in production. Because it is generally easier to catch up to the productivity frontier than to push it further, developing economies have the potential, as they close the productivity and technology gap, for much higher growth rates than leading economies. Realizing this potential requires functioning markets and at least partially favorable "intangible" factors, but there is no compelling *a priori* reason to rule out future successful economic development even in regions where recent growth prospects have been far from auspicious, such as sub-Saharan Africa.

---

[2]See, e.g., Aghion and Howitt, 1998; Barro, 1997; Grossman and Helpman, 1991; Maddison, 1995.

The historical successes of much of Western Europe, Japan, and more recently the "Asian Tigers" all point in this direction. The potential for developing economies to catch up to economic leaders does not mean that success can be taken for granted, nor should one be surprised that overall progress is overlaid with short-term business cycle fluctuations and even financial crises. Rapid growth can aggravate small imperfections that can ultimately lead to crisis. However, the key lesson is the importance of maintaining a long-term perspective when facing short-term downturns and fluctuations: the developing world *can* catch up, although the gap to the most developed countries will not be entirely closed by 2100.

Successful cases of industrialization and catch-up all appear to be characterized by a similar dynamic pattern. After a particular development threshold is passed, economic growth accelerates, passes through a maximum, and then decreases once the industrial and infrastructure base of an economy is firmly established. Historical data support Rostow's (1980) thesis that "the poor get richer and the rich slow down." This conceptual model has been incorporated into the time path of economic development underlying the three cases (*Figure 4.3*).

Many alternatives have been proposed for the conditions that trigger the "take-off" into economic development. Recent research (Lucas, 1988; Romer, 1986; UN, 1997; World Bank, 1997b) places less emphasis on physical endowments, such as land, minerals, and energy. More emphasis is given both to intangible factors – such as appropriate institutional arrangements, the market orientation of an economy, trade, and education – and to traditional explanatory variables – such as technology availability and transfer, savings rates, and other factors influencing labor productivity. These "intangible" social and institutional factors are, however, difficult to quantify for a scenario exercise and for the most part must be dealt with qualitatively. They are assumed to be generally favorable in Cases A and C, and somewhat less so in Case B.

With these cautions in mind, let us turn to the future economic growth patterns reflected in the three cases of this study. Common to all three are the following features:

- Within the time horizon of the scenarios, all countries and regions manage to successfully take off into a period of industrialization and accelerating economic development.

- In each case, GDP per capita follows the historical pattern of an initial accelerating growth, a peak, and then an eventual decline once the economy's industrial base becomes established. Growth rates for high-income economies (OECD) decline gradually in all three cases. Because of already high income levels, future OECD growth rates are below historical rates (see *Table 4.1*). Long-term growth rates of 2% per year or lower appear modest only at first

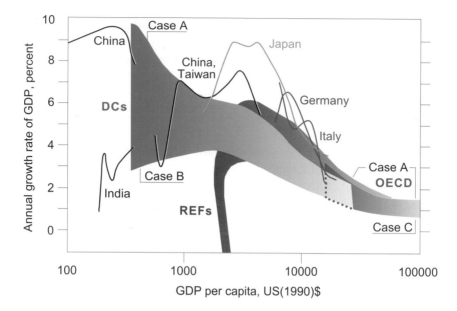

**Figure 4.3**: Economic growth rates, GDP in percent per year, versus degree of economic development, GDP per capita in US(1990)$, illustrating "takeoff" into industrialization, peak, and declining economic growth rates of high-income economies. Historical data for selected countries (as a 10-year moving average) and range of economic growth rates assumed in the three cases are indicated for the three macroregions: DCs, REFs, and OECD countries.

glance. They have to be contrasted with resulting per capita income levels. For example, even in the conservative Case B, per capita income in PAO reaches US$50,000 by the middle of the 21st century and US$75,000 by the century's end.

- Where there are currently no signs of an economic takeoff, as is the case in AFR where GDP per capita is stagnating, peak economic growth is assumed to correspond to the period of maximum demographic transition (i.e., maximum decline in population growth rates).

- Short-term economic growth rates reflect data from 1991 to 1997, plus the latest World Bank (1997a) forecasts to 2006. For the economies in transition, particularly FSU, this leads to a pronounced decline during the early 1990s, as was anticipated in this study's scenarios when they were originally formulated five years ago.

Beyond these basic commonalities, the three sets of scenarios – Cases A, B, and C – diverge in terms of economic growth patterns.

In the three Case A scenarios, the timing and speed of economic restructuring and development follow the pattern of the most successful historical examples of industrialization. This is a future characterized by free trade and favorable geopolitics. GWP increases by an average of 2.6% per year to 2050 and by 2.2% per year thereafter. The result is a fivefold expansion by 2050 and a 15-fold expansion by 2100. By 2050, average GWP per capita is US$10,000, and by 2100 it exceeds US$25,000 per capita. Average family incomes are over US$30,000 by 2050 and around US$70,000 by the end of the 21st century, based on a decline in average family size to 3.2 persons in 2050 and 2.7 persons by 2100, as is characteristic of central population projections (MacKellar *et al.,* 1995). Thus, by any standards, Case A represents a wealthy world in which the current distinction between "developed" and "developing" countries will no longer be appropriate.

The single Case B scenario also reflects significant economic expansion but incorporates more cautious expectations about geopolitics and international trade. This is a more fragmented world than in Case A, resulting in more heterogeneous patterns of economic development. For example, economic development in CPA is rapid and impressive, while that in AFR and SAS remains painfully slow. GWP again expands significantly, to US$75 trillion in 2050 and US$200 trillion in 2100. But the pace is slower than in Case A (2.1% per year), and regional disparities are larger. Through 2020 the pace of economic growth (2.2% per year) is also slower than in the original Case B of *Energy for Tomorrow's World* (WEC, 1993). This reflects recent setbacks and slow economic restructuring in EEU, FSU, and many developing countries.

The two Case C scenarios contain strong normative policy elements that distinguish them from Cases A and B. They reflect aggressive efforts to advance international economic equity and environmental protection. They include massive direct resource transfers from North to South and stringent environmental taxes, or other levies and grants, that are paid principally by OECD countries and transferred to the developing world. Such transfers constitute a positive-sum game that fuels economic growth in the developing countries and leads to GWP which exceeds that in Case B but falls short of that in Case A. GWP in Case C reaches US$75 trillion in 2050 (as in Case B) and US$220 trillion in 2100 (compared with US$200 trillion in Case B). A summary of the three cases is given in *Figure 4.4.*

It should be emphasized that in none of these cases can economic developments be taken for granted. All assume, to various degrees, numerous domestic and international policies that promote the free exchange of goods and technology, a favorable investment climate, and high levels of education and social protection for the work force. These are the necessary foundation for the formidable productivity increases of labor and capital needed to realize the high levels of economic growth reflected in all three cases. There is a risk that the energy sector could become a

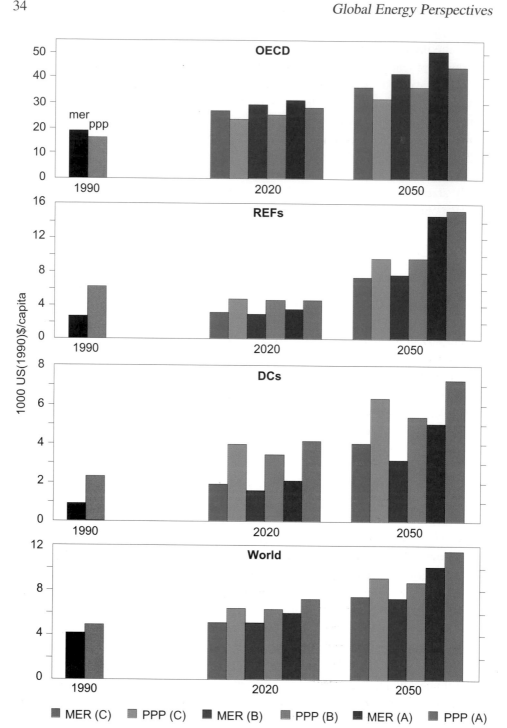

**Figure 4.4**: GDP per capita, in US(1990)$1,000, in 1990 and for the three cases in 2020 and 2050, in $GDP_{mer}$ and $GDP_{ppp}$.

bottleneck. It will have to compete for much-needed investment funds in a period of increasing competition where substantial funds are also needed for the health, communication, and transport infrastructures that rapidly growing economies require, for the industrial investments expanding economies need, and for the needs of rapidly growing urban areas and "megacities." The energy sector will also face greater competition for the policy attention it has traditionally enjoyed but will now have to share with a multitude of economic and development concerns. Without appropriate RD&D and investment efforts in the short to medium term, the energy sector also risks lagging behind overall productivity growth and becoming a possible long-term bottleneck for successful economic and social development.

## 4.3.  Energy Intensity Improvements

Chapter 3 describes regional differences in energy intensities and briefly mentions the historical development in currently industrialized countries of decreasing aggregate energy intensities over time. The causes of those decreases are many and complex. They include, first, technological improvements in individual energy end-use and conversion components – for example, a more efficient stove or a more efficient power plant. They also include structural shifts in energy systems, such as a shift from coal-fired electricity generation to gas-fired combined cycle plants. They include interfuel substitution at the level of energy end use, like the replacement of fuelwood with electricity or LPG (liquefied petroleum gas). They include economic shifts from more to less energy-intensive activities. They include changing patterns of energy end uses and, ultimately, changing lifestyles.

Not every change in every one of those categories represents a decrease in energy intensity. But taken together, the overall trends are persistent and pervasive. They are incorporated into all three cases presented in this report. In this section, we present the historical evolution of energy intensities that underlie all three cases and the future development of these trends for each case individually. We do not attempt a detailed dissection of all contributing factors to improvements in energy intensities. In the next section, we discuss technological change in general and how it is incorporated into the scenarios.

In all scenarios, economic development outpaces the increase in energy use, leading to substantial reductions in energy intensities (see *Box 4.1*). As individual technologies progress, as inefficient technologies are retired in favor of more efficient ones, and as the structure of the energy system and patterns of energy services change, the amount of primary energy needed per unit of GDP – the energy intensity – decreases. In some developing regions, the intensity of commercial energy can increase initially as traditional and less efficient energy forms are replaced by commercial energy, but the intensity of total energy decreases in these cases as

**Box 4.1: Energy intensity improvements in Cases A, B, and C**

Energy intensity improves in all three cases as a result of economic development and improved technology. Improvement rates in Cases A and B are primarily a function of economic growth. The higher the per capita GDP growth, the higher the energy intensity improvement rate. Case C goes further to incorporate additional vigorous efficiency improvements through demand-side management and economic instruments, including substantial increases in energy prices and taxes. This is not to say that such measures might not turn out to be necessary in the other two cases. Energy intensity improvement rates for all three cases are summarized in *Table 4.2*.

**Table 4.2:** Three scenarios of energy intensity improvements, primary energy per $GDP_{mer}$, in percent per year.

|        | Case | | |
|--------|------|------|------|
|        | A | B | C |
|        | 1990 to 2050 | 1990 to 2050 | 1990 to 2050 |
| OECD   | −1.1 | −1.1 | −1.9 |
| REFs   | −2.0 | −1.7 | −2.2 |
| DCs    | −1.6 | −1.2 | −1.9 |
| World  | −1.0 | −0.8 | −1.4 |

In comparing our energy intensity improvement rates with those of other studies and earlier WEC cases, important definitional and measurement issues have to be kept in mind. These measurement issues are illustrated for Case B for three of our 11 world regions – NAM, CPA, and SAS – giving energy intensity improvement rates (percent per year) to 2020 for total primary energy (TPE) and commercial primary energy (CPE) for $GDP_{mer}$ and $GDP_{ppp}$, respectively. For OECD regions, differences are insignificant.

**Table 4.3:** Energy intensity improvements, 1990 to 2020 (Case B), for three regions, in percent per year.

| Region | $TPE/GDP_{mer}$ | $TPE/GDP_{ppp}$ | $CPE/GDP_{mer}$ | $CPE/GDP_{ppp}$ |
|--------|------|------|------|------|
| NAM    | −1.2 | −1.3 | −1.2 | −1.3 |
| CPA    | −2.2 | −0.7 | −1.6 | −0.1 |
| SAS    | −1.0 | −0.3 | 0.3  | 1.0  |

The dynamics of energy intensity improvements change drastically for developing regions as exemplified by CPA and SAS. (The generally higher energy intensity improvements for CPA are the result of the much higher short-term economic growth rates for CPA; higher GDP growth leading to faster turnover of capital stock yields faster energy intensity improvements in the scenarios.) Thus considering the evolution of the total primary energy use per $GDP_{mer}$ yields a challenging pace of energy intensity improvements. Conversely, the intensity of commercial primary energy use per unit of $GDP_{ppp}$ remains roughly at 1990 levels (i.e., commercial energy consumption in Case B is projected to increase at the same rate as the $GDP_{ppp}$ for CPA).

well. With all other factors being equal, the faster the economic growth, the higher the turnover of capital and the greater the energy intensity improvements.

These long-term developments are reflected in the scenarios and are consistent with historical experience across a range of alternative development paths observed in different countries. *Figure 4.5* illustrates the range of historical energy intensity improvements for four representative countries and FSU. Energy intensities are measured both in terms of total energy divided by GDP and as commercial energy divided by GDP. *Commercial* energy intensities increase during the early phases of industrialization as traditional and less efficient energy forms are replaced by commercial energy. When this process is completed, commercial energy intensity peaks and proceeds to decline. This phenomenon is sometimes called the "hill of energy intensity." Reddy and Goldemberg (1990) and many others have observed that the successive peaks in the procession of countries achieving this transition are ever lower, indicating a possible catch-up effect and promising further energy intensity reductions in developing countries that have still to reach the peak. In the USA, for example, the peak of commercial energy intensity occurred during the 1910s and was higher than Japan's subsequent peak, which occurred in the 1970s.

*Figure 4.5* shows energy intensities for China and India measured both at market exchange rates and in terms of purchasing power parities. For both countries, energy intensities in terms of market exchange rates are very high, resembling the energy intensities of the now industrialized countries more that 100 years ago. The reason is that China's and India's GDPs are comparatively low when measured at official market exchange rates due to generally low prices in the two countries (see *Box 3.1*). Energy intensities in terms of GDP measured at purchasing power parities are generally much lower, indicating substantially higher energy effectiveness in these countries than would be calculated using official exchange rates. In terms of purchasing power parities, the peaks in commercial energy intensities may eventually prove to be lower than those experienced by the now industrialized countries.

The substantially lower energy intensity of GDP when expressed in terms of purchasing power parities ($GDP_{ppp}$) rather than at market exchange rates ($GDP_{mer}$) should be contrasted with the generally much lower energy intensity improvement rates when GDP is expressed as $GDP_{ppp}$ rather than $GDP_{mer}$. The differences can indeed be substantial. In 1990 the reported energy intensity in China was about 2.3 kgoe per US(1990)\$ for $GDP_{mer}$ with an average historical reduction rate of 2.8% per year since 1971, compared with about 0.4 kgoe per US(1990)\$ for $GDP_{ppp}$ for the same year. Since 1971 $GDP_{mer}$ has grown by 7.5% per year whereas the estimated $GDP_{ppp}$ has grown by 4.3% per year, roughly the same rate as total primary energy use. Caution is therefore needed when interpreting the apparent rapid energy intensity improvements, measured at market exchange rates, that are reported for some countries. As countries develop and their domestic prices

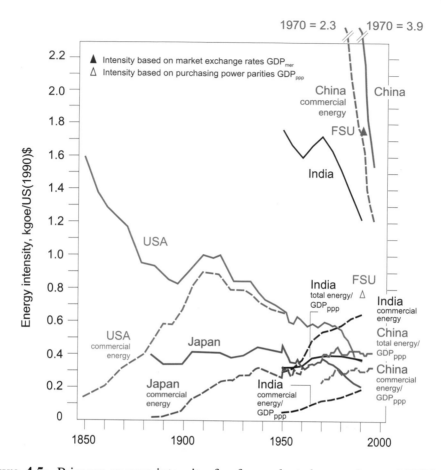

**Figure 4.5**: Primary energy intensity for four selected countries and FSU, total (solid lines) and commercial energy (dashed lines), in kgoe, per GDP, in US(1990)$. Unless otherwise specified, GDP refers to GDP$_{mer}$. For China, India, and FSU intensities based on GDP$_{ppp}$ are also given. Data sources: Nakićenović, 1987; Martin, 1988; TERI, 1994.

converge toward international levels, the difference between the two GDP measures disappears. As would be expected, energy intensity improvement rates are lower in more developed world regions.

Adding traditional energy to commercial energy reflects total energy requirements and yields a better and more powerful measure of overall energy intensity. Total energy intensities generally decline for all four countries in *Figure 4.5*. There are exceptions, including periods of increasing energy intensity that can last for a decade or two. This was the case for the USA around 1900, India during the 1960s, and China during the 1970s. Today, energy intensities are (temporarily)

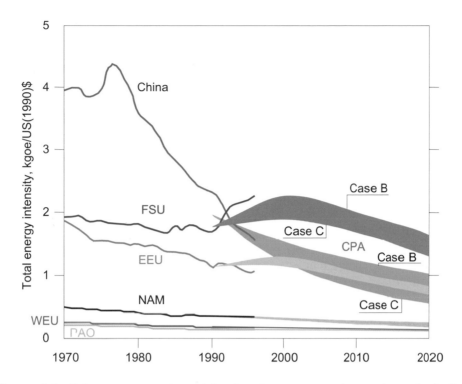

**Figure 4.6**: Primary energy intensities for six representative regions, including historical development from 1970 to 1996 and future ranges as reflected in the scenarios, primary energy divided by GDP in kgoe per US(1990)$. Historical data are shown for China instead of CPA as they are not available for the rest of the region.

increasing in the economies in transition due to economic depression (*Figure 4.6*). In the long run, however, the development is toward lower energy intensities. Data for countries with long-term statistical records show improvements in total energy intensities by more than a factor of five since 1800, corresponding to an average decline of total energy intensities of about 1% per year (Nakićenović, 1987).

Energy intensity improvement can continue for a long time to come. The theoretical potential for energy efficiency and intensity improvements is very large; current energy systems are nowhere close to the maximum levels suggested by the second law of thermodynamics. Although the full realization of this potential is impossible, many estimates indicate that the improvement potential might be large indeed – an improvement by a factor of 10 or more could be possible in the very long run (see Ayres, 1989; Gilli *et al.*, 1990; Nakićenović *et al.*, 1993, 1996). Thus, reduction of energy intensity can be viewed as an endowment, much like other natural resources, that needs to be discovered and applied.

In the six scenarios, improvements in individual technologies vary across a range derived from historical developments and current literature about future technological possibilities and structural change. Combined with the economic growth patterns of the three cases, the overall global average energy intensity reductions vary between about 0.8 and 1.4% per year. These figures bracket the historical rate of approximately 1% per year experienced by the more industrialized countries during the past hundred years. Cumulatively they lead to substantial energy intensity decreases across all scenarios. Energy intensity improvements are significantly higher in some regions, especially over shorter periods of time and when high economic growth rates result in fast turnover of the capital stock.

Because the six scenarios start with a base year of 1990 and were originally formulated five years ago, actual short-term trends in recent years can be compared with initial developments in the longer-term scenarios. *Figure 4.6* shows the range of the scenarios' energy intensity improvement rates for six regions compared with short-term historical trends. These range between vigorous improvements of up to 4% per year for CPA[3] and a (temporary) increase of energy intensities in the reforming economies of EEU and FSU. Overall, the scenario trajectories correctly anticipated short-term developments during the 1990s, especially for the reforming economies of Eurasia. All scenarios reflect a continuing process of successful economic reform and restructuring in all of Eurasia in the coming decades that leads to sustained investments in the energy sector and in economic development, and thus to long-term energy intensity improvements. The cases track particularly closely the recent developments in CPA, FSU, and EEU – the three regions with the most dynamic and heterogeneous changes in energy intensities.

*Figure 4.7* shows shorter-term historical data since 1970 for a number of the 11 regions in terms of energy intensities versus per capita GDP. These shorter-term developments are contrasted with the experience of the USA since 1800. In all cases, economic development is characterized by a reduction of energy intensities, with the energy intensities of today's DCs at levels generally comparable with those of the now industrialized countries when they had the same per capita GDP. Different countries can follow different development paths, however, and there are some persistent differences in energy intensities even at similar levels of per capita GDP. For example, PAO and WEU have consistently lower energy intensities than NAM even at the same level of development, and apparently similar differences are emerging between SAS and CPA. Energy intensities mirror economic and technological development, geography, natural resources, and lifestyles. But history also matters in the sense that the differences between "high" and "low"

---

[3]The difference in 1990 energy intensities in *Figure 4.6* reflects the difference between China and the CPA region as a whole. Historical data are shown for China as they are not available for the rest of CPA.

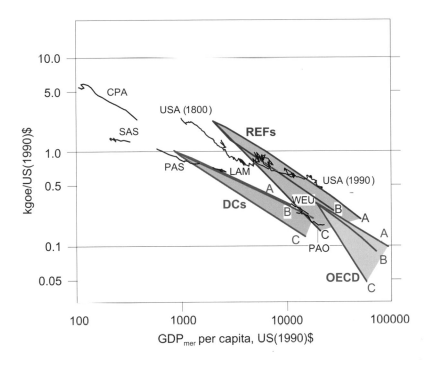

**Figure 4.7**: Energy intensities, in kgoe per US(1990)\$, as a function of degree of economic development, in GDP$_{mer}$ per capita in US(1990)\$. Historical data (black) and Cases A, B, and C (color).

energy intensities persist for decades across different regions. The range of possibilities is reflected in *Figure 4.7* by the range between Cases A, B, and C. The energy intensity improvement paths are shown for the three macroregions as it was not possible to show the individual developments for all 11 regions in such a concise manner. Energy intensities diverge within a "cone" that starts with the 1990 energy intensity of each macroregion and widens to reflect the range of possible paths.

The conclusions that can be drawn from a full analysis of historical energy intensity changes and that are incorporated in the six scenarios for the various regions can be summarized as follows:

- Aggregate energy intensities, including noncommercial energy, generally improve over time, and this is true in all countries. A unit of GDP in the USA, for example, now requires less than one-fifth of the primary energy needed 200 years ago. This corresponds to an average annual decrease in energy intensity of roughly 1% per year. The process is not always smooth, as data from the USA and other countries illustrate. Periods of rapid improvements are interlaced with periods of stagnation. Energy intensities may even rise in the early takeoff stages of industrialization, when an energy- and materials-intensive industrial and infrastructure base needs to be developed.

- Whereas aggregate energy intensities generally improve over time, commercial energy intensities follow a different path. They first increase, reach a maximum, and then decrease. The initial increase is due to the substitution of commercial energy carriers for traditional energy forms and technologies. Once that process is largely complete, commercial energy intensities decrease in line with the pattern found for aggregate energy intensities. Because most statistics document only modern, commercial energy use, this "hill of energy intensity" has been frequently discussed. Its apparent existence in the case of commercial energy intensities, however, is overshadowed by the powerful result for the secular development of the aggregate, total energy intensities – there is a decisive, consistent long-term trend toward improved energy intensities across a wide array of national experiences and across different phases of development.

- History matters. While the trend is one of conditional convergence across countries, the patterns of energy intensity improvements in different countries reflect their different situations and development histories. Economic development is a cumulative process that, in different countries, incorporates different consumption lifestyles, different settlement patterns and transport requirements, different industrial structures, and different takeoff dates into industrialization. Thus the evolution of national energy intensities is *path dependent*. In *Figure 4.5*, for example, there is an evident distinction between the "high intensity" trajectory of the USA, and the "high efficiency" trajectory of Japan.

- Despite improvements, aggregate energy intensities for developing countries remain consistently higher (especially when GDP is measured at market exchange rates) than those in industrialized countries. However, the more important relationship is that between energy intensity and economic development, measured by either $GDP_{mer}$ or $GDP_{ppp}$ per capita. This is shown in *Figure 4.7*. Energy intensities in developing countries are similar to those of industrialized ones *at similar levels of economic development (GDP per capita)*.

Given these conclusions, two "stylized facts" underlie the energy intensity trends of all three cases. First, energy intensity improvement rates are related to per capita GDP growth rates. The faster an economy grows in per capita terms, the faster its productivity growth, rate of capital turnover, and introduction of new technologies, and the faster energy intensities improve. Conversely, in cases of negative per capita GDP change, such as the recent experiences in FSU, energy intensities deteriorate. Second, there is conditional convergence among regions over time. A region at a given level of per capita GDP is assumed to achieve at least similar energy intensities as did industrialized countries when they were at similar per capita GDP levels. For example, in Case A PAS reaches US$10,000 per capita around 2040, with energy intensities that correspond to those of WEU in the early 1970s when its per capita income was at the same level.

The resulting energy intensity improvement rates for the world as a whole are 1% per year for Case A, 0.8% per year for Case B, and 1.4% per year for Case C. Such long-term improvements may appear conservative, particularly when compared with short-term energy intensity improvement rates. Part of the difference reflects the requirement in this longer-term analysis that these energy intensity improvements be sustained beyond 2020 to 2100. Another part is that the aggregate global figures mask the fact that for some regions energy intensity improvement rates are much higher than the global aggregates due to higher GDP growth rates. Energy intensity improvement rates for the three cases are illustrated in *Box 4.1* and the regional improvement rates for the three cases are presented in Chapter 7.

## 4.4.    Technological Change

Technological change is at the heart of the productivity increases and economic growth that have allowed increases in the standard of living. Increases in performance, including energy efficiency, and cost reductions (see *Figures 4.9, 4.10*, and *4.11*) have historically allowed economies not only to increase the quantity and quality of energy produced and delivered, but to do so with declining real costs. Technology is also crucial for easing the burden humanity imposes on the environment. We first discuss some general features of technological change and then describe how we have incorporated technological change into the scenarios.

Only recently have long-term energy models and scenarios begun to move beyond relatively simple representations of technological progress.[4] Five principal characteristics of the process can be identified:

- Innovation and experimentation with new technology in expectation of economic opportunities and profits;

- Innovation and experimentation as a hedging strategy against uncertainty of future developments;

- Continuous change and improvements through RD&D and actual "hands-on" experience ("learning by doing" and "learning by using");

---

[4]For a comprehensive discussion see Grübler, 1998; for modeling applications see Messner *et al.*, 1996; Grübler and Messner, 1996; Messner, 1997.

- Diffusion and substitution, that is, the gradual adoption and spread of new technologies in time and space and their interaction with existing technologies;

- Impacts (social, economic, resource, and environmental) and their feedbacks on technological change.

Basic human curiosity, ingenuity, and uncertainty about the future lead individuals and institutions to explore technological frontiers, to invent, and to innovate – where by "innovation" we refer to the first practical application of an invention. The driving forces of invention and innovation also encompass uncertainty about future economic conditions (market opportunities), political and regulatory developments (constraints), natural resources, and the severity and irreversibility of environmental impacts. Following innovation, which often takes the form of development and demonstration projects in industrial laboratories, the useful services of a new technology are typically first employed in *niche markets*. These are specialty applications where a novel technology either performs a new task that cannot be accomplished otherwise or has substantial performance advantages over existing technologies even though it is most likely more expensive (see *Box 4.2*). For the gas turbines now used pervasively for power generation, the essential precursor was the jet engine introduced in military niche markets after World War II. Jet aircraft outperformed all propeller-driven planes in a niche where there was a premium on performance (speed) and cost hardly mattered.

The first commercialization in such niche markets allows suppliers and users to "learn by doing" and "learn by using," which leads to further improvements in performance and cost (see *Box 4.2*). Use in a wider array of markets, or *pervasive diffusion*, follows as the new technology replaces older competitors. For example, after military applications jet engines found their way into commercial aircraft, first the Comet in England and then the Boeing 707 in the USA – a step that dramatically changed long-range civil aviation. In parallel, the first stationary gas turbines were developed as derivatives of jet engine technology. They were expensive and could not match the performance of steam turbines in generating electricity, but found a niche market providing peak power. Development continued, highly efficient combined cycle designs were introduced, installed capacity expanded, and today gas turbine technology is the most efficient way to generate electricity as well as the cheapest when the transport infrastructures are in place. Close to four decades have passed between the first application of gas turbines in military niche markets and their perfection to a preferred technological option for power generation, that is, their pervasive diffusion. This diffusion has been helped along the way by changing market conditions which now favor technologies that emit little pollution and have low capital costs. At the same time, partly due to this diffusion of gas turbines, the deregulation and liberalization of markets have been promoted by the availability of a low capital-cost capacity. This emphasizes the importance of uncertainty – about market conditions and diffusion in this case – which both motivates experimentation in the first place and then combines with performance and cost improvements gained through experience to turn some innovations into business and social successes, but many more into historical curiosities.

The typical resulting pattern of a technology's market share over time is an S-shaped curve.[5] At the earliest stage of commercialization, growth in a technology's market share is slow as the technology is applied only in specialized niche markets and costs are high. Growth accelerates as early commercial investments lead to compounding cost reductions and standard-setting, which leads to imitation and adoption in a wider array of settings and thus to pervasive diffusion. As the potential market is saturated and a product matures, growth in market share declines to zero. With the arrival of better competitors, the market share of the senescent technology declines.

The evolution of technological systems is as important as the evolution of individual new technologies. As the use of an individual technology, such as the automobile, expands, its evolution becomes intertwined with the evolution of many other technologies and institutional and social developments. The evolution of automobiles both affects and is affected by what happens to a host of component suppliers and their technologies. It affects and is affected by the expansion of infrastructures such as road networks and the system for refining and distributing petroleum products in gas stations. And it affects and is affected by social and institutional developments, such as settlement patterns, business adaptations to changed transportation options, and training institutions for both drivers and mechanics.

As a new technology prompts adaptations among potential component suppliers and influences related infrastructural developments, it gains advantages that will make it more difficult for subsequent competitors to catch up. A particular technological configuration becomes "locked-in" and its future development becomes "path dependent" (Arthur, 1983 and 1989) as changes build on previous ones and radical change becomes ever more difficult to implement. The same is true institutionally. The more schools, businesses, and individuals that use Microsoft (and the QWERTY keyboard), the more likely it is that the next school, business, or individual will also choose Microsoft.

Lock-in effects have two implications. First, early investments and early applications are extremely important in determining which technologies – and energy resources – will be most important in the future. Second, learning and lock-in

---

[5]See Schumpeter, 1939; Marchetti and Nakićenović, 1979; Freeman, 1983; Vasko *et al.*, 1990; Nakićenović and Grübler, 1991; Grübler and Nakićenović, 1991; Grübler, 1996.

**Box 4.2: Technological progress and learning curves**

Technology costs and performance – including energy efficiency in particular – improve with experience, and there is a pattern to such improvements common to most technologies. Initially, costs are high due to batch production modes requiring highly skilled labor. Performance optimization and cost minimization are rarely important, with the overriding objective being the demonstration of technical feasibility. When the technology seeks entry into a market niche, costs begin to matter, although the technology's ability to perform a task that cannot be accomplished in any other way is usually of central importance. Examples are fuel cells in space applications, photovoltaic systems for remote and unattended electricity generation, gas turbines for military aircraft propulsion, and drill bit steering technology.

Niche markets, however, are not sufficient to reach commercialization. Improvements need to be made in reliability, durability, efficiency (*Figure 4.9*), and, most important, cost (*Figure 4.10*). Any RD&D devoted to these objectives creates a *supply push*. Complementing the *supply push*, there must be a *demand pull* by which initial markets are sufficiently expanded to further reduce costs through economies of scale. The *demand pull* may well be policy driven. Technically feasible technologies that are not yet economically competitive might benefit from environmental or energy security policies that increase their competitors' costs. For example, other electricity generation options benefit from requirements for flue gas desulfurization in coal-fired plants, or from bans on electricity generation from natural gas that restrict combined cycle gas technology.

There are steep cost improvements during the RD&D phase. For example *Figure 4.11* shows a 20% reduction in investment costs per doubling of cumulative production in the case of gas turbines. These are followed by more modest improvements after commercialization – for gas turbines, some 10% per production doubling. Improvements continue for a while at a slower pace and then cease as the technology approaches the end of its life cycle.

This pattern of diminishing costs with increasing experience is the "learning" or "experience" curve (Argote and Epple, 1990). Its specific shape depends on the technology but is a persistent characteristic of *all* successful technologies. Modular and small-scale mass-produced products usually have steeper learning curves than do large complex technologies built in the field. This is shown in *Figure 4.11* for photovoltaics, windmills, and gas turbines, all of which display similar learning-curve effects. New technologies may also benefit from economies already achieved by older technologies. The existing transmission infrastructure, for example, can be readily used by new electricity generating technologies.

make technology transfer more difficult. Successfully building and using computers, cars, and power plants depends as much on learning through hands-on experience as on design drawings and instruction manuals. And a technology that is tremendously productive when supported by complementary networks of suppliers, repairmen, training programs, and so forth, and by an infrastructure that has coevolved with the technology, will be much less effective in isolation.

Technological progress is central to all three sets of scenarios – Cases A, B, and C – and a principal conclusion of this study is that the long-term future (after 2020) will largely be determined by the technological choices made in the next few decades. In constructing the three cases, it was therefore necessary not to treat technology as static, but to incorporate anticipated future characteristics and improvements – such as performance, cost, and diffusion – of new energy technologies such as photovoltaics, hydrogen production, and fuel cells.

We took a novel approach in which we pooled all available technology data into a single data bank, from which medians and ranges could then be extracted. The data bank used for our analysis (see Messner and Strubegger, 1991; Messner and Nakićenović, 1992; Strubegger and Reitgruber, 1995) now includes some 1,600 technologies that together cover the whole energy system from primary energy extraction to end-use conversion. For example, representative investment costs for solar systems and nuclear reactors were derived from 45 and 34 independent estimates, respectively (*Figure 4.8*). Near-term technology costs assumed for the three cases were derived from the medians of the empirical cost distributions. Lower ranges defined the scope for the future cost reductions that occur at different rates in the three cases: optimistic in Cases A and C, and more cautious in Case B (see *Figures 4.10* and *6.1*).

Cases A, B, and C incorporate technological change through different learning-curve effects for various individual and generic technologies (see *Box 4.2*), reflecting different priorities for RD&D, socioeconomic development, and energy system requirements. Technology in our cases is not a free good but instead is the result of deliberate search, experimentation, and implementation, directed by both social and political policies as well as economic opportunities and profit expectations. Through technological change, upward cost pressures facing future energy systems can be mitigated. Measured either by overall energy systems or macroeconomic costs, technology improvements can yield large returns on investments. However, they are not free, requiring RD&D, entrepreneurship, and appropriate incentives and policies to progress from the initial design stage through early implementation in niche markets to full market potential and widespread diffusion.

Within each case, there are also differences between industrialized and developing regions for technologies that are manufactured domestically. For technologies that are internationally traded, costs are assumed to be uniform across regions.

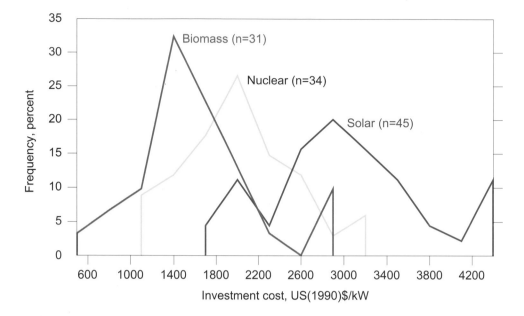

**Figure 4.8**: Range of investment cost distributions from the IIASA technology inventory for biomass, nuclear, and solar electricity generation used as input to assess costs of current and future energy systems, in US(1990)$ per kW.

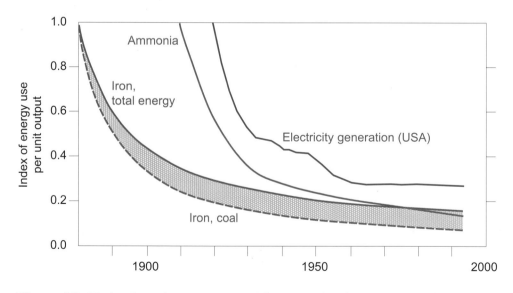

**Figure 4.9**: Technology improvements: The example of energy input for production of iron, ammonia, and electricity shown as an index with energy input at introduction equal to 1. Sources: adapted from USDOC, 1975 and consecutive volumes; Marchetti, 1978; Grübler, 1990.

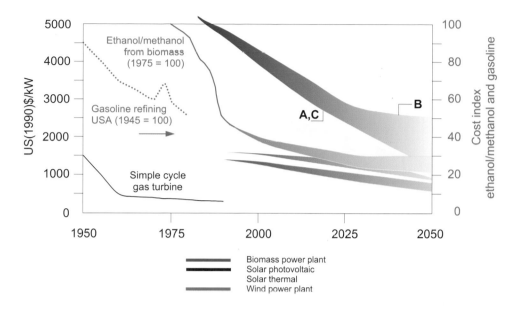

**Figure 4.10**: Technology cost improvements, in US(1990)$ per kW. Past achievements for US gasoline refining (Fisher, 1974) and Brazilian ethanol/methanol (Goldemberg, 1994) and range for biomass, solar, and wind electricity generation assumed for the three cases.

In all cases, energy options that are not technically feasible today are excluded. Nuclear fusion, for example, is excluded, while hydrogen is included as an energy carrier because it can be produced with current technologies, although not yet at competitive costs. New and emerging technologies are kept as generic as possible. Stationary fuel cells, for example, are represented as one technology: we do not distinguish between solid oxide, molten carbonate, phosphoric acid, and solid polymer fuel cells. Similarly, advanced coal-fired electricity generation is considered a single technology, with no distinction being made between integrated gasification combined cycle and pressurized fluidized bed technology. For synfuels from coal or biomass, we assume methanol as an energy vector, although in some regions ethanol may be the fuel of choice.

In Case A, there are substantial learning-curve effects for all new, and currently marginal, energy production and conversion technologies. Thus there are considerable advances in hydrocarbon exploration and extraction, renewable and nuclear electricity generation, and hydrogen and biofuel production and conversion.

For Case B, the learning-curve effects are also substantial, especially for new and environmentally desirable technologies. However, they lag on average 30% behind those in Case A, which is consistent with the less concerted RD&D efforts in Case B.

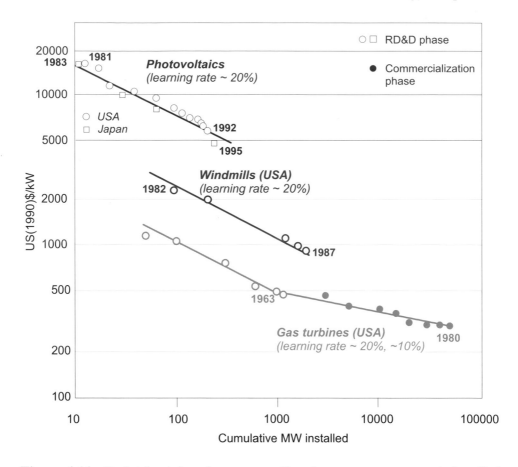

**Figure 4.11**: Technology learning curves: Cost improvements per unit installed capacity, in US(1990)\$ per kW, versus cumulative installed capacity, in MW, for photovoltaics, wind, and gas turbines. Sources: adapted from MacGregor *et al.*, 1991; Christiansson, 1995.

For Case C, learning-curve effects by design favor low-carbon fossil and renewable technologies. These technologies benefit from improvements equal to those in Case A. All other technologies develop as in Case B.

## 4.5. The Resource Base

Energy resources have obvious implications for future energy system development. In presenting the estimates of this study, we use the terminology proposed by McKelvey (1972) and Fettweis (1979). Resources are classified according to a two-dimensional matrix. One axis represents decreasing geological certainty of occurrence. The other represents decreasing economic recoverability. For energy, the term "occurrence" covers all types and forms of hydrocarbon deposits, natural uranium, and thorium in the earth's crust.

*Identified* occurrences have the highest geological assurance, followed by inferred and speculative occurrences. *Reserves* are those occurrences that are known and are recoverable with present technologies at prevailing market conditions. *Resources* are occurrences that are distinct from reserves in that they have less certain geological assurance or lack present economic feasibility, or both. Changing market conditions, innovation, and advances in geosciences can transform resources into reserves. The *resource base* is the sum of reserves and resources. It includes all potentially recoverable coal, conventional oil and natural gas, unconventional oil resources (such as oil shale, tar sands, and heavy crude), and unconventional natural gas resources (such as gas in Devonian shale, tight sand formations, geopressured aquifers, and coal seams). Quantities that are not considered potentially recoverable are classified as "additional occurrences" and are excluded from the resource base of future, potentially recoverable resources. Occurrences include methane hydrates and natural uranium dissolved in seawater. Both are known to exist in enormous quantities, but there is great uncertainty about their extent and the eventual technology and economics of their recovery.

Typically, unconventional sources differ from conventional sources by one or more of the following characteristics:

- They occur in significantly lower concentrations.

- They require unusual or extreme technological prerequisites for their recovery.

- They need complex and capital-intensive conversion for modern-day use.

- They have significant environmental implications.

Given the large quantities of conventional oil and gas reserves and resources, the actual resource dimensions of unconventional oil and gas (not to mention additional occurrences) have received relatively little attention.

*Box 4.3* summarizes recent estimates of reserves and recoverable resources of oil, gas, and coal as reported by IIASA (Rogner, 1997), IPCC (Nakićenović *et al.*, 1996), the US Geological Survey (Masters *et al.*, 1994), and WEC (1992, 1993, 1998). We also give a summary of the global nonrenewable energy resource base and compare cumulative historical and 1990 consumption levels of fossil energy and natural uranium with estimates of reserves, resources, and additional occurrences. Driven by economics, technological and scientific advances, and policy decisions, this resource base has expanded over time, and reserves have been continuously replenished from resources and from new discoveries (see Masters *et al.*,

**Box 4.3: Estimates of energy reserves, resources, and occurrences**

**Table 4.4:** Global fossil and nuclear energy reserves, resources, and occurrences, in Gtoe.

| | Consumption[a] | | Reserves[b] | Resources[c] | Resource base[d] | Additional occurrences |
|---|---|---|---|---|---|---|
| | 1850 to 1990 | 1990 | | | | |
| Oil | | | | | | |
| Conventional | 90 | 3.2 | 150 | 145 | 295 | |
| Unconventional | – | – | 193 | 332 | 525 | 1,900 |
| Natural gas | | | | | | |
| Conventional[e] | 41 | 1.7 | 141 | 279 | 420 | |
| Unconventional | – | – | 192 | 258 | 450 | 400 |
| Hydrates[f] | – | – | – | – | – | 18,700 |
| Coal[g] | 125 | 2.2 | 606 | 2,794 | 3,400 | 3,000 |
| Total[h] | 256 | 7.0 | 1,282 | 3,808 | 5,090 | 24,000 |
| Uranium[i] | 17 | 0.5 | 57 | 203 | 260 | 150 |
| in FBRs[j] | – | – | 3,390 | 12,150 | 15,540 | 8,900 |

Note: – negligible amounts; blanks, data not available.
[a] Grübler and Nakićenović, 1992.
[b] Masters *et al.*, 1994; IPCC, 1996b; OECD/NEA and IAEA, 1995; WEC, 1993.
[c] Resources to be discovered or developed to reserves. Masters *et al.*, 1994 (upper range); IPCC, 1995; OECD/NEA and IAEA, 1995.
[d] Resource base is the sum of reserves and resources.
[e] Includes natural gas liquids.
[f] MacDonald, 1990; Kvenvolden, 1988, 1993.
[g] WEC, 1993.
[h] All totals have been rounded.
[i] OECD/NEA and IAEA, 1995.
[j] FBRs = fast breeder reactors.
Sources: Rogner, 1997; Nakićenović *et al.*, 1996; Masters *et al.*, 1994; Nakićenović *et al.*, 1993; WEC, 1992; Grübler, 1991; MacDonald, 1990; Rogner, 1990; BP, 1995 and earlier volumes; BGR, 1989; Delahaye and Grenon, 1983.

**Table 4.5:** WEC estimates of reserves and ultimately recoverable resources, in Gtoe and 1,000 tons uranium (tU).

| | Proved reserves | Ultimately recoverable |
|---|---|---|
| Conventional oil | 150 | 200 |
| Unconventional oil | – | 511–595 |
| Conventional gas | 133 | 220 |
| Coal and lignite | 430 | 3,400 |
| Total | 713 | 4,331–4,415 |
| Uranium in 1,000 tU | 3,400 | 17,000 |

Sources: WEC, 1993, 1998.

**Table 4.6:** US Geological Survey estimates of reserves of conventional oil and gas, in Gtoe.

| | Identified reserves | Future discoveries with probability of | | |
|---|---|---|---|---|
| | | 95% | 50% | 5% |
| Conventional oil | 150 | 36 | 74 | 145 |
| Conventional gas | 129 | 68 | 126 | 265 |
| NGL[a] | 12 | – | 14 | – |
| Total | 291 | 104 | 214 | 410 |

Source: Masters *et al.*, 1994.
[a] NGL = Natural gas liquids.

1991, 1994). *Figure 4.12* shows the history of oil and gas reserves, for example, and increases in reserve-to-production ratios in the past four decades. Reserve additions have shifted to inherently more challenging and potentially costlier frontier locations, with technological progress outbalancing potentially diminishing returns. However, with reserve-to-production ratios above 40 years, there is little economic incentive for private sector industries to explore and vigorously develop more reserves.

We expect the resource base to continue to expand in the future. But no one can anticipate how market conditions, knowledge, exploration methods, and production technologies will continue to change. Furthermore, we cannot continue to express complex concepts such as the resource base by simple measures or single numbers. The inherent difficulty is well expressed by Adelman (1992), who describes energy resource assessment as an effort "in estimating the potentially economic portion of an unknown total."

In this study, the "resource base" column in *Table 4.4* in *Box 4.3* is taken as the availability range of nonrenewable energy sources. The "reserves" column represents the fraction of the resource base that is recoverable today, that is, can be produced with current technology under present market conditions. The resource base in turn includes those resource quantities that, with technical progress, could become economically attractive within the study horizon and that could become potentially available over the long term (i.e., beyond 2020). Because of the uncertainty about the ultimate potential of the fossil resource base, its availability is varied across the cases, and the scenarios vary from conservative (Case C), or cautious (Case B and Scenario A2), to optimistic (Scenarios A1 and A3). The "additional occurrences" listed in the far right column are not included in any of the three cases or their scenarios. They do indicate, however, that if resource exhaustion were the only criterion, the world would be unlikely to run out of nonrenewable energy sources any time before the 22nd century.

The fossil reserves as listed in *Table 4.4* amount to 1,300 Gtoe, and the fossil resource base (including reserves) is estimated at approximately 5,000 Gtoe. That is certainly sufficient for more than 100 years, even in the high-energy Case A scenarios. Thus, while the next 100 years may bring temporary and structural energy shortages, there is no expectation that we will be limited by absolute resource constraints. However, the consumption of such large quantities of fossil energy represents a considerable technological and investment challenge and would translate into cumulative carbon emissions corresponding to six to seven times the current atmospheric carbon content of about 760 GtC (or $CO_2$ concentration of 358 parts per million by volume, ppmv). Therefore, concerns about economic recoverability and the environment are more likely to cause a future shift to other sources of energy than are absolute resource constraints.

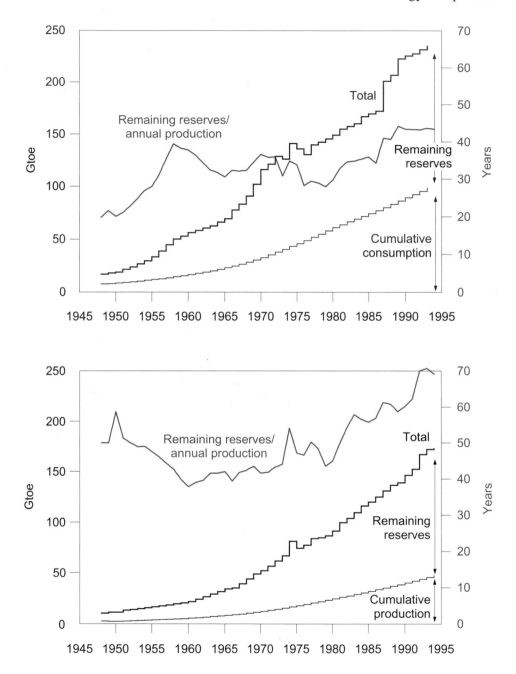

**Figure 4.12**: Technically and economically recoverable reserves and cumulative production of (*top*) conventional oil and (*bottom*) natural gas, in Gtoe. The increase in the reserve base despite growing production (i.e., the continuous replenishment of reserves from resources) is reflected in the stable or increasing reserve-to-production ratios shown in the figure.

*Box 4.3* also includes figures for uranium. As shown at the bottom of *Table 4.4*, the energy effectiveness of uranium depends greatly on the conversion technology used, but the resources are potentially extremely large. Nuclear energy's future will depend on, among other things, how current controversies concerning safety, waste disposal, and proliferation are resolved. The successful development of new technologies that resolve these issues is likely to play a much more important role in the future of nuclear power than will resource constraints.

The resource situation for renewable energies is quite different from that of fossil energy or uranium. Rather than being a (finite, albeit large) stock variable, the resources of renewables are characterized by huge – yet dispersed and diffuse – annual flows of energy available in the environment. With few exceptions (like some hydropower sites), renewable resources offer *energy densities* orders of magnitude lower than those of deposits of fossil fuels in the form of coal mines or oil fields. As an extreme illustration, the peak power requirement in Manhattan is of the order of 1.5 kW per m$^2$, with an average of 0.2 kW or 1,750 kilowatt hours (kWh) per m$^2$ per year (Bowman, 1995), which is up to 10 times larger than the mean direct solar radiation onto New York City (0.15 kW per m$^2$; WEC, 1994). Not unlike fossil energy sources, therefore, "new" renewables also require elaborate systems of energy conversion, transport, and distribution to "bridge" spatially separated areas of energy supply and demand. The limitations of renewables are therefore not constraints imposed by the magnitude of the energy flows available in the environment, which are indeed gigantic, but rather how these diffuse flows can be harnessed and converted to fuels required to provide energy services.

Renewable resources are characterized by three major advantages. First, they can provide energy services on an indefinite basis given that their use does not fundamentally disturb natural energy flows. Second, they can do so without major alterations of global geochemical cycles, being either carbon free (solar or hydropower) or carbon neutral (sustainable use of biomass). Finally, they can provide energy for numerous generations to come, and – at least in principle – for any level of future energy demand, as the natural energy flows are indeed large by any standards. The earth intercepts about 130,000 Gtoe of solar energy annually compared with 9 Gtoe total global energy consumption in 1990. *Figure 4.13* provides a schematic illustration of annual global energy flows without anthropogenic disturbances.

Clearly, only a fraction of these flows could be practically used for energy purposes. This in itself is not a problem, as all conceivable human energy needs could be provided for by diverting only a small fraction of the solar influx to energy use. Renewable potentials are limited by numerous factors such as mismatch between power densities and locations of energy supply and demand, possible land-use conflicts, adverse local environmental impacts, capital costs, or infrastructure

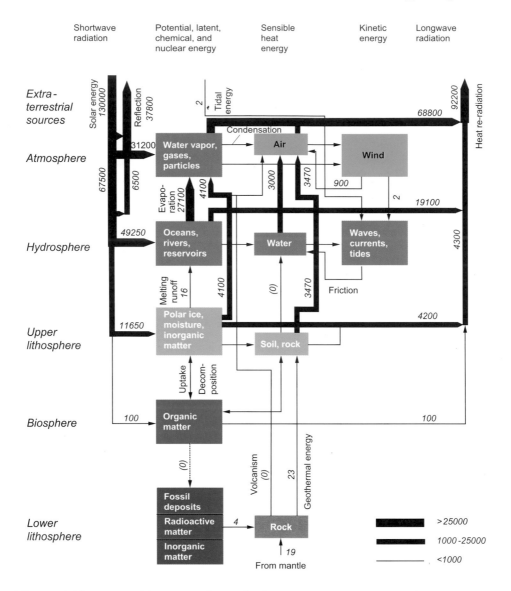

**Figure 4.13**: Renewable energy flows, in Gtoe. Source: Sørensen, 1979.

requirements. The renewable potentials that could be realized with current technologies and economics can be compared with fossil energy reserves, while the maximum technical potentials that could be developed under favorable future technological and economic conditions can be compared with the concept of the fossil "resource base."

The IPCC in its Second Assessment Report identifies global renewable potentials by the 2020s in the range of 3.2 to 5.6 Gtoe per year and the long-term technical potentials beyond 2020 at up to 100 Gtoe per year (Nakićenović *et al.*, 1996). Other estimates include the WEC *New Renewable Energy Resources* (WEC, 1994) report, which gives a range of renewable energy potentials reaching 3.3 Gtoe per year by 2020, and *Energy for Tomorrow's World* (WEC, 1993), which identifies renewable energy potentials (in the Epilogue) of up to 13 Gtoe by the year 2100, of which 10 Gtoe could be supplied by "new" renewables. These so-called new renewables do not include the traditional renewable sources – such as large hydropower – or the traditional uses of fuelwood and agricultural wastes.

Thus, renewable energy sources have the promise of meeting most human energy needs in the long term, but their actual contribution is likely to be much more modest in the shorter term. There is obviously a wide range of uncertainty on the possible timing and extent of the diffusion of renewables. Only with major, effective, and internationally coordinated policy support could developments be accelerated and significant inroads into total primary energy supply be made in the next two to three decades. Conversely, over more distant time horizons, the potentials for renewables increase significantly. Technology improvements can ultimately bring the cost of renewable energies down to competitive levels, whereas the cost of traditional (fossil) energies is likely to increase in the long term due to both the exhaustion of their technology improvement potentials and the eventual shift to lower-grade and higher-cost resources.

## 4.6.    Environment

Concerns about the environmental impacts of energy production and consumption are not new. Complaints about air pollution from burning "sea coal" date back to the 13th century in England. With the advent of industrialization and its resulting concentration of energy consumption in urban and industrial areas air and water pollution began to exceed the assimilative capacities of local environments on a large scale and became major issues. For developing countries, air pollution from industrialization has now joined traditional indoor air pollution as a major challenge. At the same time, environmental concerns in the industrialized countries have shifted more toward very long-term and global issues, most prominently climate change.

The type and extent of pollution are closely related to the degree of economic development and industrialization, as is illustrated in *Figure 4.14*. To some extent, economic development enables societies to successfully address environmental problems of both poverty (such as inadequate sanitation or indoor air pollution from traditional biomass use) and industrialization (such as sulfur emissions). But economic development and affluence can also generate new environmental problems, such as waste disposal and possible global warming.

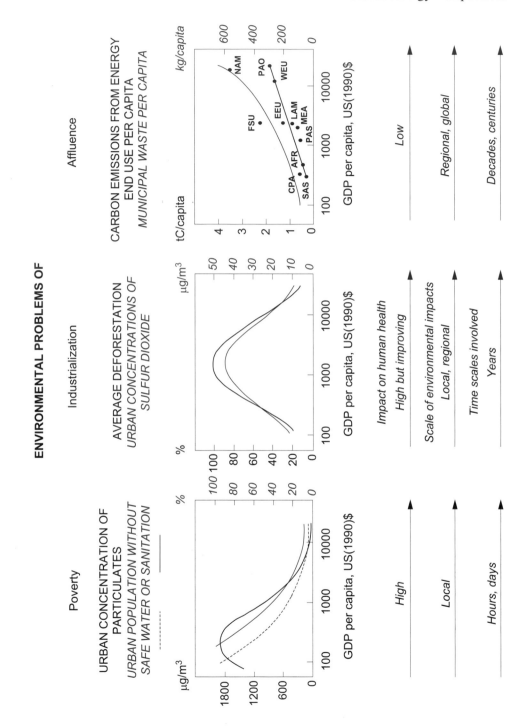

**Figure 4.14**: A typology of environmental problems as they evolve with economic development. Source: adapted from World Bank, 1992.

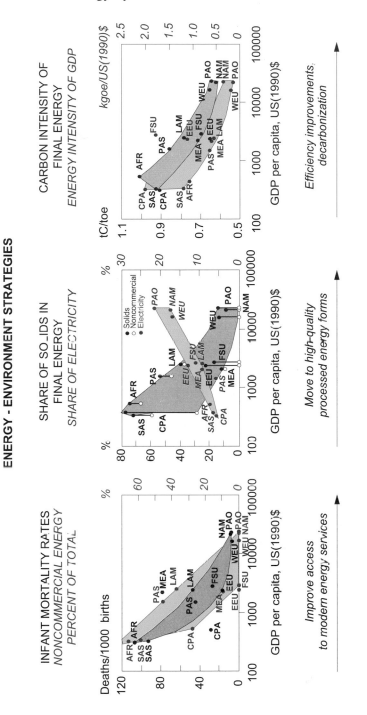

**Figure 4.15**: Energy strategies to address environmental problems of poverty, industrialization, and affluence. Improvements in energy efficiency and structural change in energy systems toward cleaner energy carriers ("decarbonization") are important generic long-term strategies to tackle environmental problems.

Local impacts, including indoor air pollution, have been and continue to be a major environmental concern related to energy use. Indoor and outdoor air quality were important in Victorian London and are equally important for many megacities of the developing world today. The cause is basically the same now as then: emission-intensive fuels, such as coal and wood, burned in inefficient open fireplaces and cookstoves. In industrialized countries, these problems have been successfully resolved in three ways: first, by the use of more efficient end-use devices (e.g., central-heating systems and fuel-efficient stoves); second, by more complete combustion within these devices; and third, by the shift to clean (often grid-dependent) fuels such as electricity and gas (see *Figure 4.15*). Compared with the historical reductions in environmental impacts through these sorts of changes in efficiencies and fuels, "end-of-pipe" environmental solutions applied to large point sources are recent phenomena.

Despite major reductions in local air pollution problems in the industrialized world, these problems remain acute in the developing world. More than one billion people in developing world cities are exposed to unacceptably high ambient concentrations of suspended particulate matter and sulfur dioxide ($SO_2$), significantly exceeding World Health Organization (WHO) guidelines (World Bank, 1992). The situation with transport-related air pollution is also dramatic in many cities of the developing world. It is estimated that people in the poorest countries are exposed to air pollution levels orders of magnitude higher than those in the industrialized countries, despite the fact that the latter are using orders of magnitude more energy. Due to cooking on open wood fireplaces, indoor air pollution in poor rural areas is more than 20 times higher than in industrialized countries (Smith, 1993). Particulate concentrations in urban areas of the developing world can be more than five times those in OECD cities (WHO, 1992). That more time is spent outdoors because of the climate and inadequate housing conditions only serves to compound the problem. Consequently, pollution exposure both indoors and in urban areas can be up to a factor of 20 higher in developing countries than in industrialized ones (Smith, 1993).

Local environmental problems should thus have first priority. As in the past, the most effective solutions will be those that are most comprehensive, relying on a mix of efficiency improvements, cleaner fuels, and "end-of-pipe" control technologies. Both efficiency improvements and the shift to cleaner fuels will yield triple dividends: lower resource use, lower overall energy system costs, and lower emissions. Past structural changes in the energy system have moved in exactly these directions, although not yet quickly enough to offset the vast expansion of human activities. This is illustrated in *Figure 4.16* where the carbon intensity of primary energy over time is taken as an indicator of other pollutants. (Sulfur and, in many cases, $NO_x$ emissions are generally higher for high-carbon fuels such as

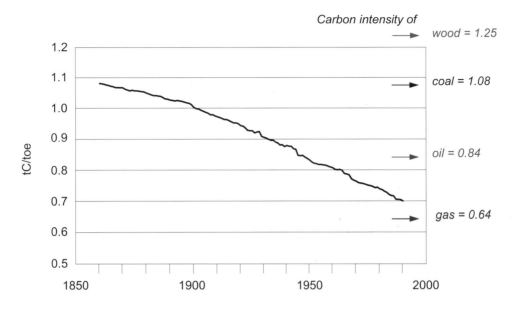

**Figure 4.16**: Carbon intensity of world primary energy mix, 1850 to 1990, in tC per toe, including emissions from unsustainable uses of fuelwood. Source: Nakićenović *et al.*, 1996.

coal.) Although the decarbonization of the world's energy system shown in *Figure 4.16* is comparatively slow at 0.3% per year, the trend is consistent and in the right direction. Measured in terms of final energy, progress has been quicker due to the increasing use of clean, grid-dependent energy carriers such as electricity, district heat, and gas.

The next chapter addresses the question of whether the trend of *Figure 4.16* can continue. The answer is "yes." Indeed, we expect that all possible energy futures covered by our three cases are capable of maintaining favorable decarbonization trends for final energy and, with the exception of Scenario A2, also for primary energy.

# Chapter 5

# Energy System Alternatives: Six Scenarios

This chapter uses many illustrations to present the key quantitative features of the energy systems that evolve in each of the six scenarios: the three Case A scenarios, the single Case B scenario, and the two Case C scenarios. The emphasis is on results supporting the central conclusion of this book, namely, that the historical trend toward ever more flexible, more convenient, and cleaner energy reaching the consumer can continue in all the cases spanned by this study – ranging across a variety of supply options, and from high to low economic growth, from aggressive to modest technological progress, and from strong to weak environmental constraints. Implications of this conclusion for financing, trade, costs, technology, the environment, and each of the major energy supply industries are elaborated in Chapter 6. However, the combination of sound and effective policy measures together with market-driven technology and price initiatives will be required to hasten the move to desired trajectories and goals.

Cases A, B, and C all reinforce the conclusion at the global and regional levels alike, that up to 2020 energy use will rise and the world will have to rely largely on fossil fuels. The purpose of this study was to take the next step, to look to 2050 and beyond in greater detail. The question was whether more options would open up after 2020 and what they might be. We initially planned on three cases to ensure that a broad range of energy perspectives would be examined. We found, however, that more possibilities needed to be considered than originally anticipated, and our three cases blossomed into the six scenarios introduced in Chapter 2.

The increase from three cases to six scenarios reflects the reality that new opportunities emerge in the long run. There is more time to develop new energy technologies, energy resources, and financial institutions, and there is more time for capital turnover and fundamental change in the energy system. By 2020, many

## Box 5.1: Defining energy systems: From primary energy to end use

This study departs from the traditional analytic separation of energy supply and demand. It treats the energy system holistically, including all stages from primary energy to the provision of energy services (see *Figure 5.1*). Energy is demanded and delivered to the points of consumption to provide energy services. It is energy services that are needed and not energy *per se*.

*Primary energy* is the energy recovered or gathered directly from nature, for example, mined coal, produced crude oil or natural gas, collected biomass, harnessed hydropower, solar energy absorbed by collectors, and heat produced in nuclear reactors. Primary energy generally is not used directly. Commercial energy is converted in refineries, power plants, and other such facilities into *secondary energy*, in the form of fuels and electricity. Such fuels and electricity can be used in a much wider range of applications, and with greater ease, than can primary energy. Secondary energy is then transported, distributed, and delivered to the point of consumption as *final energy*. Final energy is used in appliances, machines, and vehicles and is transformed to *useful energy* such as work and heat. Application of useful energy to a given task in the form of a running car engine or luminosity of a light bulb results in *energy services*. The ultimate purpose of the energy system is to provide *energy services* such as mobility (e.g., expressed in passenger-km) or illumination.

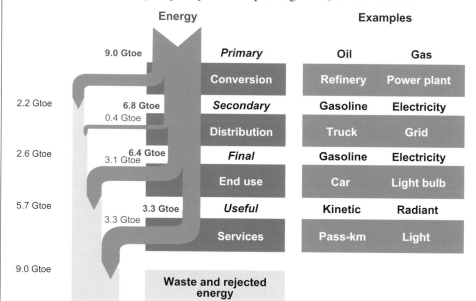

**Figure 5.1:** Global energy flows in 1990, in Gtoe. Examples are given of different stages of energy conversion, transport, distribution, and end use corresponding to the main energy flows.

Energy systems become increasingly complex as more sophisticated conversion processes are introduced in the future. From the consumer's perspective, the important issues are the flexibility, convenience, cleanliness, and cost of energy services. It is less important what the fuel or the source of that fuel is. Already electricity consumers, for example, are unaware of the actual source of their electricity. The tendency continues in all the scenarios here as more and more energy services become independent of upstream activities in the energy system.

current energy end-use devices will have been replaced by those just being introduced for commercial use today – such as new vehicles, new industrial processes,
and new heating systems. Many power plants will also have been replaced, and
others will be nearing the end of their useful lifetimes. The time horizon to 2050
and beyond means that all energy technologies and devices, and some energy infrastructures as well, are likely to have been replaced at least twice. Such turnover
creates the new supply and end-use opportunities reflected in the results presented
in this chapter, which covers the period to 2050 and beyond.

While the dynamics of change and structure of the energy system vary across
the six scenarios, the patterns of energy end use show convergent developments.
The energy system is service driven, from consumer to producer (i.e., bottom-up;
see *Box 5.1*, which also defines energy system terminology). In contrast, energy
flows are resource and conversion process driven, from producer to consumer (i.e.,
top-down). The interaction between bottom-up and top-down forces occurs at the
level of final energy. Therefore, the energy supply part of the energy system, the
"energy sector," should never be analyzed in isolation. The assessment must account for how and for what purposes energy is used in addition to considering how
energy is supplied. The dichotomy between supply and demand has been replaced
in the scenarios by a holistic approach that integrates driving forces of energy demand with energy supply and use. To facilitate the description of changes in the
structure of future energy systems and patterns of energy use, scenarios are described in terms of both primary and final energy use. Primary energy depicts the
structure of energy extraction and conversion while final energy shows the structure
of energy end use. We start with primary energy.

## 5.1.    Primary Energy in Three Cases

The six scenarios include three Case A scenarios, a single Case B, and two Case C
scenarios. *Figure 5.2* shows total global primary energy requirements for all three
cases together with population growth. Across the three cases, the primary energy
needs increase up to threefold by 2050 and up to fivefold by 2100. Case A portrays
primary energy growth rates approaching those of the historical experience since
1850. In the other two cases growth rates are substantially lower. In particular,
Case C represents a radical change, with an emphasis on energy efficiency and
conservation that results in the decoupling of energy and economic growth. The
three Case A scenarios have almost identical primary energy requirements so that
we need not distinguish among them; the same is true for the two Case C scenarios.
The variations are less than 1.5 Gtoe for primary energy in the Case A scenarios
and less than 0.1 Gtoe for the Case C scenarios.

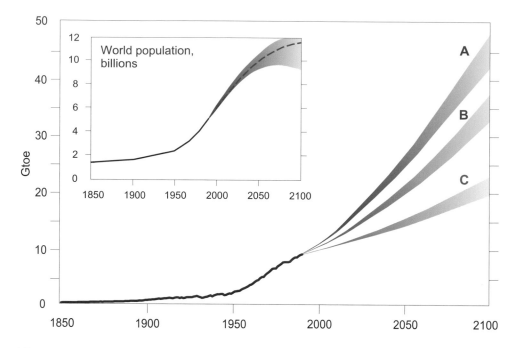

**Figure 5.2**: Global primary energy use, historical development from 1850 to 1990 and in the three cases to 2100, in Gtoe. The insert shows global population growth, 1850 to 1990 and projections to 2100, in billion people. Source: Bos *et al.*, 1992.

The increase in primary energy requirements in the developing regions is especially large. Although most of the current primary energy use is in the already industrialized North, in all scenarios it moves gradually to the industrializing South. Consequently, the most challenging undertaking will be to meet this enormous increase in energy needs and services in the industrializing parts of the world. Energy needs in the North increase modestly in Case A, grow marginally in Case B, and actually decline in Case C.

*Figure 5.3* illustrates the same global developments in a different way. It shows population and per capita primary energy use for the 11 world regions. In this figure, the width of the rectangles corresponds to population, and the height to per capita primary energy. Thus the area of the rectangles corresponds to total primary energy use. For each region, there are three overlapping rectangles. The foremost shows the situation in 1990, the next the situation in 2020, and the third the situation in 2050.

There is a large difference in the structure of primary energy use among the regions. For 1990, regions in the North show a significant degree of diversification among primary energy sources with a strong reliance on oil and gas, while regions in the South have comparatively low per capita consumption levels consisting

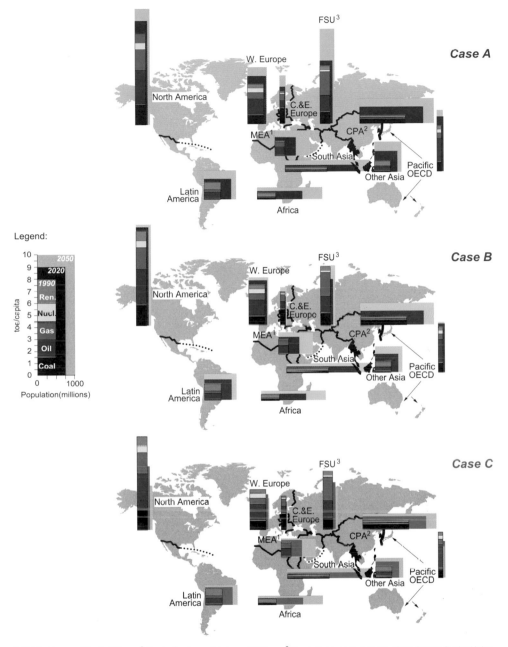

<sup>1</sup> Middle East and North Africa   <sup>2</sup> Centrally planned Asia and China   <sup>3</sup> Newly independent states of the former Soviet Union

**Figure 5.3**: Per capita primary energy use and population by world region in 1990 and for the three cases in 2020 and 2050, in toe per capita and million people. Height of the bars equals per capita primary energy, width equals population, and area equals total primary energy use.

mostly of coal and biomass. In the future this picture changes to a more balanced distribution of primary energy use among the world regions.

For the South, the figure illustrates the importance of population growth and the growth of per capita consumption in all cases. For the North, population growth is of minimal importance and per capita energy use ranges between a substantial increase in Case A and a substantial decrease in Case C.

## 5.2.    Overview of Six Scenarios

Where all six scenarios diverge is in the dynamics of energy system transformation as reflected in the contributions of individual primary energy sources – what percentage is supplied by coal, what percentage by oil, and so forth. That divergence is shown first in *Table 5.1*, which summarizes key numerical characteristics for all six scenarios. It presents a snapshot of how the scenarios will look in 2050 (see also Appendix C). *Figure 5.4* summarizes the changing structure of primary energy use across the six scenarios and compares it with historical developments. *Figures 5.5* to *5.7* then present the development over time of the structure of primary energy shares for each of the six scenarios separately.

Assumptions concerning the most salient forces driving and shaping future energy systems are varied parametrically across the scenarios to explore both differences and commonalities of alternative future energy systems. The scenarios vary in terms of four important clusters of influencing forces: with respect to future technologies in terms of penetration rates, performance, and costs; with respect to availability of energy sources, a question also closely related to technology; with respect to the geopolitical and policy issues such as trade, technology transfer, environmental regulation, and energy deregulation; and with respect to energy end-use patterns and services despite a convergence across the scenarios in the structure of final energy forms.

The high-growth Case A consists of three scenarios (A1, A2, and A3), Case B of a single scenario, and the ecologically driven Case C of two scenarios (C1 and C2). This range reflects the possibility of alternative development strategies with comparable levels of affluence and energy use. The Case A alternatives indicate that high levels of energy demand could be supplied by three basically different strategies. The differences between the two alternatives considered in Case C are less dramatic. For the intermediate Case B a single scenario was developed. In general, it is associated with more modest, perhaps also more realistic, changes so that it did not appear to be useful to consider more extreme alternatives for the development of the energy system.

*Figure 5.4* illustrates the long-term divergence in the structure of energy systems across the scenarios. Each corner of the triangle corresponds to a hypothetical

**Table 5.1**: Characteristics of the three cases for the world in 2050 compared with 1990.

| | Base year 1990 | Case (Scenario) in 2050 | | | | | |
|---|---|---|---|---|---|---|---|
| | | A | | | B | C | |
| | | (A1) | (A2) | (A3) | | (C1) | (C2) |
| Primary energy, Gtoe | 9 | 25 | 25 | 25 | 20 | 14 | 14 |
| *Primary energy mix, percent* | | | | | | | |
| Coal | 24 | 15 | 32 | 9 | 21 | 11 | 10 |
| Oil | 34 | 32 | 19 | 18 | 20 | 19 | 18 |
| Gas | 19 | 19 | 22 | 32 | 23 | 27 | 23 |
| Nuclear | 5 | 12 | 4 | 11 | 14 | 4 | 12 |
| Renewables | 18 | 22 | 23 | 30 | 22 | 39 | 37 |
| *Resource use 1990 to 2050, Gtoe* | | | | | | | |
| Coal | | 206 | 273 | 158 | 194 | 125 | 123 |
| Oil | | 297 | 261 | 245 | 220 | 180 | 180 |
| Gas | | 211 | 211 | 253 | 196 | 181 | 171 |
| Energy sector investment, trillion US$ | 0.2 | 0.8 | 1.2 | 0.9 | 0.8 | 0.5 | 0.5 |
| US$/toe supplied | 27 | 33 | 47 | 36 | 40 | 36 | 37 |
| As a percentage of GWP | 1.2 | 0.8 | 1.1 | 0.9 | 1.1 | 0.7 | 0.7 |
| Final energy, Gtoe | 6 | 17 | 17 | 17 | 14 | 10 | 10 |
| *Final energy mix, percent* | | | | | | | |
| Solids | 30 | 16 | 19 | 18 | 23 | 20 | 20 |
| Liquids | 39 | 42 | 36 | 33 | 33 | 34 | 34 |
| Electricity | 13 | 17 | 18 | 18 | 17 | 18 | 17 |
| Other[a] | 18 | 25 | 27 | 31 | 28 | 29 | 29 |
| *Emissions* | | | | | | | |
| Sulfur, MtS | 59 | 54 | 64 | 45 | 55 | 22 | 22 |
| Net carbon, GtC[b] | 6 | 12 | 15 | 9 | 10 | 5 | 5 |

Note: Subtotals may not add due to independent rounding.
[a] District heat, gas, and hydrogen.
[b] Net carbon emissions do not include feedstocks and other non-energy emissions or $CO_2$ used for enhanced oil recovery.

situation in which all primary energy is supplied by a single source: oil and gas at the top, coal at the left, and non-fossil sources (renewables and nuclear) at the right. In 1990 their respective shares were 53% for oil and gas (measured against the magenta grid lines with percentages shown on the right), 24% for coal (measured against the black grid lines with percentages on the left), and 23% for non-fossil energy sources (measured against the green grid lines with percentages at the bottom). Historically, the primary energy structure has evolved clockwise according to the two "grand transitions" discussed in Chapter 3. As shown by the black

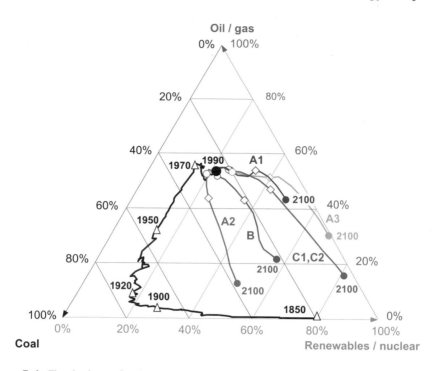

**Figure 5.4**: Evolution of primary energy structure, shares of oil and gas, coal, and non-fossil sources, in percent, historical development from 1850 to 1990 (triangles) and in scenarios to 2020 (open circles), 2050 (diamonds), and 2100 (closed circles). For an explanation of the figure see text.

line in *Figure 5.4*, traditional renewables were replaced by coal between 1850 and 1920. Coal reached its maximum market share shortly before 1920 and was then progressively replaced by oil and natural gas between 1920 and 1970. Since then, structural change in the global primary energy mix has been comparatively modest.

Because of the long lifetimes of power plants, refineries, and other energy investments, there is not enough capital stock turnover in the scenarios prior to 2020 to allow them to diverge significantly. But the seeds of the post–2020 divergence in the structure of energy systems will have been widely sown by then based on RD&D efforts, intervening investments, and technology diffusion strategies. It is these decisions between now and 2020 that will determine which of the diverging post–2020 development paths will materialize.

After 2020 all scenarios move away from their current reliance on conventional oil and gas. This transition progresses relatively slowly in Scenario A1 where oil and gas are plentiful. In Scenario A3 and Case C, it progresses more rapidly due to faster technological progress (Scenario A3) or because of energy and environmental policies favoring non-fossil fuels (Case C). In these cases the global energy system comes almost full circle by the end of the 21st century. As in 1850 at the beginning of the Industrial Revolution, it relies predominantly on non-fossil energy forms, but these are high-technology renewables and nuclear power rather than the traditional biomass fuels used 250 years earlier. In Scenario A2 and Case B, the transition away from oil and gas includes an important contribution from coal, whose long-term market share after 2050 ranges between 20 and 40%. Nonetheless, little of this coal is used directly. It is instead converted to the high-quality energy carriers (electricity, liquids, and gases) demanded by the high-income consumers of the second half of the 21st century.

The following sections discuss each of the scenarios in more detail, as shown in *Figures 5.5* to *5.7*.

### 5.2.1.  Case A: High growth

Case A is characterized by enormous productivity increases and wealth. It is technology and resources intensive. Technological and economic change, as reflected in rapid capital stock turnover, yields substantial improvements in energy intensities and efficiency. Technological change also provides greater access to energy sources, either by replenishing fossil reserves from the large resource base (see *Box 4.3*) or by developing new supply systems from non-carbon sources capable of meeting rapidly growing energy requirements. This is of particular importance in developing countries where energy needs increase rapidly despite impressive improvements in energy intensity. Case A is three-pronged, recognizing that technological change could develop along alternative "trajectories" combining clean energy carriers and various supply sources into an orderly transition to the post-fossil-fuel age. This unfolding into three different development trajectories results in three scenarios with almost identical energy end-use patterns but different energy system structures. They represent alternative combinations of technological change and energy requirements.

Scenario A1 is challenging in that it goes beyond the conventional wisdom on the availability of oil and gas after 2020. Technological change enables the vast potential of conventional and much of unconventional oil and gas resources to be used at competitive costs and without significant efficiency or environmental penalties. Fossil energy sources become phased out only toward the end of the century. There is no need to resort to exotic or speculative oil and gas occurrences or other "backstop" resources such as shales or low-quality coals.

Scenario A2 provides a contrasting and more conservative strategy with respect to both technological change and resource availability. Technological progress is more gradual, evolving along the lines of current supply technologies and the most abundant energy resources. Consequently, the development strategy is coal intensive. Large-scale and costly development programs are needed for new energy

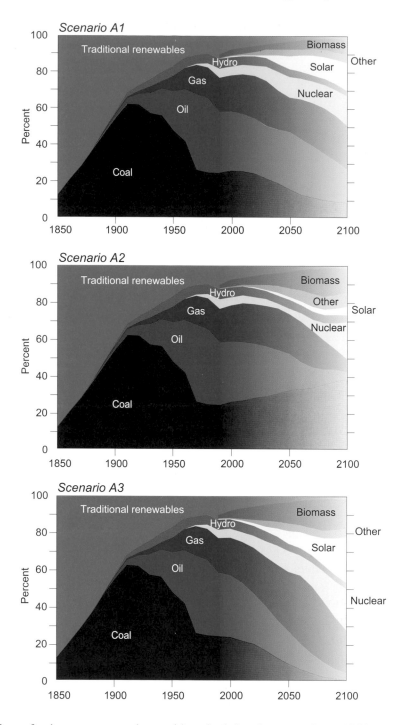

**Figure 5.5**: Evolution of primary energy shares, historical development from 1850 to 1990 and in scenarios to 2100 for Case A: The oil- and gas-intensive Scenario A1, the coal-intensive Scenario A2, and the "bio-nuc" Scenario A3.

supply lines, conversion industries, and the changing regional balance of energy supply and trade. Significant technological challenges emerge with the progressive depletion of open-cast coal deposits. Improved technology is needed to resolve the increasing costs of coal production, to reduce human risks from ever deeper coal mines, and to introduce advanced conversion of coal to methanol and other synliquids.

Scenario A3 is also "technology intensive" but with a different direction compared with the other two variants. New renewables and new nuclear combine to form a "bio-nuc" technology cluster that permits the transition to a post-fossil-fuel age along market penetration dynamics similar to those by which fossil fuels phased out traditional energy forms over the course of the 19th century. Natural gas provides the transitional fossil fuel of choice. This strategy leads to a significant degree of decarbonization of the global energy system. Global net carbon emissions amount to 9 GtC by 2050 compared with 6 GtC today, while primary energy use is nearly three times higher than at present (25 Gtoe versus 9 Gtoe). Scenario A3 is an illustration of a case in which a "rich and clean" energy future resolves some of the challenges of global warming without recourse to stringent environmental policy measures.

### 5.2.2. Case B: Middle course

Case B, with a single scenario, is based on a more cautious approach regarding economic growth prospects, rates of technological change, and energy availability. In short, the scenario is perhaps best characterized by "modest dynamics" and derives its appeal primarily because of its "pragmatic" attitude. It might also have a higher probability of occurrence than the more challenging technology- and resource-intensive Case A scenarios or the policy-intensive Case C scenarios. Overall, the Case B scenario is "reachable" without relying on drastic changes in current institutions, technologies, and perceptions of the availability of fossil fuel resources. Case B's lower energy demand implies that it can rely on fossil fuel resources to an extent that is commensurate with current estimates of ultimately recoverable oil and gas reserves. Energy supply and end-use patterns are also closer to the current situation for a longer period in Case B than in Cases A and C.

In the very long term, however, the changes become increasingly dramatic in Case B. Possible resource scarcities may need to be counterbalanced either by new fossil fuel discoveries or by the development of new energy sources. Both oil and gas are still important sources by 2100, with coal, nuclear, and new renewables at about equal shares. Nuclear and renewables eventually replace fossil energy sources. Overall, even with comparatively modest expectations of technological change, an orderly transition away from fossil fuel use is not only feasible but also appears manageable in terms of energy sector and institutional adjustments

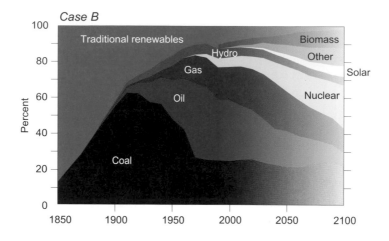

**Figure 5.6**: Evolution of primary energy shares, historical development from 1850 to 1990 and in scenario to 2100, for Case B.

extending toward the end of the 21st century. Resource constraints in Case B do not materialize because of geological scarcity of oil and gas, but rather arise due to the financing and environmental constraints of moving into progressively more remote, deeper, and dirtier fossil resources.

### 5.2.3.   Case C: Ecologically driven

The two Case C scenarios present challenging global perspectives. Ambitious policy measures accelerate energy efficiency improvements and develop and promote environmentally benign, decentralized energy technologies. In addition to vigorous control of local and regional pollutants, a global regime to control the emissions of greenhouse gases is established. The goal is to reduce carbon emission levels to 2 GtC by 2100, corresponding to one-third of current emissions or the decrease indicated in the IPCC Second Assessment Report as necessary to stabilize the atmospheric $CO_2$ concentration (IPCC, 1996a) and mitigate against undesirable climate change impacts (IPCC, 1996c).

Case C describes a challenging pathway of transition away from the current dominance of fossil sources to a dominance of renewable energy flows. By 2050 renewables account for 40% of global energy consumption, a share that increases to close to 80% by the end of the 21st century. The quality of the energy carriers delivered to end users is high in order to meet environmental constraints at the local level as well as the requirements of high efficiency end-use devices. This means that renewable energy sources are transformed into electricity, liquid, and gaseous energy carriers. Fossil fuels continue to be used as transitional fuels. Nuclear

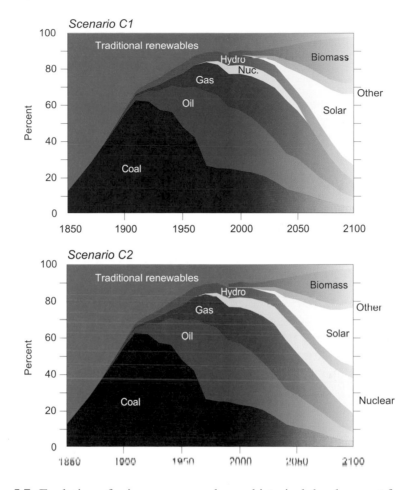

**Figure 5.7**: Evolution of primary energy shares, historical development from 1850 to 1990 and in scenarios to 2100, for Case C: Scenario C1 "nuclear phaseout," and Scenario C2 "bio-nuc."

energy is at a crossroads in Case C, and which direction it takes constitutes the main difference between the two Case C scenarios.

In Scenario C2 a new generation of inherently safe, small-scale (100 to 300 MW$_e$) reactors is developed and finds widespread social acceptability, particularly in areas of scarce land resources and high population densities. Its expansion limits the contribution from renewables.

In the other variant, Scenario C1, nuclear energy is a transient technology that is virtually phased out by 2100 together with most fossil fuels. This latter variant assumes that the nuclear industry has been unable to adapt to public concerns or to pressures for downsizing and decentralization of the energy system.

Sections 5.3 and 5.4, and the subsequent figures, highlight the differences among the scenarios with respect to the future role of fossil, renewable, and nuclear energy.

## 5.3.    Fossil Sources

In all six scenarios the peak of the fossil era has passed. Fossil energy consumption grows more slowly than total primary energy needs. The share of fossil energy sources declines immediately in Case C, after 2000 in Scenario A3, and after 2010 in Case B and Scenarios A1 and A2. The most extreme developments occur in the Case C scenarios, where fossils' share returns to the level that prevailed around 1850, that is, less than 20%. In the other cases the transition is more gradual, especially during the interim period to 2020. Oil and gas are important transitional sources of energy in all scenarios, but their share gradually declines throughout the next century. In absolute amounts, however, future oil and gas requirements are huge compared with current levels. By 2050 the highest scenarios imply increases in oil production of more than a factor of two, and of gas production by close to a factor of five. The role of coal differs across scenarios, from a revival in Scenario A2 and in Case B to a decline in the other scenarios. Direct uses of coal disappear by 2100 in all scenarios; coal is converted either to electricity, gas, or synliquids.

In Scenario A1 oil and gas maintain the highest market share of all scenarios and for the longest time. Technological advance also allows the expansion of renewables and nuclear energy. Their market share grows gradually, however. Coal's market share declines throughout.

In Scenario A2 coal's share expands steadily and by 2100 returns to its 1950s level of 40%. Most of the expansion is at the expense of oil and gas. At substantial cost, coal power plants are equipped with $SO_x$ and $NO_x$ scrubbers to reduce local environmental impacts to viable levels.

In Scenario A3 technological developments favor new renewables and nuclear energy. They expand their market shares at rates consistent with the historical dynamics by which fossil fuels replaced traditional renewables in the 19th century. Oil and coal use is reduced, and by 2100 all fossil fuels account for less than half the energy market.

In Case B, technology progresses more slowly than in Case A and coal's abundant resources give it an eventual edge over oil and gas. Coal is a backstop option *par excellence* in this scenario. Oil and gas maintain significant market shares through 2070 by moving into costlier categories of first conventional and later unconventional resources.

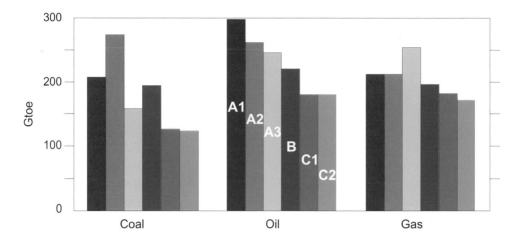

**Figure 5.8**: Cumulative fossil energy requirements, 1990 to 2050, in Gtoe.

The Case C scenarios exert minimum pressure on fossil resources other than perhaps natural gas. Although overall technical progress is the same as in the Case A scenarios, improvements in fossil fuel technologies do not go beyond conventional oil and gas production. Given the strict constraints on carbon emissions, coal use is minimized. Even in this extreme scenario, cumulative coal consumption between 1990 and 2100 equals 180 Gtoe, 40% more than cumulative coal production to date.

The scenarios are not limited by fossil resources directly, but rather were designed to reflect gradual changes in the structure of energy supply as conventional reserves and resources become more expensive and difficult to exploit. Consequently, the six scenarios lead to different cumulative energy requirements and different ultimate fossil resource consumption. *Figure 5.8* compares the cumulative coal, oil, and gas requirements of the six scenarios from 1990 to 2050.

By 2050, cumulative requirements are lower than the identified global energy reserves, estimated at about 1,300 Gtoe (see *Box 4.3* on energy reserves, resources, and occurrences). The total fossil energy needs range between 500 Gtoe (Scenario C2) and 750 Gtoe (Scenario A1). This also means that the current conventional oil and natural gas reserves of about 290 Gtoe need to be totally replenished in all cases, either from resources or by tapping into unconventional reserves, or both, as actually occurs in the scenarios. Cumulative consumption of oil and gas ranges between 350 Gtoe (Scenario C2) and 510 Gtoe (Scenario A1). Maximum cumulative coal consumption (A2) stays below half of the estimated recoverable reserves of about 600 Gtoe.

The differences in the cumulative resource requirements among the three variants of Case A are especially pronounced, ranging between 660 Gtoe and 750 Gtoe.

Case B is the least challenging from this point of view. It is designed to provide a gradual and limited transition to new and alternative sources of energy. Thus, it results in relatively high fossil energy needs of 610 Gtoe, compared with intermediate levels of total primary energy use. Fossil energy requirements are comparatively low in Case C at 470 to 490 Gtoe.

Resources *per se* do not appear to be the main global limiting factor by 2050, even for the variants of Case A. However, beyond 2050 the relative cumulative contributions of coal, oil, and natural gas spread over a significantly wider range. Thus, after 2050 regional resource scarcities may well occur, and consequently all six scenarios envisage a gradual shift away from fossil energy sources by the end of the next century. This transition may not be smooth in all cases as the cumulative energy requirements are indeed very large. For example, the three variants of Case A result in the consumption of between 1,400 Gtoe and almost 1,800 Gtoe of fossil energy by 2100. The cumulative demands on oil and gas, ranging between 900 Gtoe (Scenario A2) and 1,400 Gtoe (Scenario A1), will require substantial technological advances to ensure the replenishment of needed reserves from the resource base.

These large quantities are certainly beyond the "conventional wisdom" about global resource availability. They compare with the global fossil energy resource base of up to 5,000 Gtoe, including 1,300 Gtoe of estimated reserves (see *Box 4.3*). Thus, up to 500 Gtoe would have to be transferred from the resource base to reserves.

Although the potential magnitude of the global fossil resource base is truly gigantic (even without the speculative and exotic occurrences), it will be a serious challenge to transfer more than a small portion of these quantities to actual energy reserves. In the past, transfers from resources to reserves have kept pace with increasing production (and demand). *Figure 4.12* illustrates how oil and natural gas reserves have increased historically and compares them with cumulative consumption. It shows that the reserves-to-production ratio has varied between 40 and 70 years for gas and between 20 and 45 years for oil. Additions to reserves (from the resource base) have generally outpaced increases in annual production, leading to a gradual rise in the reserves-to-production ratio, especially for natural gas. Transferring sufficient quantities of fossil resources to reserves to keep up with future increases in consumption will be difficult but feasible with high rates of technological progress and changes in extraction economics. Such future improvements, however, are consistent with past improvements and are assumed to be feasible in all scenarios.

There are other possible obstacles to utilizing large portions of the global fossil resource base, including financing and environmental constraints. For example, a large part of the estimated resource base of 5,000 Gtoe consists of coal. This is certainly sufficient to provide enough fossil energy well beyond the next 100 years, even for the high-demand Case A scenarios, but this may not be a viable strategy. Consumption of such large quantities of fossil energy translates into cumulative $CO_2$ emissions of the same magnitude – almost 5,000 GtC, or more than six times the current atmospheric $CO_2$ concentration. A shift to other sources of energy and global energy decarbonization are likely to occur for reasons other than resource depletion.

## 5.4.    Renewables and Nuclear

In all scenarios there is a significant expansion of *renewables*, characterized by the following features. Traditional uses of renewables and traditional technologies are gradually phased out. Traditional renewables are gradually replaced by high-quality commercial energy carriers, including those from "new" renewable sources. In Case C, the phaseout of traditional renewables is somewhat slower than in the other cases due to explicit policies of rural (re)development that result in a slower shift away from traditional economic patterns. End-use conversion of traditional biomass, however, improves significantly in Case C (e.g., via fuel-efficient cooking stoves), which has the added benefit of reducing deforestation rates and improving indoor air quality.

The renewables portfolio is particularly diversified. It varies geographically in response to available resources and technologies, and the level and structure of energy demand. Diffusion is not instantaneous. It proceeds at different rates in different scenarios via a succession of specific market niches. Thus, even massive policy interventions would appear limited in their ability to shortcut the learning and experimentation required for the spread of new renewable technologies.

Compared with a number of other scenarios available in the literature (e.g., Williams, 1995; IPCC, 1996b), growth in the market potential of renewable energy is more gradual in the near to medium term, ranging between 2.3 Gtoe (Case B) and 3.3 Gtoe (Scenario A3) by 2020, compared with 1.6 Gtoe in 1990. This more modest near-term growth is counterbalanced, however, by higher growth rates in the long term, as performance improvements developed in initial niche markets translate into vigorous penetration of broader markets.

Renewables in all cases are driven by consumer demands for more flexible, more convenient, and cleaner energy. Especially after 2020 (*Figure 5.9*), renewables reach the consumer increasingly in the form of high-quality, processed energy carriers – electricity, heat (via either small-scale district heating grids or direct on-site solar collectors), biogas, and liquids (e.g., methanol or ethanol). These shifts are strongest in low-demand scenarios (e.g., Scenario C1) but also emerge in the more distant future for all other cases.

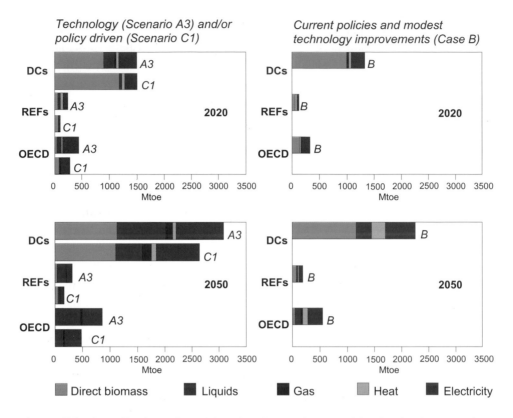

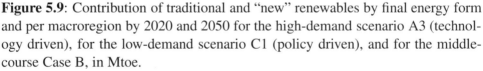

**Figure 5.9**: Contribution of traditional and "new" renewables by final energy form and per macroregion by 2020 and 2050 for the high-demand scenario A3 (technology driven), for the low-demand scenario C1 (policy driven), and for the middle-course Case B, in Mtoe.

Although all scenarios show a substantial increase in renewables, there are differences, most notably in the driving forces. These vary from improved technological and economic performance to significant policy support to the gradual depletion of low-cost fossil resources. The remainder of this section addresses these differences.

By design, Case B takes the most cautious approach to renewables, assuming gradual improvements but no radical breakthroughs. Traditional uses are gradually phased out, capital requirements constrain the expansion of hydropower, and limited technological progress yields comparatively slow penetration rates for new renewables. In 2020, traditional biomass remains the dominant category of renewables (1.1 Gtoe), followed by hydropower (0.7 Gtoe) and new renewables (0.5 Gtoe). The total contribution of renewables is 2.3 Gtoe in 2020. This reflects a gradual phaseout of traditional biomass use and a smaller contribution of large hydropower (0.7 Gtoe) compared with earlier scenarios (WEC, 1993, 1994). The smaller hydropower contribution results from a more conservative estimate of capital availability for large-scale hydropower in developing countries (Martin, 1994; Moreira and Poole, 1993).

Over the longer term, however, the contribution of renewables expands at a steady pace even in Case B. By 2050 renewables account for 4.4 Gtoe, or 22% of global primary energy use (*Figure 5.6*), and by 2100 their share is 33%, or 11 Gtoe – more than total current world primary energy use. As shown in *Figure 5.10*, traditional and new biomass and hydropower make up the bulk of the renewables in 2050, while genuinely new renewables – solar and wind – make comparatively modest contributions. In absolute terms, however, even these are significant: wind and solar together total approximately 1 Gtoe, comparable with hydropower.

Cases A and C are more bullish on renewables, as shown in *Figure 5.10*. The driving forces are either technology push (as in the high-demand "bio-nuc" Scenario A3) or policy pull (as in the low-demand Case C). In Scenario A3, renewables reach as much as 22 Gtoe (with over 8 Gtoe being supplied by biomass) by 2100, pushing the limits of alternative energy futures. Such results require further detailed analysis of potential land-use conflicts between energy and agriculture, particularly in developing countries, where most future demand growth is located and where most of the increased production of renewables will have to take place (see *Box 5.2*). The potentially severe environmental impacts on biodiversity, natural habitats, and species protection also require careful scrutiny.

*Box 5.2* presents estimates for land requirements of biomass production for an illustrative high biomass growth scenario (A2 and A3). Depending on productivity increases, land requirements range between 390 and 610 million hectares (ha) by 2050 and between 690 and 1,350 million ha by 2100. These additional land areas for both agriculture and biomass production are available in principle – particularly in developing countries, where most of these additional land requirements are concentrated – but they stretch future land requirements (and land-use changes) to their ultimate limits. Both agriculture and biomass energy are moving into a "high-tech, high-productivity" future where ecosystems are tightly managed and controlled by humans.

In view of significant ecosystem changes and regional land-use conflicts, particularly in Asia, the future of biomass is likely to be more constrained than the staggering numbers of the ultimate technical potential of biomass plantations seem to suggest. Also, biomass energy is no "energy supply panacea" for unlimited energy growth. As in the case of fossil energy, energy efficiency improvements will be instrumental in overcoming possible constraints from the available resources (renewable or not). This applies in particular to developing countries that combine high population density with vigorous energy demand growth.

**Box 5.2: Food and biomass: Coevolution or a source of land-use conflict?**

Both agricultural food production and biomass production for energy require land. To address the issue of possible land-use conflict between biomass and agriculture, we have used IIASA's Basic Linked System (BLS) of agricultural models (Fischer *et al.*, 1988, 1994) to calculate the food and agricultural land requirements of the high-growth Case A (see Appendix A). The required expansion of agricultural cropland to increase diet and to provide food for an additional 5 billion people is estimated at 250 million ha, with 200 million additional ha required in developing countries. This compares with 1,440 million ha currently used by agriculture (*Table 5.2*).

The required land area for energy biomass production in the high-growth Case A was estimated at 400 to 600 million ha by 2050 and 700 to 1,350 million ha by 2100. In an optimistic scenario (lower bound biomass land requirements), biomass land productivity is assumed to grow to 10 toe per ha per year, with two-thirds being produced in dedicated plantations and the remainder being recovered from agricultural residues and from forest management. These assumptions were based on the LESS scenario (Williams, 1995).

These additional land areas for both agriculture and biomass are available in principle (see *Table 5.2*), but they stretch future land requirements (and land-use changes) to their ultimate limits. By 2100 agriculture (1,700 million ha) and biomass (700 to 1,350 million ha) could require up to 3,000 million ha – as much land as is currently covered by forests. Land-use conflicts may become a major constraint.

For instance, in Asia (CPA, PAS, and SAS) the land required for expanding agricultural production and achieving maximum biomass use would require the entire potential arable land by 2100. In view of significant ecosystem changes and regional land-use conflicts, the future of biomass in all likelihood will be constrained, particularly in areas of high population density such as Asia, and is most likely to evolve along the lines of Cases B and C.

**Table 5.2:** Land use in 2050 and 2100 compared with 1990, for selected world regions, in million ha.

| | Current use | | | Additional land use | | | Potential |
|---|---|---|---|---|---|---|---|
| | | | | Agric. | Biomass[a] | Biomass[a] | arable |
| | Forests | Pasture | Agric. | (2050) | (2050) | (2100) | land[b] |
| ICs[c] | 1,770 | 1,190 | 670 | 50 | 70–100 | 150–350 | – |
| Africa[d] | 630 | 700 | 150 | 95 | 110–180 | 140–340 | 990 |
| Asia | 600 | 880 | 470 | 33 | 160–250 | 260–340 | 500 |
| Latin America | 890 | 590 | 150 | 72 | 50–80 | 140–320 | 950 |
| DCs[e] | 2,120 | 2,170 | 770 | 200 | 320–510 | 540–1,000 | 2,440 |
| World | 3,890 | 3,360 | 1,440 | 250 | 390–610 | 690–1,350 | – |

[a] For maximum biomass scenarios (A2 and A3), range corresponds to 4 to 6 toe/ha land productivity by 2050 and to 6 to 10 toe/ha by 2100. Lower bounds also assume that 80% (2050) and 67% (2100) of biomass will be produced in plantations. Higher bounds assume 100% plantation biomass.
[b] FAO (UN Food and Agriculture Organization) estimate (Alexandratos, 1995).
[c] Industrialized countries, including OECD, EEU, and FSU.
[d] Including MEA.
[e] Developing countries.

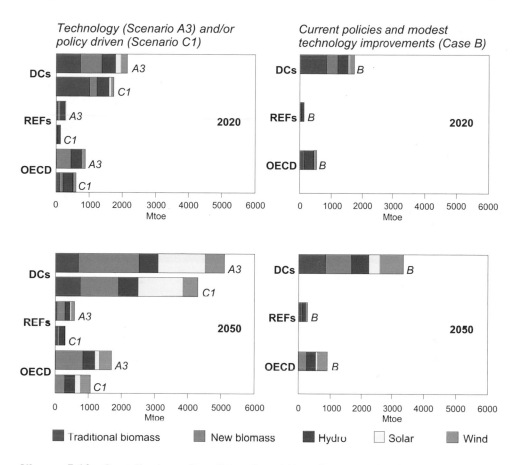

**Figure 5.10**: Contribution of traditional and "new" renewables by energy source and per macroregion by 2020 and 2050 for the high-demand Scenario A3 (technology driven), for the low-demand Scenario C1 (policy driven), and for the middle-course Case B, in Mtoe.

The changing geography of renewables is best illustrated by Scenario A3. In 1990, industrialized countries consumed only 16% of global biomass but two-thirds of "modern" uses of renewables, principally hydropower. By 2050, in Scenario A3, industrialized countries would use only one-third of global "modern" renewables. Thus, while industrialized countries now dominate modern uses of renewables and have taken the lead in early niche markets for "new" renewables such as wind-, solar-, and biomass-fueled district heating, modern renewables must ultimately be transferred to the rapidly growing South. The large-scale, post–2020 transfer reflected in the scenarios should not be seen as simply selling off-the-shelf technology from North to South. The variety of site-specific conditions and the need for diverse

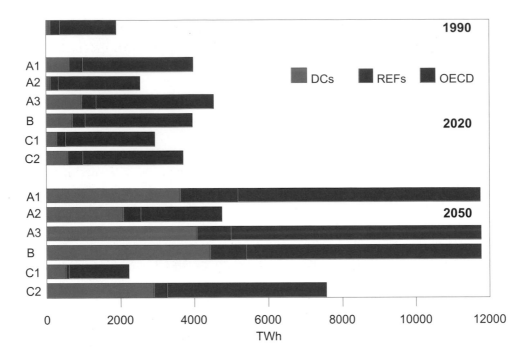

**Figure 5.11**: Nuclear electricity generation in the three macroregions in 1990, and in the six scenarios in 2020 and 2050, in terawatt hours (TWh).

energy carriers to match demand and supply will require local adaptations, local technology improvements, and the development of local technological capabilities.

Uncertainty and changing fortunes characterize the long-term role of *nuclear* energy in the six scenarios. *Figure 5.11* shows nuclear power generation in the three macroregions for the years 1990, 2020, and 2050. Until 2020, all scenarios experience growth, but variation across the scenarios is considerable. The scenario with the lowest nuclear power generation path (in absolute terms), Scenario A2, foresees a capacity increase of some 20%. In contrast, Scenario A3 requires more than a doubling of current generating capacity by 2020. The resulting nuclear installed capacity in the scenarios ranges between 410 gigawatts electric (GW$_e$) and 740 GW$_e$.

This initial divergence in the scenarios mirrors the substantial uncertainties in the future of nuclear energy that result from the present lack of public acceptance due to concerns about reactor safety, the proliferation of weapons-grade fissile materials, and the final disposal of spent fuel and fission products. The socio-political controversy about the actual need for nuclear power and the associated risks has significantly complicated licensing procedures and, because of public opposition to the handling and disposal of radioactive materials, to date, no long-term sites exist (Häfele, 1990). In addition, many utilities are no longer willing to order new nuclear plants, principally because of construction delays and interruptions leading to high costs and financial risks (COSEPUP, 1991).

Nevertheless, at the end of 1997, 437 nuclear power plants were connected to the electricity grid, with a total installed capacity of some 352 GW$_e$ providing approximately 17% of global electricity supply. Another 36 reactors, corresponding to an additional 27 GW$_e$ of installed capacity, are currently under construction (IAEA, 1998). Once these capacities become operational, 33 countries will be using nuclear power for electricity generation. Construction starts of nuclear plants peaked at the end of the 1960s, and have since slipped from 37 to 5 or fewer plants per year. New capacity orders are being placed primarily in Asia and to a lesser extent in FSU and EEU.

Although the momentum of global nuclear power expansion has slowed considerably in recent years (Beck, 1994), there are important differences with respect to nuclear developments at the national and regional levels. Some countries have imposed complete nuclear moratoria (e.g., Austria) or quasi-moratoria (e.g., USA, Germany, Switzerland, and Sweden), while others have vigorously advanced nuclear power for electricity generation (e.g., France, Japan, Belgium, and South Korea). Apparently, societies and cultures perceive the virtues and risks of nuclear power differently and, as for most technology configurations, there is no homogeneous and globally accepted solution.

Today it is not clear how and by which technologies the problems currently facing nuclear energy may be resolved. What actually happens will depend on how safety, waste disposal, and proliferation concerns are resolved, and whether the greenhouse debate adds increasing importance to nuclear energy's "carbon benignity." Consequently, after 2020 completely different nuclear futures may unfold varying from an almost fivefold expansion between 1990 and 2050 (Scenarios A1, A3, and Case B) to a 20% decline over the same period (Scenario C1).

The share of nuclear power in global primary energy supply appears to be concentrated in two narrow and scenario-dependent "clusters." One cluster reflects the current outlook for nuclear power, that is, the continuation of country-specific energy policies and investment behaviors; here, nuclear energy achieves market shares of 4 to 6% of primary energy by 2020 and about 4% by 2050. Scenarios A2 and C1 are in this cluster, though for different reasons. Whereas Scenario C1 uses nuclear as a temporary bridge to a long-term nuclear-free future, Scenario A2 presents a pragmatic wait-and-see strategy over the medium term, thus keeping open the option of accelerated use in the longer run. All other scenarios fall into the second cluster, where nuclear contributes 11 to 14% to primary energy supply by 2050. Here, Case B realizes the highest nuclear share, although the highest absolute electricity generation occurs in Scenarios A1 and A3.

Beyond 2050, the relative contribution of nuclear energy to primary energy supply envelops a wide span of potential futures, from a phaseout in Scenario C1 to an almost 24% share in Case B. In the latter case, nuclear assumes the role of the long-term "swing" supplier. Both fossil fuel and renewable energy technologies experience modest improvement rates; their future development, therefore, cannot unfold at the rate observed in the other scenarios requiring higher "backup" from nuclear.

The range of nuclear futures in the scenarios is linked to the deployment of specific nuclear technologies. In Cases A and B, it is assumed that new improved nuclear reactor lines are developed, replacing current generating stations. More sophisticated fuel cycles and advanced reactors including breeding may be part of these scenarios. In Case C, however, future nuclear technologies consist of small to medium-size reactors (100 to 300 $MW_e$), relatively simple fuel cycles, and drastically improved operating and safety features such as the 100 $MW_e$ high-temperature reactor currently under development in South Africa. Smaller reactors could also lead to nuclear power applications other than electricity generation. For example, small reactors can be used for district and process heat generation, desalination, hydrogen production, oil extraction, or energy supply of conventional loads in remote areas.

It is interesting to compare the paths of nuclear energy in developing countries with those in the OECD region. In 1990, nuclear energy in developing countries was 7% of OECD use. This situation changes drastically in all six scenarios. Until the mid-21st century, nuclear power is vigorously deployed in developing countries. In the scenarios with high nuclear shares, the installed capacity in developing countries is twice the current global capacity. Clearly, if the current use of nuclear energy in developing countries is already creating concerns about proliferation, as illustrated by reactions to the growing "club" of countries with nuclear weapons, we urgently need international agreements, enforceable regulations, and comprehensive material accounting that would allow the world to deal with the implications of up to one terawatt electric ($TW_e$) of installed nuclear capacity in developing countries by the mid-21st century and much more thereafter in a scenario of high nuclear growth.

The resource base for uranium, as described in Chapter 4 (See *Box 4.3*), equals 260 Gtoe, assuming a typical once-through fuel cycle. In comparison, the cumulative uranium requirements for the three scenarios with a high contribution of nuclear energy require either larger resources or fuel reprocessing and higher conversion rates. But even if natural uranium resources were larger, there seem to be only two major long-term routes for the large-scale development of nuclear fission power. One is breeder reactors, which appear quite problematic in the foreseeable future because of many technological and proliferation issues. The alternative would be significant new developments in fuel cycles and burn-up rates as well as substantial progress in the economic recovery of low-grade uranium, which is abundant in the earth's crust and oceans.

We have not included nuclear fusion explicitly in the study. Had this been done, fusion would have assumed or shared the role played by breeder reactors. However different the two may be from an engineering point of view, there is little to distinguish them in the scenarios. Both are large, centralized, carbon-free producers of electric power that generate radioactive waste and have a practically infinite resource base. Therefore, in our attempt to treat future technologies as generically as possible, breeding and fusion reactors, from today's viewpoint, look very much alike.

## 5.5.    Electricity

Electricity and hydrogen are the only end-use fuels capable of providing all energy services without harmful emissions. Hydrogen currently plays an insignificant role, serving a limited number of niche markets. In the absence of very stringent environmental policies, hydrogen is unlikely to play more than a marginal role before 2050. Electricity is an important energy carrier today, and its contribution increases in all six scenarios. Thus we focus on the role of electricity in the different scenarios.

Up to 2020, there are two developments common to electricity generation in all scenarios: the decline of oil products and the increase in absolute terms of all other established fuels (see *Figure 5.12*). In Case A, capacity increases by over 50%. That expansion, plus the replacement of older facilities, results in efficiency increases. Average thermal efficiency of coal-fired electricity generation increases to about 40%. For natural gas, the increase is even larger, to about 50%. This is primarily the result of investments in combined-cycle turbine technology. In contrast, in Case C, coal-fired capacity remains near 1990 levels and new technologies enter the market only as obsolete facilities are retired. The slower turnover of the capital stock leads to lower overall efficiency gains for coal-fired electricity, even though nameplate technology improvements are identical in Cases C and A.

Non-fossil technologies increasingly make inroads into both new capacity and replacement markets. By 2020, renewable technologies have grown out of their initial niche markets and established themselves as genuine competitors, especially in decentralized markets. The penetration of solar, wind, biomass, and geothermal is highest in the high-growth Case A (where technology develops rapidly) and in Case C (where environmental policies develop rapidly). In Case C, however, policy-driven requirements for costly carbon, sulfur, and nitrogen mitigation

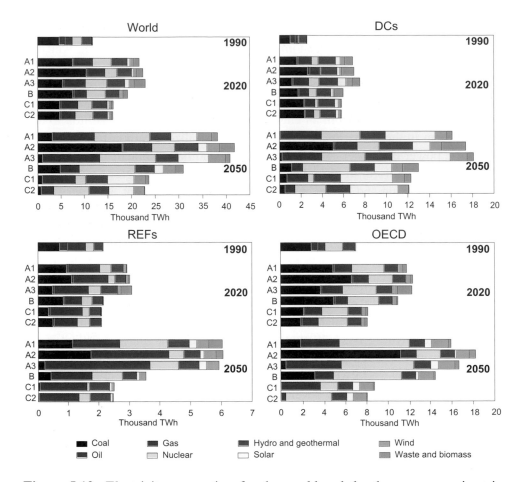

**Figure 5.12**: Electricity generation for the world and the three macroregions in 1990, 2020, and 2050, in TWh.

increase electricity costs and induce extensive cost-effective end-use efficiency improvements. Thus end-use efficiency improvement, rather than large supply restructuring and capacity expansion, is the principal characteristic of Case C up to 2020. Except for Scenario A2, the fossil fuel share of electricity generation decreases from 62% in 1990 to 45 to 55% in 2020. The expansion of nuclear and hydropower is chiefly responsible. In the high-coal Scenario A2, the fossil share increases slightly to 64% of electricity generation.

*Figure 5.12* shows electricity generation for the three macroregions. Renewables expand fastest in DCs, where the need for capacity expansion is the highest. In general, the preferred renewable source is hydropower, followed by the "new" renewable sources: biomass, solar, wind, geothermal, and waste. In Case C, biomass is needed largely for methanol and ethanol production and therefore does not make notable inroads into electricity generation. In REFs, the fuel of choice is natural gas. Only in Case B, where technological progress is slower, does gas fail to become the dominant utility fuel. In the OECD region, the shape of the electricity sector continues along current trends, with the bulk of production coming from coal, gas, and nuclear energy. Wind is by far the leading new renewable source of electricity, but in relative terms it makes hardly a dent in the electricity supply menu.

In all scenarios the electricity supply structure after 2020 is best characterized by supply diversification. With the exception of perhaps coal in Scenario A2, no single technology or fuel truly dominates global electricity generation. The period to 2050 corresponds roughly to a full replacement of the utility capital stock and is also sufficient to allow renewable technologies to move further down the learning curve and penetrate the base-load market. Thus, by 2050, the existing capital stock has been replaced, the menu of replacement options has more than doubled, and there is much more diversity among scenarios. However, the differences evident in 2050 essentially depend on the technology path established prior to 2020. Thus, RD&D priorities and the investment decisions of the next two decades will largely determine the economic effectiveness of future electricity generating options. The differences between scenarios in 2050 do not provide opportunities for switching strategies late in the game, hence the need for action starting *now*.

By 2050, electricity from renewables, excluding hydropower, increases to between 18% and 37% of electricity generation, with a median value of 26%. Electricity from fossil fuels declines to a median share of approximately 30% (58% in Scenario A2 and 15% in Scenario C2). Essentially this is all gas and coal, as oil virtually disappears as a utility fuel. Although coal boasts higher reserves, natural gas outperforms coal in all scenarios except the coal-intensive Scenario A2. Thus, the scenarios suggest that unless investments in renewables and in gas exploration and development are constrained by limited capital, coal is unlikely to remain the staple fuel of electricity generation. Nuclear power emerges as a robust option in all scenarios. Even in Scenario C1, the absolute output of nuclear is slightly higher than in 1990. Still, in the high-technology Cases A and C, solar, wind, and hydro jointly generate more electricity than nuclear.

At the regional level, *Figure 5.12* indicates that electricity generation by 2050 is less diversified and more scenario dependent. OECD relies principally on nuclear power, except for the coal-intensive Scenario A2 (in which coal is the preferred fuel), and the nuclear phaseout Scenario C1 (in which natural gas is preferred). In REFs, natural gas is the dominant fuel in all scenarios. Only in DCs is there no consistently dominant source. Given a five- to sevenfold increase in this region, electricity generation commands the maximum possible contribution from all sources. One trend appears robust across all scenarios: whenever natural gas, coal,

## Box 5.3: Calculating final-to-primary energy efficiency

In this study primary energy use is calculated based on the net calorific values of fuels. For non-fossil electricity we apply the widely used "substitution equivalence" method, by which primary energy equivalents are calculated assuming a conversion efficiency of 38.6% (WEC, 1993). Other conventions also exist. The "direct equivalence method" assumes that primary energy equals 100% of the direct heat value of the electricity produced. The "engineering method" calculates primary energy based on the actual efficiency of the energy system. Brisith Petroleum and the International Energy Agency (IEA) combine methods, assuming efficiencies of 33% (i.e., the substitution equivalence method) for nuclear electricity, and 100% (i.e., the direct equivalence method) for hydropower.

The choice of method makes little difference for systems with little energy conversion beyond electricity generation based on the steam cycle. For elaborate energy systems the difference is significant. In 1990, for example, the difference between methods is small. The global average final-to-primary energy efficiency is 72% using the substitution method and 74% using the direct equivalence method. But in 2100 in the high-technology Scenario A3, calculated final-to-primary efficiencies vary from 78% using the direct equivalence method to 65% using the engineering method to only 55% using the substitution equivalence method (*Figure 5.13*). Caution is therefore advised in comparing the results of this study with those of others. One reason we focus on final energy is the increasing dependence, as systems become more complex, of primary energy calculations on the choice of the accounting convention. The very concept of primary energy becomes increasingly problematic, particularly as renewable energy forms gain importance.

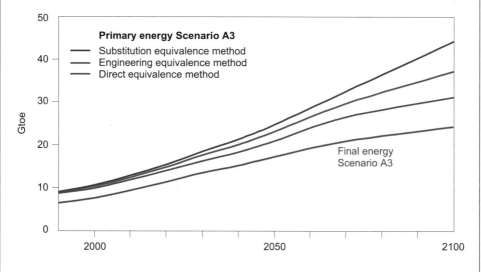

**Figure 5.13:** Final energy and three alternative methods for calculating the primary energy equivalent of non-fossil energies for Scenario A3, in Gtoe.

or nuclear power face constraints, renewable technologies, especially solar, play the role of the swing supplier.

Given the enormous increase in electricity generation as well as the increasing contribution from non-fossil sources in the scenarios, an increasing fraction of primary energy goes into the electricity sector. Exactly how much is to a large extent a definitional issue (see *Box 5.3*). Given increasing conversion and complexity of energy systems as described in the scenarios, the traditional focus of energy studies on primary energy may increasingly be inappropriate.

Perhaps the most important message for electricity generation, however, lies less in what happens on the supply side than in what might happen on the demand side. The Case C scenarios provide the clearest examples. There, capacity buildup occurs more slowly than might have been anticipated from the growth in energy service demands. The reason is that it is more economical to invest in end-use efficiencies (e.g., the introduction of heat pumps) and interfuel substitution (e.g., from electricity to district heat) than in new capacity. Though this effect is less pronounced in other scenarios, an increasing trend is expected favoring those who successfully switch from selling just kilowatt hours of electricity to selling energy services. We also expect the utility industry to become flatter, particularly as renewable technologies enter the generating market. First, the number of independent producers is likely to increase. Second, production will probably become more decentralized. And third, deregulation and common carrier rights will divide production, transmission, and distribution into separate businesses. The increased number of producers may well include users of both electricity and process heat who use industrial co-generation to generate their own electricity. Overall, the increase in small scale decentralized generating capacity may discourage investments in grid expansion.

## 5.6. Final Energy

Despite all the variations described in this chapter so far, the pattern of final energy use is remarkably consistent across all cases and scenarios. Equally consistent is the continuing trend toward energy reaching the consumer in ever more flexible, more convenient, and cleaner forms. *Figure 5.14* presents the same conclusion graphically. The figure shows the small variation among Cases A, B, and C. Variations among the three Case A scenarios, and the two Case C scenarios, are even smaller.

As shown in the figure, all three cases reflect a continuing pervasive shift from energy used in its original form, such as traditional direct uses of coal and biomass, to elaborate systems of energy conversion and delivery. This shift continues in all cases, leading to ever more sophisticated energy systems and higher-quality energy

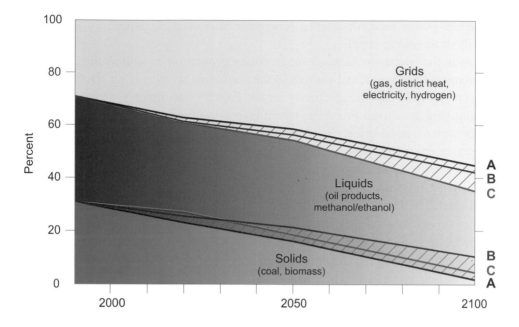

**Figure 5.14**: World final energy by form, in percent, as solids, liquids, and grids. Overlapping shaded areas indicate variations across Cases A, B, and C.

carriers. A second profound transformation is the increasing degree to which energy is delivered by dedicated transport systems, such as pipelines and networks. This development enhances trade possibilities and promotes similar end-use patterns across regions with fundamentally different primary energy supply structures. Finally, changes in final energy patterns reflect the changes in economic structure presented in the scenarios. As incomes increase, the share of transport and residential/commercial applications also increases (*Figures 5.15* and *5.16*).

These converging final energy patterns yield substantial quality improvements in the final energy (and resulting energy services) delivered to the consumer. Quality improvements are measured by two indicators: fuel-mix-induced efficiency gains and carbon intensity of final energy. The efficiency of final energy use improves as the final energy carrier portfolio changes in the direction of higher-quality fuels. The effect is an improvement via interfuel substitution of 20 to 30% in our cases. The actual end-use efficiency gains are of course much larger, as they are mostly driven by technological change in end-use devices (cars, light bulbs, etc.). The main points are that more efficient end-use devices will require higher-quality fuels and there is a high degree of congruence across all six scenarios. Thus, while primary energy supply structures and resulting carbon intensities diverge between the cases, those of final energy converge. The decarbonization trend of final energy is also faster across all cases compared with primary energy.

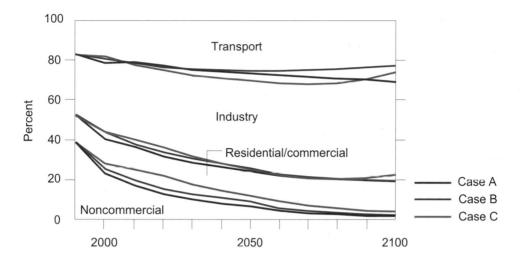

**Figure 5.15**: Final energy by sector, developing countries, in percent.

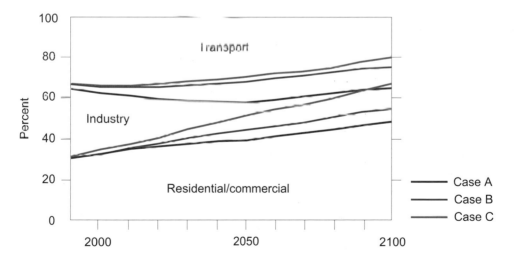

**Figure 5.16**: Final energy by sector, OECD countries, in percent.

The principal difference between Cases A, B, and C is that the shift away from direct uses of coal and biomass is less complete in Case B than in the more technologically challenging Cases A and C. In Case B, limited direct use continues, although with improved, more environmentally benign technologies. In Cases A and C, direct use of coal and biomass is phased out after 2050; otherwise the final energy structures are similar. The share of electricity increases in all three cases, as do the shares of new energy carriers such as methanol and hydrogen. The role of oil diminishes as it becomes reserved mostly for transportation. Methanol and ethanol, and later hydrogen, are also used as transport fuels. By 2050, the share of noncommercial final energy generally decreases to less than 6%, compared with 16% today. By 2100, it is virtually gone. Industrial and transport energy shares generally grow, largely due to an enormous increase in industrial production and mobility in the developing countries.

*Figures 5.15* and *5.16* contrast the final energy structures in the developing countries and the OECD. In developing countries, final energy needs for transport and industry grow faster than those in the residential and commercial sector, and the share of noncommercial energy uses declines. In the OECD region, however, residential and commercial needs grow faster than those for transport and industry. With high levels of affluence and leisure, new services and new activities emerge that shift final energy requirements away from production. Demographic changes associated with aging and single-person households reinforce the trend.

Interestingly, of the three macroregions, the OECD region (*Figure 5.16*) shows the greatest variation among cases. In Case C, final energy needs in the residential and commercial sector increase to more than half of all final energy after 2050. Mobility and production are replaced by communication and services, resulting in lower material and energy intensities. This leads to significant differences across regions and scenarios in the end-use devices (heating systems, cars, etc.) that are used and in *how* they are used (i.e., lifestyles), even when differences in total final energy demands are small. An illustration is given in *Box 5.4*, which contrasts the high-growth Case A for LAM with the ecologically-driven Case C for WEU. Both regions have a strong tradition of detailed analyses of energy end use and associated lifestyle changes (Goldemberg *et al.*, 1988; Schipper and Meyers, 1992; IEA, 1997b).

## Box 5.4: Energy use and lifestyles

Different levels of energy use imply differences in economic activities and in consumer lifestyles. Even when levels of energy use are similar, lifestyles may differ, as illustrated here for Case A in LAM and Case C in WEU in 2050. In both, per capita final energy use is about 1.7 toe, but there are important differences *how* energy is used (see *Table 5.3*).

In LAM, Case A reflects high affluence and consumption largely following the historical experience of OECD countries. Energy end-use patterns and lifestyles in LAM in 2050 are not too different from those in 1990 WEU considering their similar levels of per capita GDP$_{ppp}$. Case C for WEU is very different. In 2050 Case C WEU, comfort and mobility are high, at 100 m$^2$ of floor area per capita and some 20,000 km of travel per capita per year. But energy use is low because efficiency is high: for example, the specific fuel use of buildings per square meter of floor area improves by a factor of four, and most of the increase in mobility is provided by public transport including high-speed rail and aircraft (perhaps the first market niche for hydrogen fuels) that complements the high information and communication flows of a unified Europe.

Per capita car travel in 2050 is similar in Case A for LAM and Case C for WEU (6,000 to 7,000 passenger-km per year). But this similarity masks differences in technologies used and especially *how* they are used. In Case A for LAM, car mobility is based on a car ownership rate of 0.4 vehicles per capita and about 7 liters of gasoline per 100 km driven. In Case C for WEU, cars are rented or leased rather than owned: small vehicles using 3 liters per 100 km for urban commutes; larger (5 liters per 100 km) vans for weekends and holidays. Car manufacturers become providers of "cradle-to-grave" car services, ensuring the high recycling rates underlying Case C's decline in industrial energy use.

**Table 5.3:** Per capita final energy use in Case A for LAM and in Case C for WEU (toe per capita, roman) with illustrative activity indicators (italics).

|  | LAM 1990 | WEU 1990 | LAM ~2050 (A1) | WEU ~2050 (C1) |
|---|---|---|---|---|
| *GDP/capita, US(1990)$* | *2,500* | *16,000* | *8,500* | *33,000* |
| *GDP$_{ppp}$/capita, US(1990)$* | *4,700* | *13,200* | *10,300* | *27,000* |
| Total final energy, toe/capita | 1.10 | 2.50 | 1.70 | 1.70 |
| *Floor space, m$^2$* | *10* | *40* | *30* | *100* |
| Final energy, residential and commercial |  |  |  |  |
| Appliances | 0.05 | 0.10 | 0.10 | 0.30 |
| Space conditioning | 0.50 | 0.70 | 0.25 | 0.50 |
| *Value added, US(1990)$* | *800* | *5,300* | *3,700* | *9,200* |
| Final energy, industry |  |  |  |  |
| Light, drives, etc. | 0.05 | 0.20 | 0.15 | 0.15 |
| Process heat | 0.20 | 0.60 | 0.30 | 0.20 |
| Feedstocks | 0.05 | 0.20 | 0.15 | 0.10 |
| *Total passenger-km* | *4,700* | *10,400* | *10,000* | *20,000* |
| *(Cars)* | *(2,100)* | *(7,400)* | *(7,000)* | *(6,000)* |
| *Total ton-km* | *2,000* | *3,400* | *4,000* | *3,000* |
| *(Trucks)* | *(900)* | *(2,400)* | *(3,000)* | *(1,500)* |
| Final energy, transport |  |  |  |  |
| Passengers | 0.15 | 0.40 | 0.40 | 0.20 |
| Goods | 0.10 | 0.30 | 0.20 | 0.25 |

Notes: All numbers are rounded. Activity data for 1990 based on Criqui, 1998; IEA, 1997b; Sathaye *et al.*, 1989; Schäfer, 1995; Schäfer and Victor, 1997; and Schipper and Meyers, 1992. National data can vary substantially from the regional averages reported here.

# Chapter 6

# Implications

Having presented the cases and how they unfold into six energy system scenarios, we now turn to their implications. The following sections discuss implications in terms of investments and financing, international trade, energy costs, technology, energy industries, and the environment.

## 6.1. Investments and Financing

Capital investment is crucial for energy development. Both the overall development of energy systems and structural changes result from investments in plant, equipment, and energy infrastructures. Because adequate and affordable energy supplies are critical for economic growth, any difficulties in attracting capital for energy investments can slow economic development, especially in the least developed countries where some two billion people have yet to gain access to commercial energy services. And while energy investments account for only a small share of the global capital market, the availability of the capital needed for a growing global energy sector cannot be taken for granted.

Capital markets have been growing faster than total GDP for quite some time, and this trend is unlikely to change. Present annual global energy investments are approximately 7% of international credit financing of about US$3.6 trillion[1] (Hanke, 1995). With capital markets growing relative to GDP, and assuming largely stable future energy investment ratios, capital market size does not appear to be a limiting factor for energy sector finance.

Estimates of global capital requirements for energy development are often derived from back-of-the-envelope calculations based on aggregate energy investment indicators for several major energy-consuming countries that are then extended to

---

[1]Unless specified otherwise, all monetary values in this section are expressed in constant 1990 US dollars, US(1990)$.

the rest of the world. These estimates tend to be highly influenced by present market realities and short-term market expectations and necessarily incorporate a number of assumptions about the relationship between income growth and energy requirements. For example, if energy intensities are assumed to increase, capital requirements will, other things being equal, differ significantly from scenarios where energy intensities decline. Investments are likely to grow faster than GDP in the former case and slower than GDP in the latter. Capital estimates also depend greatly on the assumed costs of different technologies, including infrastructures, and the projected energy mix. As a result, comparisons among estimates of future investment requirements must recognize that each reflects a set of assumptions consistent with a specific energy–economy–environment scenario. The capital requirements presented below for the six scenarios of this study are no exception.

Investment requirements for each scenario are the result of detailed bottom-up cost calculations for the entire energy sector, extending from resource extraction (e.g., coal mining and oil exploration) through development and production to delivery of energy products to final consumers. Each technology – an oil platform, gas pipeline, liquefied natural gas (LNG) terminal, electricity generating plant, district heat grid, etc. – is characterized by a set of techno-economic parameters, one of which is investment cost in terms of US dollars per unit of installed capacity, for example, US$/kW (see Chapter 4 and *Figure 6.1*). In the long term, specific investment costs are not static. Innovation and technological learning tend to lower such costs, and future energy sector investment requirements will depend greatly on the degree to which innovation and learning improve specific investment costs, efficiencies, emissions, and other performance characteristics. Environmental regulation and resource depletion, on the other hand, tend to increase specific investment costs. In the past, innovation has more than compensated for depletion and often for environmental regulation as well. The extent to which this trend continues in the future varies among the scenarios (see Sections 2.2 and 4.4). *Figure 6.1* shows the ranges of specific investment costs assumed for several key energy technologies.

To put the energy sector investment needs, calculated in the scenarios, into context, it is helpful to first compare current worldwide energy investments to overall economic activity. During the early 1990s, global energy capital expenditures, on average, have varied between US$240 and 280 billion per year. This amounts to just over 1% of GWP. Investments are most capital intensive in the power sector, which includes generation, transmission, and distribution. Power sector investments currently account for approximately 0.7 to 0.8% of GWP (EIA, 1998a, 1998b; IEA, 1998). Upstream operations in the coal, oil, and gas sectors account for another 0.3 to 0.4% (Beck, 1995; IEA, 1998). Even if non-electric downstream investments are included, total energy sector investments in the 1990s still do not exceed 1.5% of GWP. This reflects the fact that since the mid-1980s growth in

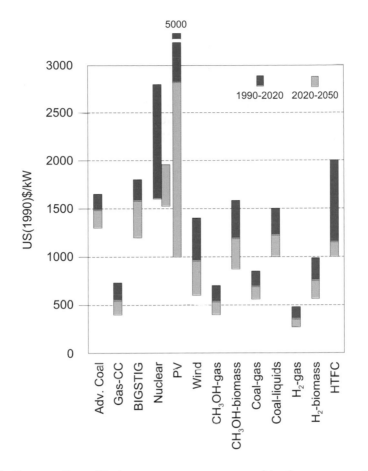

**Figure 6.1**: Range of specific investment costs assumed in the scenarios for several key energy technologies, in US(1990)$ per kW installed capacity. For abbreviations see list of acronyms.

electricity generating capacity in the OECD region has been slower than growth in actual electricity generation.[2] The faster growth in actual generation is due to considerable improvements in load factors – in part as a result of demand-side management and the appearance of independent power producers. Once load factor improvements level off, capacity and generation expansion will occur at similar rates and power sector investments as a percentage of GWP might become larger. Still, total global energy sector investments are unlikely to exceed 2% of GWP even

---

[2]Between 1985 and 1995 generating capacity in North America expanded by 2.6% per year on average while electricity generation increased by 3.9% per year. The corresponding rates for the European Union are 1.1% for capacity expansion and 2.0% for generation (EC, 1997).

under extreme conditions (e.g., rapid capacity expansion worldwide together with stringent environmental constraints).

Energy investments are thus a relatively small share of total GDP, although that share varies greatly among countries and between different stages of economic development. Regional deviations can be several times the global averages presented above. For example, recent power sector investments in Africa equal some 3% of GDP, and in China 5 to 6% of GDP (IMF, 1995). Although current average savings rates are much higher – at about 22% of GDP globally, 21% in the industrialized countries, and 24% in the developing countries[3] – it remains to be seen whether all regions will be able to mobilize the needed share of savings for energy investments. Estimates for competing infrastructural investment requirements in Southeast Asia, for example, have been reported by Desai (1996) for the period 1995 to 2004 at US(1994)$41 billion for telecommunication, US(1994)$100 billion for transportation, and US(1994)$20 billion for water and sanitation systems, compared with US(1994)$77 billion estimated for the power sector.

Within the context of economic growth, savings, and the size of capital markets, the enormous capital requirements for the energy sector in Cases A, B, and C, as shown in *Table 6.1*, are less intimidating. As in the past, regional deviations from these averages can be as high as a factor of two to three during the next decades, particularly for large energy exporters and rapidly developing countries. But on balance, the fraction of GDP that goes to energy investments in the scenarios remains relatively stable, and comparable with historical averages, from about 1 to 1.5% of GWP. It has also been emphasized that energy investments have positive spillover effects on the rest of the economy beyond the energy sector (IEA, 1998). These benefits include infrastructure improvements, access to advanced technology, know-how transfer, job creation, increased tax revenues, and improved competitiveness, especially if private sector involvement releases scarce public funds for other public investment needs that have limited attractiveness to private investors (e.g., education and health care).

Among all energy sector capital projects, oil investments remain among the most attractive. The cost of replacing or expanding production capacity for oil is a lower percentage of expected gross revenues than for other primary energy sources. Oil's advantage is likely to continue, especially in light of the increased demand for transportation services in the scenarios, and the existence of a versatile premium market will ensure the availability of necessary investment capital. For electricity, however, new capacity and infrastructures are capital-intensive investments with long amortization periods and are thus less attractive in a rapidly changing market environment characterized by deregulation and privatization.

---

[3]In the reforming economies, recent GDP declines have been matched by reduced savings, keeping the savings rate relatively constant at about 20% (IMF, 1995).

Hence, the real challenge in raising funds for energy investments will be ensuring adequate rates of return and reducing the perceived risks to investors. Returns in the energy sector do not always compare well with those of other infrastructure investments. Between 1974 and 1992, for example, electricity projects supported by the World Bank realized average rates of return of 11% per year, whereas urban development and transport returns were 23 and 21%, respectively (Hyman, 1994). Also important is the allocation of funds within the energy sector. Rate-of-return considerations discriminate against smaller-scale, clean, and innovative energy supplies, and against investments in energy efficiency improvements. Economic risk considerations make large-scale capital-intensive investment projects with long amortization periods such as nuclear power unattractive, especially to private investors who demand a risk premium on returns. Market size and product mobility often favor investments in oil exploration over, for example, natural gas or energy conservation.

In the past, in many countries much of the energy sector has been publicly owned, and in most developing countries substantial international funding has supplemented limited domestic capabilities. The share of private sector capital has usually been less than 20%. More recently, growing public and private debt in industrialized and developing countries alike has made energy sector financing, with its long amortization periods, more difficult. Privatization has become the accepted political remedy. A second development increasing the likely dependence of energy investments on private capital is stagnation in international development finance, despite an increase in international credit financing from 5% of GWP (or about US$175 billion) in 1973 to 17% (or about US$3.6 trillion) in 1993 (Hanke, 1995). Although energy financing therefore must increasingly come from the private sector, government policies can make a difference by restructuring subsidies that reduce investment risks consistent with long-term development targets, by encouraging energy prices that reflect real costs, by separating monopoly activities from potentially competitive activities (the existence of state industries is an impediment to private investment), by facilitating unrestricted access to markets, and by maintaining a stable political climate that reduces investment risks and broadens access to international capital markets. Especially in developing countries, capital market reform and a clearer separation of government and industry activities are prerequisites for attracting private sector investment both domestically and internationally. New financing mechanisms such as build-own-operate (BOO), build-own-transfer (BOT), and production sharing agreements (PSA) with unhampered financial transfers across borders need to be institutionalized. Nonetheless, the bottom line for energy investments is clear: returns must at least match opportunity costs.

**Table 6.1**: Cumulative investments in energy supply by region, 1990 to 2020 and 2021 to 2050.

| Energy investments | Case | | | | | |
|---|---|---|---|---|---|---|
| | A1 | | B | | C1 | |
| | 1990-2020 | 2021-2050 | 1990-2020 | 2021-2050 | 1990-2020 | 2021-2050 |
| Cumulative, in trillion US(1990)$ | | | | | | |
| OECD | 6.8 | 8.2 | 6.5 | 8.6 | 4.1 | 3.4 |
| REFs | 2.6 | 4.0 | 1.9 | 3.2 | 1.7 | 1.7 |
| DCs | 5.0 | 11.3 | 3.9 | 10.6 | 3.7 | 9.0 |
| World | 14.3 | 23.4 | 12.4 | 22.3 | 9.5 | 14.1 |
| As share of GDP, in percent | | | | | | |
| OECD | 1.0 | 0.7 | 1.0 | 0.8 | 0.6 | 0.4 |
| REFs | 7.7 | 3.1 | 6.4 | 4.2 | 5.1 | 2.2 |
| DCs | 2.2 | 1.4 | 2.0 | 2.0 | 1.7 | 1.4 |
| World | 1.5 | 1.1 | 1.4 | 1.3 | 1.1 | 0.8 |
| Per unit of primary energy, in US$(1990)/toe | | | | | | |
| OECD | 45 | 44 | 46 | 52 | 34 | 33 |
| REFs | 46 | 44 | 42 | 52 | 33 | 34 |
| DCs | 32 | 35 | 27 | 39 | 28 | 39 |
| World | 40 | 39 | 37 | 45 | 31 | 37 |

*Table 6.1* shows the cumulative energy sector capital requirements for Cases A, B, and C according to traditional definitions of energy investments. Thus it includes capital for production capacity, for transmission and distribution infrastructures, and for complying with environmental standards. It does not include investments in traditional end-use technologies such as furnaces, appliances, and vehicles, as they are traditionally counted as durable consumer goods or business investments.[4] Especially for Case C, which relies on accelerated rates of energy efficiency improvements, these investments in end-use technologies would be substantial and could approach the investments in energy supply proper. The fact that the performance of end-use technologies plays such an important role in all three cases in this study is a strong argument in favor of new approaches to evaluating energy sector investments. Integrated resource planning, for example, has begun to extend the traditional energy sector perspective to take into account investments in end-use technologies. Approaches that assess both supply and end-use options, and both expansion and conservation will be increasingly essential in all the futures represented by the three cases.

---

[4]For a discussion of new end-use devices linking both energy supply and demand, see Section 6.4.

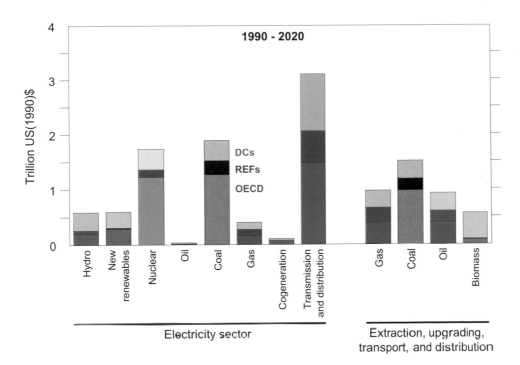

**Figure 6.2**: Breakdown of global cumulative energy sector investments for Case B from 1990 to 2020, in trillion US(1990)$.

For each of the three cases, the results shown in *Table 6.1* are based on the specific investment cost estimates and learning-curve effects discussed in Chapter 4 (see also *Figure 6.1*). Looking first at the cumulative capital requirements from 1990 to 2020, the range across the cases is between US$10 and 14 trillion. The developing region's share rises from today's 25 to 30% to between 32 and 39%. Looking at energy investments as a share of regional GDP, the reforming countries rank the highest, diverting 5.1 to 7.7% of regional GDP to the energy sector. They are burdened with slow initial economic growth and at the same time they need to replace obsolete energy infrastructures, and it is likely to be extremely difficult to attract the needed capital to the energy sector. Developing countries invest 1.7 to 2.2% of GDP in the energy sector, and OECD region investments are the lowest at 0.6 to 1.0% of GDP. By and large, it takes a greater effort to build up an energy infrastructure than to expand and maintain an existing one.

To illustrate what goes into the cumulative capital requirements of *Table 6.1*, *Figure 6.2* breaks down the different components for Case B. The figure shows that investment in electricity generation is dominated by the OECD, especially for the expansion of nuclear and coal-fired capacity. New renewable and hydropower investments are concentrated in developing regions. Given the current economic unattractiveness of most new renewables, this reflects substantial learning-curve effects. More than US$250 billion are invested in the development of a biofuel production infrastructure. The trend toward convenient grid-dependent final energy forms is mirrored by both the investment volume and the regional breakdown in the figure.

From 2020 to 2050, capital requirements in *Table 6.1* grow substantially in absolute terms, but still more slowly than GDP. This is true in all scenarios. It reflects, first, the shift from supply-side investments (shown in *Table 6.1*) to end-use technology and infrastructure investments (not included in *Table 6.1*). Detailed estimates of the latter are not possible, but they are likely to double the investment requirements. Second, the declining share of GDP going to energy investments reflects continued progress along technological learning curves throughout the energy system.[5] Finally (and most important), capital requirements grow slower than GDP because of the energy intensity improvements in all three cases beyond the effects of technological change, for example, due to economic, structural, or lifestyle changes.

The figures in *Table 6.1* are lower than some estimates in the literature and are higher than others. For example, WEC (1993) estimated the global capital requirement for 1990 to 2020 at US$30 trillion. This included efficiency-improvement-related investments of approximately US$7 trillion (which are excluded here), but excluded learning-curve effects in lowering future investments (which are included here). Once both corrections are made, the figures are comparable. The International Energy Agency (IEA, 1998) and the Energy Information Administration (EIA, 1998b) estimate an average of US$130 to 150 billion in power sector investments through 2020, which is below the range in the scenarios (US$220 to 350 billion). The different investment requirements reflect how the IEA and EIA assumptions about the evolution of the underlying generating capacity mix differ from those in our scenarios. By 2020, for example, both IEA and EIA assume no real increase in nuclear capacity worldwide, whereas in this study's scenarios there is an increase.[6] Also, EIA assumes a much more modest contribution from renewables, particularly in developing countries. In the IEA and EIA analyses electricity generation fueled by natural gas takes up the slack created by the lower nuclear and renewable capacities. As electricity generation from natural gas has considerably lower capital costs, overall capital requirements are lower. Conversely, gas infrastructure investments need to be included in the analysis, as is done in our approach.

---

[5]Had these been excluded, the capital requirements of the electricity sector in OECD would have been 8 to 15% higher for 1990 to 2020. In developing regions, the impact would have been greater, an increase of 25 to 40%, due to heavy investments in new renewables.

[6]IEA's estimates are based on an average capital cost of US$937 per $kW_e$ of generating capacity independent of fuel and technology and without accounting for necessary investments in a corresponding expansion of transmission and distribution capacity.

Finally, IEA and EIA energy demand estimates are shorter-term than those of this study. The longer-term scenarios need to incorporate foresight, taking into account resource needs over the entire study horizon. Therefore, natural gas in the long term needs to be reconciled with short-term return on investment considerations which favor higher gas use in the power sector.

Annual capital requirements for energy investments rise from less than US$240 billion in 1990 to US$370 to 570 billion by 2020, and to US$520 to 880 billion by 2050. A large share of this investment would probably need to be externally financed. Hyman (1994) calculates that a third of the global capital spending based on WEC (1993) estimates as electricity needs would be externally financed. This implies that a large share of total energy investments would also need to come from international capital markets or development assistance. These investment requirements compare with total funds transferred to developing countries in 1990 of about US$140 billion, a total debt service for these countries of about US$150 billion, and total official development assistance from the OECD countries of about US$50 billion (Grubb *et al.*, 1993).

## 6.2. International Trade

How the global economy and hence the energy system develop in the future will greatly depend on future prospects of free trade. Cases A, B, and C incorporate different expectations. Case A reflects a future moving aggressively toward free trade among all regions, although temporary tariff protection for infant industries, especially in developing regions, does remain an option. Geopolitical supply security considerations are waived in favor of a truly cooperative world. Case B is less cooperative. It incorporates a number of continuing trade barriers, although these do not greatly affect energy trade. For Case C, trade constraints reflect the sustainable development objectives central to the case. Free trade is permitted in commodities, technologies, and energy forms that meet Case C's high environmental standards. Those that do not are strictly regulated.

An analysis of the trade flows implied by the scenarios reveals a general decline in the share of primary energy (equivalent) that is traded. Currently, 17.6% of global primary energy is traded among regions as they are defined in this study (see *Box 1.1*). This figure is in close agreement with the true country-by-country figure for 1990 of 18.5%, indicating that our regional disaggregation reflects actual energy trade patterns quite well. Crude oil and oil products are currently dominant, accounting for 78% of global energy trade; coal accounts for 13%, and natural gas for 9%. By 2050, the percentage of primary energy that is traded declines to a level of between 11% and 16%. Absolute volumes continue to increase – by a factor of 1.7 to 2.5 for Case A and a factor of 1.7 for Case B. The increase in Case C is much lower at between 10% and 40%. Energy trade in Case C is limited primarily to sustainable energy forms (e.g., biomass, methanol, or ethanol) and actually shrinks beyond 2050.

The overall geopolitical shift in energy use from North to South across all scenarios is also reflected in energy trade. In 1990, OECD imports accounted for 84% of international energy trade. By 2020, the OECD shares drop to between 55% in Case C and 65% in Case B, and by 2050 the range is between 10% in Case C and 34% in Scenario A2. This geopolitical shift is likely to erode the current position of OECD countries as the dominant energy buyers. Conversely, import security concerns that traditionally have been strong in import-dependent Western Europe and Japan will increasingly be shared by developing regions. Concerns in terms of absolute import needs will also grow in developing countries compared with the OECD.

Crude oil and oil products remain the most traded energy commodities through 2050. The spread is quite large, ranging between 77% in Scenario A1 and 33% in Scenario C1. After 2050, methanol, piped natural gas, and LNG become the key traded energy commodities (see *Figures 6.3* and *6.4* for Scenario A3 and Case B, respectively). Electricity is an important component of regional energy trade and is thus considered in the scenarios, but does not play an important role in global energy trade.

In general, the global energy trade pattern shifts from primary energy to secondary energy, which improves trade flexibility and thereby lowers geopolitical concerns. For example, methanol and hydrogen can be produced from a number of primary sources ranging from coal to biomass. Biofuel and eventually hydrogen production leave more value added in the exporting regions than the export of primary energy. Exporting secondary energy forms becomes a staple source of income for a number of developing regions.

*Figure 6.5* illustrates one consequence of the decrease in the share of energy that is traded. It shows the international oil price versus exports over time for MEA. Each of the dashed lines represents a constant level of revenues. The historical trajectory of oil prices and exports from the region has moved counterclockwise through very volatile changes in total revenue and prices.[7]

In general, prospects for oil exporters are bright in the long run, and, at least through 2050, oil revenue is unlikely to be below US$140 billion per year for MEA. But there are differences among the three cases. In Case C, environmental policies to reduce fossil fuels cause declining exports, but rising prices keep revenue constant. In Cases A and B, technological change and the speed at which reserves are

---

[7]Prior to 1990, the revenues in the figure are equal to quantity times "averaged" price. After 1990, revenues are calculated from model-determined shadow prices of internationally traded oil for any given year. Revenues and calculated shadow prices are therefore higher than if calculated with average prices. For further discussion of shadow prices and costs, see Section 6.3.

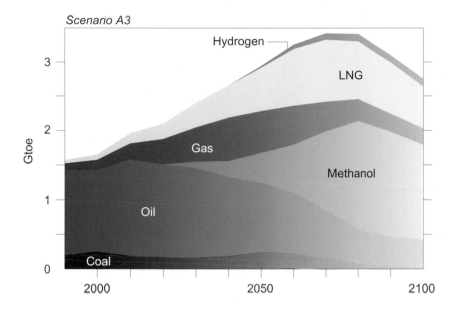

**Figure 6.3**: Global energy trade, Scenario A3, 1990 to 2100, in Gtoe.

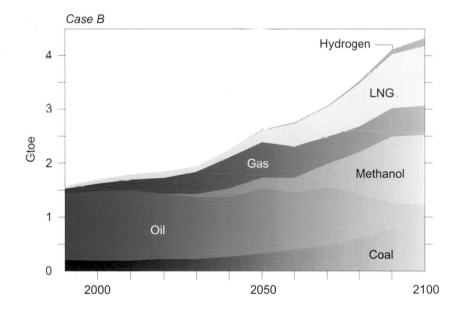

**Figure 6.4**: Global energy trade, Case B, 1990 to 2100, in Gtoe.

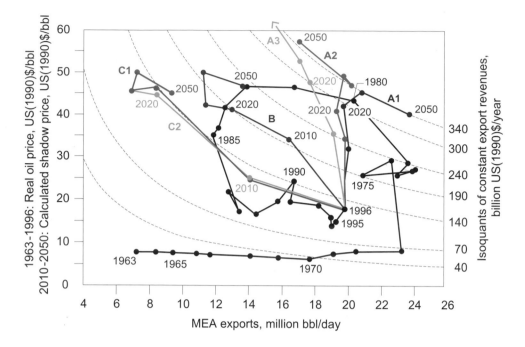

**Figure 6.5**: Oil export quantities and revenues for MEA, historical development from 1963 to 1996 and in scenarios to 2050.

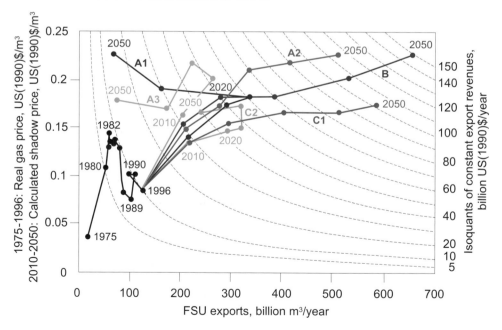

**Figure 6.6**: Natural gas export quantities and revenues for FSU, historical development from 1975 to 1996 and in scenarios to 2050.

replenished determine prices, export volumes, and revenues. In Case A, greater technological progress than in Case B enables higher exports at slightly elevated prices, and long-term revenues may exceed US$300 billion annually. The slower the rate of technological change, the more important the price component becomes in revenue generation. Export volumes slip as reserves are replenished more slowly, prices rise, and revenues vary as a function of the scenario-specific oil substitution possibilities. Long-term export revenues for the region exceed US$300 billion in all Case A scenarios and are at least around US$200 billion in Case B, and are thus substantially higher than at present.

The other major potential exporter besides MEA is FSU, where natural gas will be the principal energy export. *Figure 6.6* shows export quantities, prices, and revenues for FSU gas exports for all six scenarios. Gas exports increase in all scenarios from 88 Mtoe in 1990 to a relatively narrow range between 250 and 270 Mtoe in 2020.[8] Exports increase further to 2050, except in Scenarios A1 and A3 where regional demand and relative prices make the gas more valuable within the region. By 2050 exports range between 65 Mtoe (Scenario A1) and 610 Mtoe (Case B), where revenues could reach US$150 billion. Export revenues are likely to be at least US$50 billion per year by 2020 and exceed US$100 billion by 2050 in three out of six scenarios. These values are 5 and 10 times the 1990 revenues, respectively.

## 6.3.    The Costs of Supplying Energy

One question is always asked of any study on future energy developments: What about energy prices? Mindful of the generally poor track record of past energy price forecasts (see Schrattenholzer and Marchant, 1996), we consider scenarios based on physical rather than monetary indicators to be more appropriate for exploring the very long term (i.e., several decades ahead). Inferred prices and elasticities based on cost information in the scenarios have been used as *ex post* consistency checks.

The price of energy comprises many components: costs for establishing and maintaining the production, conversion, and transport and distribution infrastructure of energy supply; profit margins; a whole host of levies such as royalties and taxes raised at the points of energy production or use; and, finally, consumers' willingness to pay for quality and convenience of energy services. Also, given the importance of energy and the vast volumes traded, prices are influenced by a whole range of additional factors, from inevitable elements of speculation all the way to geopolitical considerations, all of which can decouple energy price trends from any underlying physical balance between supply and demand. Taxes are especially

---

[8]For FSU natural gas, 1 billion m$^3$ per year equals 0.935 Mtoe.

significant. In a number of OECD countries, up to 80% of the consumer price of gasoline is taxes (OECD, 1998b), and differences between countries are enormous. In 1997, 27% of the price of gasoline in the USA was taxes, compared with 78% in France. Even among large oil producers (and exporters) taxes vary substantially. In Mexico taxes are 13% of gasoline prices, but in Norway they are 75% (OECD, 1998b). Given pressures to reduce national deficits, reduce traffic congestion, and limit local air pollution and global warming, we believe that energy taxes are more likely to rise in the future than to fall, but any more specific projection would be well beyond the scope of our analysis.

Because long-term behavior of energy prices is nearly impossible to anticipate, and given the poor track record of past energy price forecasts, we constructed the scenarios on the basis of energy system costs, not prices, plus physical constraints such as resource availability and technological efficiencies. After the fact, however, we can draw inferences about how prices in the scenarios might behave, *other things being equal*, and that is the purpose of this section.

The energy models used to quantify the six scenarios (see Appendix A) calculate both average energy costs and so-called shadow prices. The average costs calculated by the models are reliable and straightforward indicators of cost behavior within the scenarios. The disadvantage is that they do not include other crucial components of the price formation mechanism and grossly underestimate possible future prices. Marginal costs, on the other hand, are equal to prices when supply and demand are at equilibrium. The trouble is, however, that the shadow prices in the models do not exactly correspond to marginal costs, although they are often interpreted as marginal costs. Shadow price estimates should therefore be seen for they are – model-derived indicators of energy system constraints. They are not scenarios of future energy prices.

*Figure 6.7* presents shadow price and average cost trends for oil, gas, coal, and renewables for the six scenarios. The costs shown in *Figure 6.7* are the costs for each of the energy forms as they are traded on the international market – oil, gas, coal, and renewables in the form of biofuels. These trade costs do not include regionally specific cost components such as pipelines, refining capacity, and transport and distribution costs. Such costs are accounted for at the level of the 11 world regions. The shadow prices in *Figure 6.7* are the same as those used to calculate possible export revenues for MEA and FSU in *Figures 6.5* and *6.6*. Both average costs and shadow prices of energy trade are normalized to 1990 indices.[9] Because average costs are lower than shadow prices, the 1990 indices in *Figure 6.7* for average costs are less than one.

---

[9]For comparison, in 1990 representative average prices were US$25 per barrel for oil, US$3 per million British thermal units (MMBtu) for gas, and US$40 per ton for coal (BP, 1997). For renewables no large-scale international trade exists. For an example of a representative price consider Brazilian ethanol at US$65 per barrel (Goldemberg, 1996).

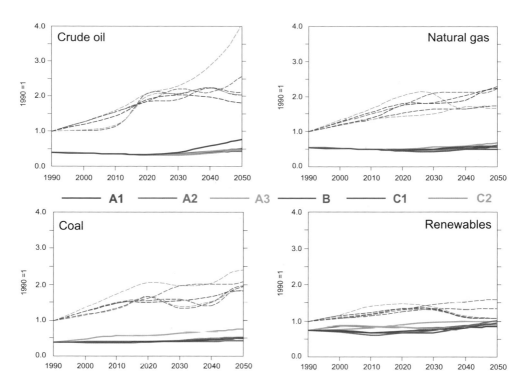

**Figure 6.7**: Trends in average costs (solid lines) and shadow prices (dotted lines) for globally traded energy forms – oil, gas, coal, and renewables (biofuels) – shown as an index with average shadow price in 1990 equal to 1.

The shadow prices in *Figure 6.7* can be interpreted as a possible upper bound on energy price trends. We interpret the average costs in *Figure 6.7* as a lower bound on energy price trends. As noted above, model calculations are more stable and reliable for average costs than for shadow prices. Setting prices below average costs for any length of time would put producers out of business.

The good news from *Figure 6.7* is that, given the scenarios' combination of adequate energy resources, free trade, and future cost reductions through improved technology (see Section 4.4), average costs rise only very gently over the next two to three decades. The overall increase, the rates of change, and their volatility are all greater for shadow prices. Overall, increases and differences among scenarios are greatest for oil, suggesting that uncertainty and volatility in oil markets are unlikely to diminish.

The second important conclusion from *Figure 6.7* is that shadow prices generally remain lowest for energy sources that are the focus of investment and technological RD&D and prove highest when anticipatory investments are lacking. Rising costs therefore contribute to a larger degree to energy intensity improvements in Case B than in the other scenarios with a more dynamic technology outlook. For oil, for example, shadow prices increase most in Scenario A3 due to high demand growth combined with limited conventional oil resources and no investments in unconventional oil. Average costs stay low, however, because the scenario increasingly uses (cheap) oil substitutes. Oil shadow prices are lowest in Scenario A1 where technological progress makes it possible to tap large amounts of unconventional oil throughout the 21st century. Relatively low demand for oil in Case C keeps shadow prices low initially. Around 2020 environmental constraints cause shadow prices to increase, but any subsequent upward pressure is balanced by lower demand than in other scenarios and by improvements in the alternatives to oil.

For gas, shadow prices increase gradually, with higher medium-term costs in Scenario A3 where gas is the preferred transitional fuel in the shift to post-fossil fuels. In this case, required investments in developing new gas fields and pipelines result in a "bulge" of shadow prices around 2030 that could translate into increased price volatility. Otherwise, large gas resources and the timely development of alternatives keep shadow prices and average costs rather flat, *provided* near- to medium-term investments promote the necessary technological improvements to ensure access to resources and required infrastructure buildup.

For coal, shadow prices rise when coal demand is high, as in Scenario A2 and Case B, and are comparatively flat when alternatives are more available, as in Scenario A1 and Case C. They rise most in Scenario A3, where social externalities are incorporated into the economics of coal mining.

New renewables are in their infancy compared with oil, gas, and coal. Currently, demand is low and the potential for cost improvements is large. As a result, shadow prices rise only very gradually with time. Keeping costs low, however, requires upfront investments and RD&D, as is best illustrated in the high-growth, "bio-nuc" Scenario A3. This scenario has the fastest initial cost increases of the six scenarios due to higher upfront investments. These stimulate technological learning and cost reductions so that by 2050 shadow prices in Scenario A3 drop below those in other scenarios where no similar "learning by doing" takes place.

The overall message of *Figure 6.7* is that average energy system costs in the scenarios are almost flat through 2020, and shadow price increases are on average less than 1.5% per year (oil in Scenario A3 is the only exception). As noted above, energy price formation includes much more than just energy costs. But to the extent that costs determine prices, real energy price increases corresponding to the scenarios are between zero and about 1.5% per year – reassuringly gentle. To the extent that prices determine costs, that is, low prices in competitive markets tend to promote cost reductions, the scenarios suggest the need for continuous productivity and efficiency improvements. There are two major caveats. The first is that each scenario incorporates early investments that, for specific energy sources, lead to technological improvements, resource expansion, and reduced costs. In the absence of such near- to medium-term investments, long-run costs will become much higher as today's conventional resources are depleted. Without technological improvements, energy supplies will be constrained by geology, and those constraints will drive price increases. *Figure 6.7* suggests that such increases would not be felt before 2020, but without early investments looking beyond the current generation of energy facilities, technologies, and fuels, costs will rise more quickly after 2020. The second major caveat is that gradual long-term averages mask the potential for price volatility suggested by the ups and downs of shadow price curves in *Figure 6.7*. Prices are potentially volatile even without considering all the non-cost components of energy prices. *Figure 6.7* suggests that the potential volatility will increase as the current generation of energy facilities are retired by 2020 and beyond.

Thus while prices matter greatly in the short term, technology matters in the long term. The ultimate question is whether geology or strategic investments in technological progress will determine long-term costs and prices.

## 6.4.  Technology

Technology is the key determinant of economic development and is essential for raising standards of living and for easing the burden humanity imposes on the environment. Technological progress is central to all three cases, A, B, and C. It is based on human ingenuity and is thus a man-made and effectively renewable resource as long as it is properly nurtured. But it has a price. Innovation, especially the commercialization of novel technologies and processes, requires continual investments of effort and money in RD&D. In turn, technology diffusion depends on both RD&D and steady improvements through learning by doing and learning by using.

Innovation and technology diffusion require both that opportunities are perceived and that the entrepreneurial spirit exists to pursue them. Long-term scenarios cannot forecast future technological "winners," but they can indicate areas of technological opportunity. Where these are consistent across a range of different scenarios, we conclude that they are especially good candidates for long-term strategic technology investments. The most promising possibilities suggested by the scenarios are shown in *Figure 6.8* and *Box 6.1*.

*Figure 6.8* illustrates for 2020, 2050, and 2100 the global market potential in the scenarios for four classes of energy technologies: new end-use energy devices, power plants, synfuel production, and energy transport, transmission, and distribution infrastructures. For each of the four classes, the minimum, maximum, and

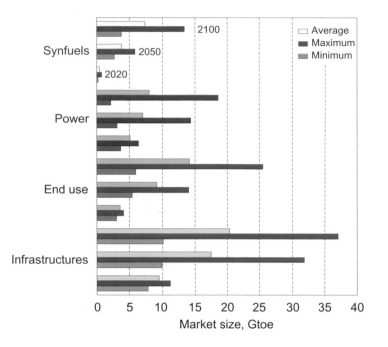

**Figure 6.8**: Global market potentials for four classes of energy technologies: power plants, synfuel production, new end-use energy devices, and energy infrastructures. Minima, maxima, and averages across the six scenarios are shown for 2020, 2050, and 2100, in Gtoe.

average market potential for the six scenarios are shown. Market potential is measured in terms of the aggregated (secondary) energy that is delivered.[10] A more detailed breakdown is shown in *Box 6.1*.

Across the wide variation in possible energy developments depicted in the six scenarios, the importance of energy infrastructures grows persistently. Even in the low-demand scenarios of Case C, energy infrastructures deliver at least 10 Gtoe per year by 2050. By the end of the century they average 20 Gtoe per year across all six scenarios, reaching close to 40 Gtoe per year in the highest growth scenarios. The markets for power sector technologies also grow substantially, with a wide spread between the maximum and minimum scenarios. By 2050, the range is between 3 Gtoe per year and 14 Gtoe per year. Part of this spread relates to uncertainties about demand growth, but part of the spread arises from energy end-use innovations in the form of new, on-site decentralized electricity generation technologies such as photovoltaics or fuel cells. The potential for end-use technologies in the long

---

[10]For end-use technologies delivering energy services (like mobility) rather than energy, the figure shows final energy input. The figure focuses on new end-use energy technologies and does not include traditional appliances and heating systems.

### Box 6.1: The market for future energy technologies

The market potential in the 21st century for energy technologies in the form of infrastructures, power plants, synfuel production, and decentralized end-use devices is large. Ranges for 2020 and 2050 are illustrated below. Mindful of the dangers of trying to "pick winners," and consistent with the aggregate representation of technology in our energy models, only generic technologies are listed. We do not distinguish between solid oxide, molten carbonate, phosphoric acid, and solid polymer fuel cells, for example. Opportunities for hybrid and transitional technologies are also wide open – on-board steam reforming, for example, or partial oxidation could provide hydrogen for fuel cell vehicles while continuing to use existing oil distribution infrastructures.

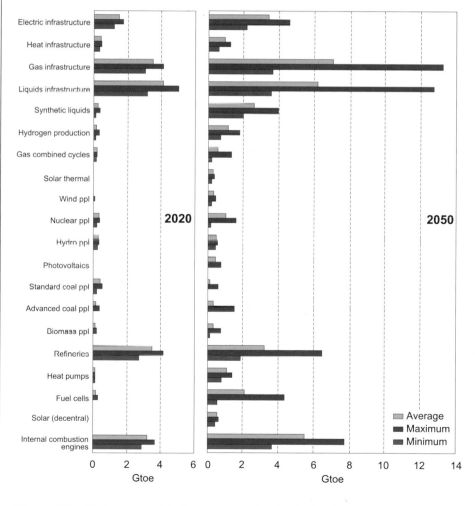

**Figure 6.9:** Market potentials for energy technologies by 2020 and 2050, in Gtoe. For explanation see text; ppl = power plants.

term outgrows that of the power sector. The most important customers for energy technologies would no longer be a limited number of utility managers but rather millions of energy consumers worldwide. Synfuels also emerge in the long term as a major technology market. An orderly transition away from conventional oil and gas translates into large technology markets for synliquids, syngas, and, in the long term, hydrogen produced from both fossil fuels (coal and natural gas) and renewables (biomass). By the end of the 21st century the synfuels market could be at least 4 Gtoe per year, comparable to the current oil market.

As noted at the beginning of this section, technological progress has a price – continual investments in RD&D. And all of the technological improvements that occur in the scenarios and that are reflected in the expansion of all categories shown in *Figures 6.8* and *6.9* presume steady RD&D investments. To state the overall conclusion reflected in *Figures 6.8* and *6.9* more broadly, RD&D in clean and efficient energy technologies and infrastructures is the only real hedging strategy for a future full of uncertainties and surprises.

Given the importance of strategic investments in RD&D, it is a cause for concern that RD&D expenditures are currently declining. The decline is not restricted to any one country or group of countries. It appears to include all industrialized countries, and it is these countries that account for most of the world's RD&D. Private RD&D is declining along with public RD&D, with private sector investments in energy-related RD&D, for example, falling by nearly a third in the USA in the past five years (Yeager, 1998). Evidently, upfront RD&D expenditures are increasingly viewed as too expensive in markets where maximizing short-term shareholder value takes precedence over longer-term socioeconomic development and environmental protection.

The scenarios allow two additional, more specific conclusions about investment strategies. First is the importance of improving and expanding infrastructures for clean grid-dependent energy forms. For energy infrastructures, the backbone of the energy system, the impact of technological change can be particularly slow because of the slow rate of capital turnover. Benefits also tend to be long-term and include significant positive spinoff effects, or externalities, from which a private investor cannot profit directly. On the other hand, such infrastructures have the advantage that their value is largely independent of which energy sources dominate future energy systems – the electricity grid can transmit electricity from coal as easily as electricity from renewables or nuclear power. Gas pipelines can transmit natural gas, syngas, and in the longer term also hydrogen, e.g., in the form of hythane. But because short-term incentives for such needed infrastructural investments are low, these investments are an area where creative government intervention is appropriate and public-private partnerships desirable. From a long-term global perspective, the improvement and expansion of energy grids in Eurasia should be given high priority.

The second more specific conclusion concerns the importance of end-use technologies. Although the traditional focus of energy technology policy has been on energy supplies, a clear result of this study is that long-term improvements in energy system costs, efficiency, and environmental impacts will increasingly be determined by technological innovations at the level of energy end use. Generic technologies that are at the interface between energy supply and end use (e.g., gas turbines, fuel cells, and photovoltaics) are therefore a robust bet for enhanced RD&D and diffusion efforts, and could become as important as today's internal combustion engines, electric motors, and microchips. Over the long term much of today's traditional, centralized energy conversion may even shift toward the end user. For example, hydrogen-powered fuel cell vehicles could, when parked, be a source of clean residential electricity. Instead of moving electrons, hydrogen molecules would be moved. In this case the importance of the energy infrastructure does not diminish, but instead grows.

The conclusion that the point of final energy use is where we expect far-reaching technological improvements to occur has two additional implications. First, it weakens the argument for extensive RD&D investments in large, sophisticated, "lumpy," and inflexible technologies such as fusion power and centralized solar thermal power plants. Improvements in end-use technologies, where millions, rather than hundreds, of units are produced and used, are more amenable to standardization, modularization, mass production, and hence exploitation of learning-curve effects. Second, institutional arrangements that govern final energy use and supply are critical. Deregulation and liberalization of electricity markets can create incentives in this direction as service packages are tailored to various consumer preferences and especially as traditional consumers can sell electricity back to the grid. But there are also concerns that liberalization will discourage longer-term RD&D investments by emphasizing short-term profits (see Section 7.6).

The fact that increased attention should be devoted to technological innovation focused on energy grids and energy end use does not in any way mean that the importance of primary energy and centralized conversion energy technologies will diminish. Fossil fuels, particularly natural gas, are central to energy supply during the next few decades in all scenarios, and the more efficiently and cleanly we can use them the better. In the short term, the message of the scenarios is that supply-related RD&D should be focused on developing clean fossil technologies, increasing efficiencies, and improving geological understanding of conventional and unconventional resources, particularly gas and oil. For the long term, RD&D on energy supplies should anticipate the grand transition from fossil to non-fossil energy sources. For example, experience with synthetic fuels from fossil resources will speed the eventual diffusion of synthetic fuels from non-fossil resources. A range of non-fossil technologies should be explored along with new and inventive conversion cycles and novel combinations of energy sources and fuels. The emphasis should be on multipurpose technologies that can be adapted to different energy sources and energy carriers. The objective is to reduce the risk of becoming locked-in to technologies and energy sources that turn out to be less advantageous as uncertainties about the future resolve with time.

Finally, the consistent emphasis on clean forms of final energy suggests that in the long term meeting even high levels of end-use energy demand can be consistent with regional and global environmental protection. For the near term it is important to accelerate the diffusion of existing environmental technologies – scrubbers, unleaded gasoline, and better stoves in rural areas of developing countries. But for the benefit of the longer term, it is important that near-term solutions do not lock in existing fossil-based technologies and slow the introduction of cleaner fossil and non-fossil alternatives. The challenge is to speed the diffusion of current clean technologies while increasing the flexibility of the energy system to explore, incorporate, and profit from yet cleaner and more efficient alternatives.

## 6.5.    Implications for Energy Industries

All three cases reflect substantial growth for all energy industries through at least 2020. This growth is based on the assumption incorporated in all cases of increasingly open markets in which subsidies favoring established industries are removed and a more "level playing field" becomes a reality. A more level playing field will create substantial reshuffling within and among energy sectors. The result is likely to be characterized by one or more of the following labels: intersector integration, energy cascading, organizational flattening, service orientation, global competition, common carrier or third-party transport, and integrated resource planning. These factors will have a much larger impact than the production profiles summarized in Chapter 5. New business opportunities will emerge, centered primarily on the conversion of non-oil sources to liquid end-use fuels and, in the distant future, to hydrogen. Coal, natural gas, biomass, nuclear power, and solar energy may all find new markets in synfuel, methanol, and hydrogen production. In the longer run, the present separation into coal, oil, and natural gas industries will probably be less distinct.

Although the next two to three decades offer ample opportunity for all the present energy sectors to do extremely well, the cases show diverging prospects after 2020, in which different energy industries embark on often mutually exclusive development paths. *Figures 6.10* and *6.11* illustrate the divergence in contributions of fossil and non-fossil energy sources, respectively, by showing the minimum and maximum production of coal, oil, natural gas, biomass, new renewables, and nuclear energy across the six scenarios.

### 6.5.1. Coal

The future of the coal industry changes across the scenarios, ranging between boom and bust. This is due in part to the potential range of technological progress in competing industries and in part to environmental policy. An almost certain impact on the coal industry stems from consumer preferences for clean and convenient end-use fuels. A common trend in all scenarios is the long-term virtual elimination of coal as a direct end-use fuel and its replacement by coal-derived electricity and synfuels.

That coal has the largest conventional reserves of all fossil fuels is both its strength and its weakness. Coal's growth prospects rise whenever oil and gas reserve replenishment falls behind expectations or the outlook for nuclear turns bleak. In those cases, coal readily fills the gap in electricity generation and, after 2020, in the production of synliquids. Likewise, coal gasification can eventually partially replace dwindling natural gas supplies. The coal sector, however, must increasingly be integrated into liquid and gaseous fuel production and delivery systems, and must place a growing emphasis on services and service businesses. In Case B, approximately 100 Mtoe of coal-based liquids need to be produced by 2020, growing substantially to 660 Mtoe by 2050.

If technological progress and exploration allow the full development of oil and gas resources, coal's future role is likely to be smaller. Even in this case, coal remains an important energy source, especially in the developing parts of the world with large coal resources. Coal's outlook will also worsen if environmental constraints are substantially tightened. Policies to reduce $SO_x$ and $NO_x$ will increase costs while policies to reduce greenhouse gas emissions could hurt coal twice – coal production contributes significantly to methane releases, and coal is the fuel with the highest specific $CO_2$ emissions. In view of the large availability of coal and growing energy needs, the scenarios suggest that over the next century more coal will be used than has been used cumulatively to date. This will require continued, stepped-up efforts to improve safety in mining and there will be a need to develop clean technologies for coal conversion, because coal in the future will be marketed as electricity, gas, and liquids.

### 6.5.2. Oil

The oil industry appears to have a long future ahead of it. As long as the industry achieves a smooth transition from conventional to unconventional sources, oil products will continue to benefit from their versatility, high energy density, and existing

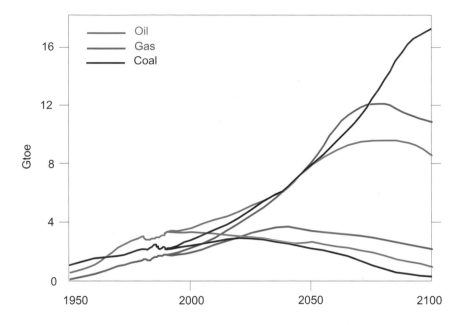

**Figure 6.10**: Minimum and maximum annual coal, oil, and natural gas production, 1950 to 2100, primary energy in Gtoe.

distribution infrastructure. There is likely to be a continuation of the trend since the mid-1970s toward concentration in the transportation and feedstock markets.

In scenarios with higher rates of technological progress, oil production becomes less geographically concentrated. By and large, producers in the Middle East continue producing inexpensive conventional oil. They have little incentive to move to unconventional oil, which is therefore first developed outside the area. This creates a risk that producers in the Middle East might temporarily take advantage of their inexpensive reserves to set the market price and to cause uncertainty and slow the capital-intensive buildup of unconventional production capacity elsewhere. However, our scenarios indicate that a cooperative approach maximizes long-run benefits to Middle Eastern and other producers alike.

If technological change is insufficient to allow the economic production of unconventional oil sources, then the Organization of the Petroleum Exporting Countries (OPEC) will quickly regain the ground lost earlier. The downstream sector may be required to integrate liquid products derived from coal, natural gas, and biomass as early as the first decades of the 21st century. In fact, in the absence of technical progress, oil could be the only sector facing declining production rates after the 2020 period. In the worst case, oil production declines to 80% of 1990 levels by 2050 (Case C). Still, current reserve-to-production ratios combined with modest technical change are sufficient to guarantee the oil industry a secure niche until at least 2050.

### 6.5.3. Natural gas

In the long term, natural gas has the best prospects among the fossil sources in the scenarios presented here: gas production at least doubles between 1990 and 2050, and could increase by a factor of four to five. Natural gas benefits from the commercialization and diffusion of highly efficient and economically attractive conversion technologies. Other important factors in the success of natural gas include its cleanliness, its largely underground invisible local distribution infrastructures, its high quality as a feedstock for low-cost methanol production and hydrogen, its ability to be used directly as an end-use fuel or as a primary energy source for electricity generation, the advent of new liquefaction technology, and its complementarity to major oil markets.

Still, bright prospects in the long term are no cause for complacency now. The realization of this future will require aggressive exploration and resource development. A gas pipeline infrastructure needs to be put in place on a continental scale in Asia. Future upstream gas operations, transmission, and distribution must further reduce leakage. Otherwise, the obvious environmental advantages of gas may be offset by methane releases to the atmosphere, especially if production and use increase severalfold. Deregulation will increase organizational flattening – at the expense of present monopolies – leading to increased operational independence, responsibility, and risk for production, transmission, distribution, and load management. In essence, deregulation will foster a shift from selling primary energy to selling energy services. Interregional natural gas trade in the scenarios expands by factors of three and two (medians) for piped natural gas and LNG, respectively. Such trade volumes require the simultaneous construction of long-distance pipelines and an infrastructure for cryogenic production, storage, regasification, and shipping. Institutionally, natural gas must liberate itself, at least temporarily, from its past as the stepsister of the oil sector. Gas exploration, production, processing, and transmission are distinctly different from the equivalent processes for oil. In many instances, the application of oil technology has hampered natural gas development. At the same time, some integration may be required to further reduce the environmentally adverse and economically doubtful practice of flaring associated gas from oil production. Finally, natural gas needs to develop new markets, especially as a staple source for transportation fuel. Three possibilities are included in the scenarios: compressed natural gas, LNG, and conversion to methanol. By 2020, up to 70 to 90 Mtoe of methanol are generated from natural gas, with a growth potential of up to 330 Mtoe by 2050.

### 6.5.4.  Renewables

In all scenarios there is a significant expansion of renewables along classic diffusion patterns of slow initial growth followed by more rapid expansion. Without upfront investments in RD&D and niche market applications, bullish future markets will not materialize. Diffusion proceeds via a succession of specific market niches, and it is difficult to use policy interventions to bypass the process. However, despite a slower start for renewables than in other studies, our outlook is optimistic in the long term. The renewables portfolio is extremely diverse across sources and regions, with the central challenge being that of matching the diverse sources to consumer demands for increasingly flexible, convenient, and clean energy. This bullish outlook is also reflected in *Figure 6.11* in the quite small medium-term variation of biomass and new renewable production across the six scenarios compared with nuclear energy in the same figure and fossil sources in *Figure 6.10*.

The OECD currently dominates uses of "modern renewables." Because of its technology, RD&D capability, and capital availability, the OECD needs to take the lead in developing and introducing "new" renewables. This provides the niche markets needed for technological learning, improvements, and reduced costs. Subsequently, massive technology transfers to developing countries will be required. For developing countries, the first priority is to make current uses of renewables, such as biomass, more sustainable. Doing so will involve efficiency improvements that bring ecological benefits – for example, by reducing deforestation – as well as health and social benefits through reduced indoor air pollution. The next priority is the development of endogenous technological capabilities to assimilate, modify, and tailor new renewable technologies to the needs of developing countries. In the long term, the market for renewables – like all other energy markets – will be in the South. The key issues are how feasible technology transfer will be, and what financial resources the South will be able to access to import the relevant technologies commercially and develop endogenous capacities.

### 6.5.5.  Nuclear

For nuclear energy, the range of possibilities is extremely broad in the scenarios. *Figure 6.11* compares nuclear energy's maximum and minimum development paths across the six scenarios with those of biomass and new renewables. The range for nuclear extends from more than a fivefold increase by 2050 in Cases A and B, through Scenario C2's new generation of small-scale nuclear technologies, to Scenario C1's phaseout of nuclear power altogether in the second half of the 21st century.

What actually happens will depend on when and how proliferation, waste, and safety concerns are resolved, and whether the climate issue will add additional

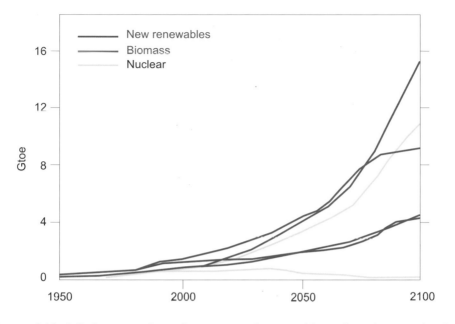

**Figure 6.11**: Minimum and maximum annual renewable and nuclear production, 1950 to 2100, primary energy equivalent in Gtoe.

weight to nuclear's characteristic as a zero-carbon option. For example, scenarios with the highest nuclear contributions (A1, A3, and B) translate into annual grid connections of up to 70 $GW_e$ of installed capacity per year by 2050. Although such additions are enormous compared with current construction rates, additions of some 30 $GW_e$ per year were realized in the mid-1980s. The realization of such construction rates requires a fundamental change in the public's perception of nuclear power. The nuclear industry must develop and demonstrate not only new, convincingly safer, possibly smaller-scale reactor designs incorporating "inherently safe" and "walk-away" features, but also efficient and "robust" fuel cycles and waste disposal solutions. These objectives are crucial, especially because a substantial share of the new capacities (even in the low nuclear scenarios) is located in developing countries. Extensive technology transfer, education, and training in the developing countries are essential prerequisites. Also required are substantial reductions in capital costs, initially through shorter construction times and ultimately through standardization and series-type production. Policymakers are called upon to design, implement, and enforce global nonproliferation mechanisms.

The range of different nuclear futures explored in the six scenarios is wide indeed, reflecting the current uncertainty surrounding its use. This uncertainty is unlikely to disappear in the near future. One conclusion, therefore, is that a precautionary nuclear strategy advances a wide range of possible technology developments, ranging between small-scale decentralized technologies and large integrated technologies. This strategy would ensure the preservation of the necessary know-how and engineering capability should a new nuclear age emerge. It would also be compatible with less ambitious prospects as it would not "lock-up" enormous resources and infrastructure investment in current technologies.

## 6.6.    Environmental Impacts

There are three major environmental impacts of concern, all arising from airborne emissions: first, extensive indoor and urban air pollution; second, sulfur and nitrogen emissions and their potential contribution to regional acidification; and third, greenhouse gas emissions, in particular $CO_2$, and their potential contribution to global warming. This section addresses each in turn.

### 6.6.1.    Local environmental impacts

Developing countries currently face numerous environmental problems. Two important generic categories can be identified (see Section 4.5; WEC, 1995c; World Bank, 1992). The first, which might be called "pollution from poverty," includes poor sanitation, lack of clean drinking water, and high levels of indoor air pollution – affecting women and children in particular – due to inefficient biomass cooking stoves. It also includes deforestation from overgathering of fuelwood and high ambient concentrations of particulate matter in urban areas. The second category might be termed "modern pollution." Its source is dense motorized traffic, resulting in high urban concentrations of very small diameter particulate matter ($PM_{10}$), lead, ozone, and volatile organic compounds (VOCs). Rural environments in Africa exemplify the first category; Mexico City, the second. Cairo exemplifies both categories in large measures. The result is a formidable challenge for the developing world. Rapid development and urbanization mean that problems that historically were dealt with one at a time in much of the industrialized world must all be addressed simultaneously in today's developing regions.

The aggregate long-term analysis presented here cannot address site-specific environmental issues in detail, but it can offer some illustrations. As discussed in the next section, airborne emissions of sulfur in Asia, for example, will increase in the coal-intensive Scenario A2 up to a factor of three over the next decades if unabated. Already today ambient concentrations of $SO_2$ in Beijing and Shanghai are twice the recommended WHO standards. Concentrations of suspended particulate matter exceed WHO guidelines by factors of three to four. Air quality in other megacities of the developing world is equally bad (WHO, 1992). Increasing emissions by an additional factor of three would pose a serious threat to both human health and local ecosystems. The historical London smog episodes with a few thousand casualties could reoccur in Beijing or Calcutta on an even larger scale.

While complete solutions to such problems will require creative and varied actions in many local settings, we can draw several general conclusions from the current study:

- Improving conversion efficiencies in end-use devices has a key role to play in reducing indoor air pollution and conserving traditional resources such as fuelwood.

- Also important are structural shifts away from traditional energy end-use patterns and energy carriers toward more efficient modern conversion technologies and cleaner energy carriers, including LPG stoves to replace wood stoves, and fluorescent lights to replace kerosene lamps.

- There needs to be a long-term shift toward energy services provided through clean, grid-dependent fuels. Such systems include heat cascading and district heating and cooling systems, electricity, gas, and eventually hydrogen.

All of these measures result in significant environmental improvements at the point of energy end use and reflect the central message of this study: patterns of energy end use and infrastructures converge across the scenarios, despite diverging energy supply structures.

Local environmental impacts are closely related to regional and global impacts, and progress at different levels often goes hand-in-hand. The coal-intensive Scenario A2 would have significant environmental impacts at all levels. Conversely, the Case C scenarios, which combine vigorous energy efficiency improvements with structural changes toward clean energy carriers, perform well on all environmental accounts. More generally, while scenarios differ with respect to environmental impacts, in all cases environmental objectives at the local, regional, and global levels are consistent with one another.

Much of the significant aggregate environmental improvement common to all scenarios is, however, long term and ultimately technology and infrastructure (i.e., capital) intensive. In the meantime, local solutions must be developed for local problems. These solutions require "ingenuity" at least as much as capital. They include better housekeeping measures such as the local redesign of end-use devices (like the Kenyan Ceramic Jiko initiative for the Nigerian cooking stove project); solving infrastructural bottlenecks such as the scarcity of LPG bottles in many areas of India; gradual environmental investments in large point sources (for example, dust precipitators); the addition of limestone to coal in boiler combustion; the use of low-sulfur crude oil in refineries; and the gradual elimination of tetraethyllead as an antiknock additive.

**Table 6.2**: Range of sulfur emissions in megatons sulfur (MtS) in 2020 and 2050 compared with 1990. Lower values correspond to the ecologically driven Scenario C1, higher values to the coal-intensive, high-growth Scenario A2. Values in parenthesis give emissions for a hypothetical Scenario A2 without any sulfur control measures.

| | 1990 | 2020 | 2050 |
|---|---|---|---|
| Africa and Middle East[a] | 4 | 3–6 (~10) | 4–11(~15) |
| Americas[b] | 13 | 3–11(~20) | 1–4 (~40) |
| Asia[c] | 18 | 22–35(~50) | 15–42(~80) |
| Europe[d] | 24 | 6–11(~30) | 2–14(~40) |
| World | 59 | 34–61(~110) | 22–64(~175) |

[a] AFR and MEA.
[b] LAM and NAM.
[c] CPA, PAO, PAS, and SAS.
[d] EEU, FSU, and WEU.
All figures are rounded.

### 6.6.2. Sulfur emissions and regional acidification

In addition to their immediate, local environmental impacts on human health, vegetation, and materials, $SO_2$ and $NO_x$ emissions also have broader regional impacts. Of principal concern is acid deposition. Acid rain is characterized by far less uncertainty than is the possibility of global warming discussed later in this section.

IIASA's RAINS model of regional acidification was used to calculate sulfur deposition levels in Europe and Asia (Amann *et al.*, 1995) and to compare them with "critical loads" – defined as the maximum deposition levels at which ecosystems can function sustainably.[11] Impacts of sulfur emissions on food production in Asia were assessed using an improved version of IIASA's BLS of agricultural models (Fischer *et al.*, 1988) and drawing on the analysis of Fischer and Rosenzweig (1996).

*Table 6.2* summarizes the range of sulfur emissions in the scenarios at regional and global levels. Sulfur emissions are lowest in the ecologically driven Case C, with its high environmental standards. Conversely, emissions are generally highest in the high-growth, coal-intensive Scenario A2 and, in some cases, in the middle-course Case B. For comparison, *Table 6.2* also shows a hypothetical Scenario A2 without any active sulfur abatement measures. This scenario was used as an example of a "worst-case" scenario for assessing acidification impacts.

In the hypothetical Scenario A2 with no sulfur abatement, sulfur emissions in Europe increase by about one-third over the next 30 years and sulfur deposition levels exceed 16 grams of sulfur per square meter ($gS/m^2$) per year in large parts

---

[11] RAINS (Regional Acidification INformation and Simulation) is described in detail in Alcamo *et al.* (1990) and Amann *et al.* (1995). See also Appendix A on methodology.

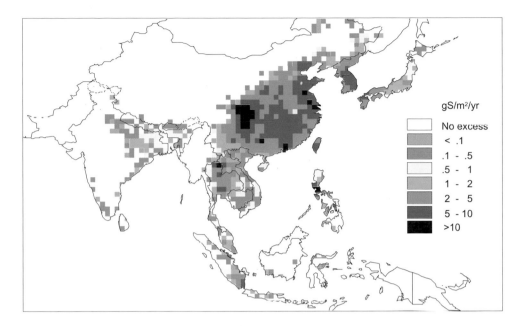

gS/m²/yr

No excess
< .1
.1 - .5
.5 - 1
1 - 2
2 - 5
5 - 10
>10

**Figure 6.12**: Excess sulfur deposition above critical loads in Asia for an unabated Scenario A2, in gS per m² per year. Source: based on Amann *et al.,* 1995.

of Central, Western, and Northern Europe. This level of deposition is very high compared with the requirements of the Second Sulphur Protocol to the Convention on Long-Range Transboundary Air Pollution (UN/ECE, 1994). This protocol calls for reduction measures to lower maximum excess deposition to below 3 gS per m² per year.

In the rapidly growing economies of Asia the situation is even more dramatic. For the same hypothetical unabated Scenario A2, $SO_2$ emissions in Asia nearly triple by 2020. Over the shorter term *actual* emissions may even run ahead of those in the unabated Scenario A2, as current national energy projections (Green *et al.,* 1995) total more than the regional projections in the scenario. However, while national projections better reflect immediate policies and plans, they are not necessarily as consistent across countries (in terms of imports, exports, and prices) as are global scenarios.

For an unabated Scenario A2, ambient air quality in South and East Asia deteriorates significantly in both metropolitan and rural areas, with sulfur deposition reaching twice the worst levels ever observed in the most polluted areas of Central and Eastern Europe. Deposition in 2020 exceeds the critical loads (see Hettelingh *et al.*, 1995) for most of the region, as shown in *Figure 6.12*. Of critical importance is that, for economically important food crops in Asia, Scenario A2's unabated emissions would cause critical loads to be exceeded by factors of up to 10. As

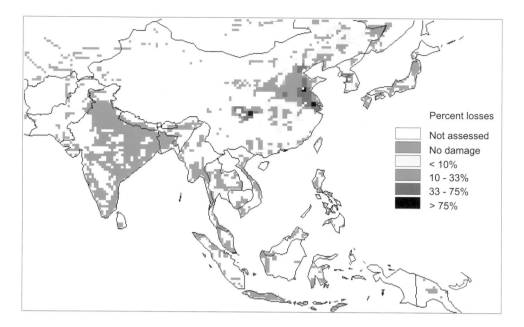

**Figure 6.13**: Crop production losses in Asia from sulfur emissions in an un-abated Scenario A2, in percent. Source: based on Fischer and Rosenzweig, 1996; Nakićenović *et al.*, 1998.

a result, severe losses in crop production could occur over large areas of Asia, as illustrated in *Figure 6.13*.

Given these results, the six scenarios described in this report all assume sulfur control measures. For North America and Europe, the scenarios reflect the most recent legislation to further reduce sulfur emissions (in particular the 1990 Amend-ments to the US Clean Air Act, and the Second European Sulphur Protocol). For developing economies, particularly in Asia, the scenarios assume gradually phased-in sulfur control measures, particularly for large point sources located in or close to major urban centers. After 2005, any new coal-fired generation in the scenarios incorporates only advanced coal technology, including scrubbers.

Sulfur controls are most stringent in the ecologically driven Case C. They are phased in more gradually in Case B and Scenario A2. Nevertheless, *global* sulfur emissions remain relatively constant at 1990 levels even in the most pessimistic scenario. Emission increases in Asia, where controls are expanding only gradually, are roughly balanced by sharp emission declines in North America and Europe as a result of tighter recent sulfur legislation.

Further emission reductions beyond those incorporated into the scenarios would be feasible. For Scenario A2, sulfur emissions in Asia in 2020 could be further reduced by some 20% with a full application of advanced technologies extending beyond the electricity generation sector. But, investment requirements would be substantial, corresponding to up to 10% of the total energy sector investments in Asia between 1990 and 2020. By 2020 sulfur control costs could represent up to 0.4% of regional GDP in Asia. This compares with the 0.2% of GDP required in Europe to comply with the Second Sulphur Protocol (Amann *et al.,* 1995).

These results suggest that in Asia concerns about sulfur emissions, and their potential regional impact on food security, will take precedence over global, long-term environmental issues such as climate change. Emission increases in Asia are therefore unlikely to remain unchecked, especially after 2020. Combined with continued sulfur emission reductions in the OECD countries, global sulfur emissions are therefore unlikely to increase substantially. This suggests that near-term progress in tackling the local and regional environmental problem of sulfur emissions may exacerbate the global environmental issue of climate change. With reduced sulfur emissions the historically important cooling effect from sulfate aerosols will ultimately disappear.

### 6.6.3. Carbon emissions and climate change

For each of the six scenarios, the level of energy use and the structure of energy supply, as presented in Chapter 5, determine future carbon emissions. *Figure 6.14* shows the results in terms of both gross and net carbon emissions from fossil fuels.

"Gross" fossil carbon emissions in a given year include all $CO_2$ associated with fossil energy resources extracted and used in that year irrespective of the conversion process chosen and whether the $CO_2$ is really emitted to the atmosphere. Thus, gross fossil emissions can be calculated by multiplying the amounts of coal, oil, and natural gas used as primary energy by their respective carbon emission factors and adding the results (see Nakićenović *et al.,* 1996). We report gross fossil carbon emissions here primarily to allow for reproducibility and comparability with other studies that use (the simpler measure of) gross emissions.

"Net" fossil carbon emissions refer to $CO_2$ released immediately through burning fossil fuels. These emissions are particularly relevant for estimating the atmospheric carbon cycle and the contribution of $CO_2$ to global warming. To calculate net fossil emissions we begin with gross fossil emissions and deduct $CO_2$ associated with non-energy purposes (feedstocks) where carbon is stored for extended periods in materials such as plastics and lubricants. Although in some statistics such carbon is considered to be "emitted" at the point that feedstocks are converted to those materials, our calculations count emissions if and when such products are subsequently incinerated as municipal waste. Next we exclude from net fossil emissions $CO_2$ that is "scrubbed" during electricity generation and synthetic fuel production and subsequently stored permanently (e.g., in depleted gas fields), as

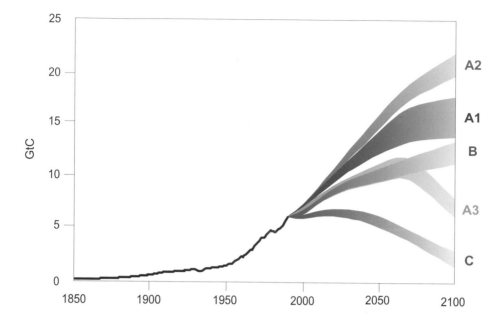
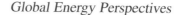

**Figure 6.14**: Global carbon emissions from fossil fuel use, 1850 to 1990, and for scenarios to 2100, in GtC. For each scenario, the range shows the difference between gross and net emissions (see text for explanation).

occurs in Case C, or that is sequestered through reinjection for enhanced oil recovery (as occurs in Scenario A1). Carbon dioxide reinjection for enhanced oil recovery already takes place in the USA, where the $CO_2$ is piped into Texas from Colorado. Reinjection occurs on a wider scale in Scenario A1.

The difference between gross and net fossil emissions is illustrated in *Figure 6.14*. Generally, the difference becomes larger over time as oil and gas are increasingly reserved for premium markets, that is, feedstock uses. The difference is particularly large in Scenario A1, with its heavy reliance on unconventional oil production. There, up to 2 GtC per year are sequestered through reinjection of $CO_2$ for enhanced oil recovery (and without any recourse to climate policies).

Both gross and net fossil carbon emissions as shown in *Figure 6.14* exclude $CO_2$ from burning biomass. To the extent that biomass use is unsustainable – that is, the $CO_2$ released during burning is not sequestered by subsequent vegetation (tree) regrowth – the resulting $CO_2$ emissions are included in estimated $CO_2$ emissions from deforestation as discussed below. Unfortunately, there are currently no reliable data for a bottom-up estimate of unsustainable biomass emissions. For 1990, total fuelwood use (unsustainable and sustainable) in developing countries (446 Mtoe) accounted for about 0.6 GtC, compared with the estimated 1.6 ±1 GtC released from land-use changes and deforestation in the early 1990s (IPCC, 1996a).

In the longer term, all scenarios relying on new renewable energy sources require that all biomass be used sustainably. Carbon dioxide emissions and uptake therefore balance out globally, although not necessarily for each region. A region producing methanol from biomass for export, for example, includes negative $CO_2$ emissions due to vegetation uptake. The positive $CO_2$ emissions from burning the exported methanol fuel show up on the balance sheet of the importing region. Thus regional carbon budgets take into account international trade in biofuels (and carbon). As they are reported here (see Appendix C), net carbon emissions of a biofuel-exporting region may thus become very small, and in extreme cases even negative.

As shown in *Figure 6.14*, gross and net energy-related carbon emissions vary substantially among the scenarios. The range of emissions is particularly large in the three Case A scenarios. In Scenario A1, they reach 14 GtC per year in 2100.[12] In the coal-intensive Scenario A2, they reach 20 GtC per year. In the "bio-nuc" Scenario A3, as a result of significant structural change in energy supply, they come to only 6 GtC per year, roughly the level of emissions today. The difference is that energy consumption in Scenario A3 in 2100 is five times greater than in 1990, with approximately the same level of emissions. Case B's emissions are very close to those of Scenario A3 up to about 2050 but then increase to nearly twice the Scenario A3 level by 2100. As described earlier, the two Case C scenarios are constrained to stabilize global emissions at 1990 levels by the mid-21st century in order to meet a $CO_2$ emission ceiling of some 2 GtC by 2100. As such, only the Case C scenarios describe a long-term emission path leading to stabilization of atmospheric $CO_2$ concentrations, consistent with the stated objective of the UN Framework Convention on Climate Change (FCCC, see also Section 6.6.4).

In terms of cumulative carbon emissions between 1990 and 2100, the Case C scenarios result in less than 540 GtC of cumulative net emissions from energy use. Cumulative net emissions for the other scenarios are 1,210 GtC for Scenario A1, 1,490 GtC for Scenario A2, 910 GtC for Scenario A3, and 1,000 GtC for Case B.[13]

It should be emphasized that emissions in the six scenarios are in most cases below levels of typical "baseline" or "business-as-usual" scenarios developed within the climate community. Only in Scenario A2 are cumulative (1990 to 2100) carbon emissions above those in the IPCC's IS92a reference scenario (Pepper *et al.*, 1992; Leggett *et al.*, 1992). The generally lower emissions in the scenarios presented here are due to technological improvements in the energy sector that are incorporated when the analysis is done in detail. From this perspective, typical baseline scenarios appear more as contrived, special cases than as potentially likely outcomes. They

---

[12]Unless specified otherwise, all numbers in this section refer to net emissions.

[13]Cumulative gross emissions vary from below 630 GtC in the Case C scenarios to 1,140 GtC in Case B to between 1,070 (Scenario A3) and 1,630 GtC (Scenario A2) in Case A.

combine optimism about high economic growth with general pessimism about technological change and resource availability, except for coal production. We believe that the scenarios presented here describe more consistent possible futures because they match high economic growth with technological changes that enlarge the resource base (particularly in the case of clean conventional oil and gas), that improve alternative energy supply sources, and that permit structural changes toward clean energy carriers.

Overall, the range of efficiency improvements in the three cases is in line with alternative long-range energy and emission scenarios as shown in *Figure 6.15*. The range of alternative scenarios is derived from the evaluation report of the IPCC emission scenarios (Alcamo *et al.*, 1995). As mentioned above, the detailed energy sector outlook presents a future characterized by continuing improvements and structural change in the energy sector, not technological and structural stagnation. As a result, the carbon intensity of primary energy use (*Figure 6.16*) improves across all scenarios. Only the coal-intensive Scenario A2 presents a picture similar to the high-emission scenarios developed within the climatic science community.

The atmospheric concentrations and the potential warming that might result from the scenarios' net carbon emissions were calculated using a carbon cycle and climate model developed by Wigley *et al.* (1993, 1994). For comparability with other studies, non-energy-sector anthropogenic emissions were taken from the IPCC IS92a scenario (Pepper *et al.*, 1992). By and large, this is a worst-case scenario, particularly concerning tropical deforestation. Non-energy emissions in the IS92a scenario correspond to another 130 GtC from land-use changes, deforestation, and cement manufacture over the 1990 to 2100 time period. In addition to $CO_2$, which turns out to be the most important greenhouse gas, the calculations include methane ($CH_4$), nitrous oxide ($N_2O$), nitrogen oxides ($NO_x$), carbon monoxide (CO), volatile organic compounds (VOCs), chlorofluorocarbons (CFCs), halocarbons, and sulfur emissions, thus incorporating all important direct and indirect greenhouse gases. Sulfur contributes a modest cooling effect via sulfate aerosols. Due to the increasing levels of sulfur control in all scenarios, however, this effect is largely a transient one, particularly in Case C and Scenario A3. The results for atmospheric $CO_2$ concentrations and possible global mean temperature change resulting from the scenarios are summarized in *Figure 6.17*.

Rising $CO_2$ emissions in Cases A and B lead to central estimates for $CO_2$ concentrations between 550 and 750 ppmv in 2100. This compares with concentrations of 280 ppmv around 1800 (the beginning of the fossil-fuel age) and current concentrations of 368 ppmv. In Case B and in Scenario A1, $CO_2$ concentrations approach 600 and 650 ppmv, respectively, by 2100. The concentrations in the "bio-nuc" Scenario A3 are lower, reaching 550 ppmv in 2100, and in the coal-intensive Scenario A2 they are higher, reaching 750 ppmv.

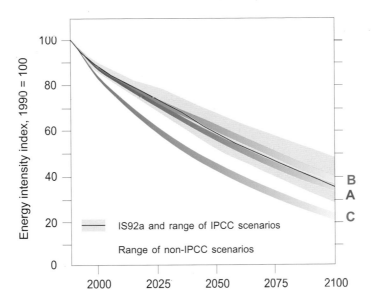

**Figure 6.15**: Global primary energy intensity shown as an index with 1990 equal to 100.

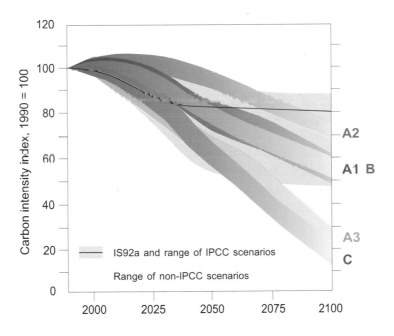

**Figure 6.16**: Global carbon intensity of primary energy supply shown as an index with 1990 equal to 100.

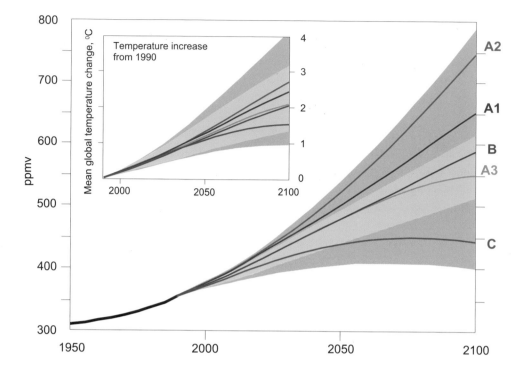

**Figure 6.17**: Atmospheric $CO_2$ concentrations, in ppmv, historical development from 1950 to 1990 and in scenarios to 2100. Insert shows global mean temperature change compared with 1990, in degrees Celsius. The (substantial) model uncertainties are also indicated.

Thus, all scenarios except Case C approach the doubling of preindustrial $CO_2$ concentrations that is the basis for most climate model calculations. And again in all scenarios except Case C, concentrations continue to rise throughout the 21st century. Based on current knowledge, an increase of $CO_2$ concentrations to 600 ppmv by the end of the 21st century could lead to an increase in the mean global temperature of about 2.5 degrees Celsius (°C) and a sea-level rise of up to half a meter. As indicated in *Figure 6.17*, however, the scientific uncertainties of such estimates are substantial.[14] In fact, the uncertainty range of possible global mean temperature changes from Case B is so large as to encompass the central estimates of all the other scenarios from Case C all the way up to (and above) Scenario A2.

---

[14]Uncertainty ranges are obtained by varying the climate sensitivity (i.e., the mean global temperature increase estimated for a doubling of $CO_2$ concentrations) from 1.5°C to 4.5°C. This is the range used in the IPCC Second Assessment Report (IPCC, 1996a), although that report assigns no probabilities to either the high or low bound. If one also accounts for the uncertainty in the global mean aerosol forcing of ±50%, the uncertainty range around the estimated change in global mean temperature in 2100 is amplified slightly.

Progress continues to be made both in terms of basic scientific understanding and models of possible future climate, but substantial uncertainties remain. For the time being, it appears premature to base near-term energy policies directly on the output of the current generation of climate models. Rather, decisions should be based on contingency considerations favoring policies that will permit maximum energy system flexibility in the future once scientific uncertainties are reduced. In addition, the six scenarios illustrate the congruence of many energy–environmental policies. Efficiency improvements and changes toward cleaner energy conversion and end-use structures not only improve local and regional environments, but also mitigate against undesirable global changes. Consequently, they draw together the urgent local and regional environmental priorities of developing and transitional economies, and the concern about potential climate change evident in many industrialized countries and in countries particularly exposed to potential climate change risks.

The Case C scenarios are the only ones in which $CO_2$ concentrations stabilize by 2100, reflecting their ambitious reduction profile, from 6 GtC of emissions in 1990 to 2 GtC in 2100. After peaking at about 450 ppmv around 2080, $CO_2$ concentrations slowly begin to decline as excess $CO_2$ is absorbed by natural sinks. Note that even with its ambitious reduction measures, Case C's $CO_2$ concentrations rise by up to 90 ppmv over the next 100 years. This increase is about equal to the concentration increase from the onset of the Industrial Revolution until today (from 280 to 368 ppmv). Thus even in Case C, some climate change appears inevitable: perhaps 1.5°C (with an uncertainty range between 1°C and 2.5°C) in increased global mean surface temperature compared with the 1990 level, and about 30 cm (with an uncertainty range between 20 cm and 50 cm) in mean sea-level rise above 1990 levels. This illustrates both the legacy of our past dependence on fossil fuels and the considerable lead times required for an orderly transition toward a zero-carbon-based sustainable energy economy.

The present calculations are consistent with the most recent findings from the IPCC Second Assessment Report, in particular the WG I carbon cycle model intercomparison (IPCC, 1995; IPCC, 1996a). The calculations reported here also include the cooling effect of sulfate aerosols. The effect is smaller than estimated in our 1995 report (IIASA–WEC, 1995) due to improvements in modeling regional sulfur emissions and correspondingly lower levels of global and hemispheric sulfate aerosols.

The cooling effect of sulfate aerosols would become significant only under two conditions: first, a strong radiative forcing of stratospheric sulfate aerosols (its magnitude continues to be debated), and second, high continuous fluxes of sulfur emissions. To quantify this effect – which has been suggested by some as a strategy to offset global warming from fossil-fuel-intensive energy scenarios – we analyzed the case of entirely unabated sulfur emissions from Scenario A2, similar to the case initially analyzed for acid rain impacts. In this hypothetical scenario, sulfur emissions increase nearly fivefold by the end of the 21st century. Assuming a high cooling effect, this would reduce the estimated radiative forcing of Scenario A2 to bring it roughly in line with the estimated forcing of Scenario A3 (also a high-growth case, but based on continuing and steady decarbonization). Thus, not only would a strategy of "cooling the greenhouse" with unabated sulfur emissions produce no better results than would structural changes toward clean energy supplies, it would also be of little use. By the time significant cooling effects became noticeable (beyond 2050), acidification would have surpassed by at least one order of magnitude the critical loads for both human health and important food crops.

Despite significant uncertainties continuing to surround the climate change issue, all six scenarios of this study confirm that the energy sector is indeed a major stakeholder. In high-growth scenarios, the energy sector alone would account for between 65% (Scenario A3) and 80% (Scenario A2) of all radiative forcing changes due to anthropogenic activities, including deforestation, agriculture, and CFC production and use. Even in the ambitious policy scenarios of Case C, the energy sector – despite drastic actions to improve energy efficiency and move aggressively to zero-carbon fuels – would still account for 45% of long-term changes in radiative forcing.

### 6.6.4.   Emissions reduction and the Kyoto Protocol

The Kyoto Protocol was agreed to in December 1997 by the Third Conference of the Parties to the FCCC. It addresses greenhouse gas emissions and specifies emission limits for nearly all countries included in Annex I of the Convention, that is, the OECD countries (as of 1992) and countries in transition to market economies (UN/FCCC, 1992). The specified limits would reduce the Annex I countries' average annual greenhouse gas emissions between 2008 and 2012 to about 5% below 1990 levels (UN/FCCC, 1997). Limits are calculated in terms of the "aggregate anthropogenic carbon dioxide equivalent emissions" of the six key greenhouse gases not controlled by the Montreal Protocol on Substances that Deplete the Ozone Layer – $CO_2$, $CH_4$, $N_2O$, halocarbons, perfluorocarbons, and sulfur hexafluoride. The limits refer to net emission changes including increased removals of greenhouse gases from "direct human-induced land use change and forestry" (see UN/FCCC, 1997; IGBP TCWG, 1998).

The Kyoto protocol specifies different emission limits for different countries. In most cases it requires reductions. The EU as a whole must reduce emissions by 8%, the USA by 7%, and Japan by 6%. Most of the countries in transition (some of which are allowed to choose a base year other than 1990) must reduce emissions by between 5% and 8%. In a few cases, increases are allowed. Iceland is allowed to increase emissions by 10%, Australia by 8%, and Norway by 1%. Limits for Russia, Ukraine, and New Zealand equal their 1990 emissions.

The protocol also permits Annex I countries to trade "emission reduction units" with each other. That is, if one Annex I country's emissions are below its limit, it may sell or barter the difference to another Annex I country that would otherwise be over its limit. Under the "clean development mechanism" defined in Article 12 of the protocol, Annex I parties may also apply "certified emission reductions" toward meeting their limits. Certified reduction units can be accrued through joint projects with non–Annex I countries that reduce emissions outside Annex I countries.

Here we compare the net energy-related carbon emissions calculated in the scenarios with the Kyoto limits. The comparison excludes non-energy related carbon emissions, the other five Kyoto greenhouse gases, and greenhouse gas sinks, but these are currently small compared with energy-related carbon emissions. Total Annex I emissions from cement in 1990 were 0.1 GtC, for example, compared with 4.1 GtC from energy (Grübler and Nakićenović, 1994).

With these caveats, the principal result from comparing the scenarios with the Kyoto limits is that in Case C all regions with Annex I countries (NAM, WEU, EEU, FSU, and PAO) are well below their Kyoto limits in 2010 and heading toward yet lower emissions thereafter. Case C is thus clearly in compliance with the protocol. None of the other scenarios is directly in compliance, but Case B and Scenario A3 come close. This is illustrated in *Figure 6.18*, which shows net energy-related carbon emissions for Annex I and non–Annex I regions against the Kyoto limit for Annex I countries. Although total *global* emissions increase in Case B and Scenario A3 by 1.4 to 1.5 GtC, they only increase in the Annex I regions by about 10 megatons of carbon (MtC) in Case B and 110 MtC in Scenario A3. The Kyoto target requires a *reduction* of some 200 MtC compared with 1990 Annex I emissions. Thus, in the aggregate the total Annex I reductions needed for full compliance in 2010 are about 210 MtC in Case B and 310 MtC in Scenario A3, which amount to only 5 to 7% of global emissions in 2010. They are well within the range of uncertainty inherent in such long-term analyses, particularly as we have not estimated the potential expansion of carbon sinks or possible reductions in other greenhouse gases and non-energy sources. With trading of emission reduction units among Annex I countries, Case B and Scenario A3 might therefore also comply with the Kyoto limits.[15]

What makes compliance a possibility in Case B and Scenario A3 is that emissions decrease initially in FSU and EEU as a result of economic recessions in the early 1990s. In 2010 EEU emissions are below their Kyoto limits by about 25 MtC

---

[15]At the time of this writing, the rules governing emissions trading (as well as certified emission reductions, greenhouse gas sinks, and equivalences between different greenhouse gases) are still under negotiation.

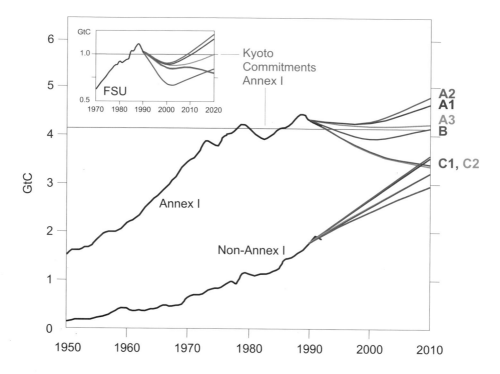

**Figure 6.18**: Net energy-related carbon emissions, in GtC, historical development from 1950 to 1990 and in the six scenarios to 2010. Also shown is the agreed aggregate Annex I emission limit of the Kyoto Protocol. The insert shows emissions in FSU compared with 1990. The Russian Federation's and Ukraine's 1990 emissions equal their Kyoto limits.

in Scenario A3 and 40 MtC in Case B. In the Russian Federation and Ukraine, 1997 emissions were some 310 MtC below 1990 levels. In the FSU the gaps are about 120 MtC in Scenario A3 and 280 MtC in Case B by 2010, a difference sometimes referred to as the "Russian bubble." The label purposely emphasizes that the bubble is temporary and will most likely disappear as energy consumption in Russia increases with economic recovery. By itself the Russian bubble does not lead to compliance with the Kyoto Protocol. But with international emissions trading the bubble can go a long way in that direction.

Studies of possible carbon trading indicate a wide range of possible prices that depend on assumed trading arrangements, emission scenarios, and other factors. Results range from as little as US$10/tC to more than US$100/tC. In this study we have focused on costs rather than prices (see Section 6.3). But we have also looked at alternative measures for reducing $CO_2$ emissions including command-and-control emission limits and taxes. In both cases, the value of marginal emissions was in the range of US$100/tC. Actual trading prices might be fundamentally different, but this result can be used for orientation. Assuming that the "bubble" is traded at half this value, revenues would equal about US$15 billion per year, not far from the US$30 billion per year in revenues from FSU gas sales (see *Figure 6.6*). Thus, the value of the Russian bubble could be substantial. One possibility is that emission reduction units could be sold to Europe, North America, and Japan with the proceeds invested in further development throughout Eurasia. A particularly important potential candidate for investment is a new gas transport infrastructure connecting the Siberian and Caspian regions with China and other rapidly developing countries in Asia. This would reduce long-term greenhouse gas emissions and air pollution in Asia and promote a cleaner environment in general. It has also been suggested that Russia might simply "bank" the bubble for after 2010, when its emissions will rise above its 2008 to 2012 Kyoto limits. This could be a riskier strategy for Russia as continuing negotiations could *de facto* eliminate the value of emission reduction units that are banked rather than sold.

Compliance with the short-term emission limits agreed to in the Kyoto Protocol is not, however, the same as meeting the ultimate objective of the UN/FCCC (UN/FCCC, 1995), that is, to stabilize "greenhouse gas concentrations in the atmosphere at a level that would prevent dangerous anthropogenic interference with the climate system" (UN/FCCC, 1992). The scenarios also need to be evaluated against this longer-term objective.

While there is not yet agreement on what "dangerous" means, the Second Assessment Report of the IPCC (IPCC, 1996d) concluded that the level at which stabilization occurs is governed more by total accumulated anthropogenic $CO_2$ emissions than by the way emissions change over time. In the first approximation, it is the cumulative emissions that matter and not the exact emissions path. The IPCC therefore analyzed cumulative emissions between 1990 and 2100 of 650, 1,000, 1,200, and 1,300 GtC, which would lead to stabilization around 450, 550, 650, and 750 ppmv, respectively. This presumes annual emissions drop well below 1990 levels by, respectively, about 40, 100, 140, and 200 years from now. In comparison, cumulative net carbon emissions are 540 GtC in Case C, 1,000 GtC in Case B, 1,210 GtC in Scenario A1, 1,490 in Scenario A2, and 910 GtC in Scenario A3. Thus only Case C is consistent with eventually stabilizing the atmospheric $CO_2$ concentration below 450 ppmv. Delaying implementation of Case C to 2010 does not affect long-term $CO_2$ concentrations. Following Case B to 2010 and only then moving to Case C would add less than 20 GtC to Case C's cumulative (1990 to 2100) emissions of 540 GtC. Also in this case, Case C's $CO_2$ concentrations stabilize below 450 ppmv. Scenario A3 is consistent with stabilization below 550 ppmv if emissions continue their decline after 2100. In all other scenarios $CO_2$ concentrations do not stabilize but continue to increase during the 22nd century.

The Kyoto Protocol is only the first step in a long process. Its short-term commitments are not sufficient in themselves to ensure the long-term emission reductions necessary for stabilization. In particular, the current commitments do not include the developing (non–Annex I) countries, whose share of global emissions will continue to increase as shown in *Figure 6.18*. In 1990 non–Annex I countries accounted for 30% of global net energy-related carbon emissions. By 2020 their share increases from 46% (Scenario A2) to 54% (Case C), and by 2050 ranges between 55% (A2) and 70% (Case C). Accomplishing the emission reductions reflected in Case C would require that timely and tight limits be negotiated efficiently for commitment periods beyond 2008 to 2012, and that limitations are agreed to by all countries, North and South. Accomplishing stabilization along an alternative path, represented by Scenario A3, might also be possible, for example, if revenues from emissions trading under the Kyoto Protocol are invested prudently in post-fossil technologies and infrastructures. Explicitly dedicating revenues from emissions trading to such environmentally friendly purposes might also make both emissions trading and resource transfers more politically acceptable in the short term.

# Chapter 7

# Energy System Alternatives: Eleven Regions

## 7.1.    Introduction and Summary

This chapter complements Chapter 5's global analysis in two ways. First, it summarizes the range of energy-related concerns across regions as identified in a detailed poll of WEC regional experts. This sets the stage for a subsequent discussion of regional perspectives and a "bottom-up" interpretation of the global scenario analysis. Second, the chapter presents separate scenario results for each of the 11 regions. (Additional details are given in Appendices C and E.) Against this quantitative background, each section of the chapter then attempts a qualitative interpretation of the global scenarios from a regional perspective.

There is a natural divide between long-term "top-down" global analyses and (usually) short-term "bottom-up" national and regional assessments. The two alternatives have basically different, but complementary, objectives, levels of detail, and temporal and spatial scales. Global, long-term analyses provide a holistic "big picture" of future possibilities with emphases on the major causes of change and on consistency – for example, that two countries are not assumed to simultaneously import the same barrel of oil. National or regional analyses usually have shorter time horizons and greater detail, and focus on national or regional development aspirations and short-term "next steps" in pursuit of those aspirations. Regional assessments can inform global analysis about the wide range of local and regional circumstances that, in their aggregate, will translate into alternative global developments. In turn, global analyses can inform national and regional assessments about the boundary conditions in which regional development strategies can expect to be productive. They provide a "big picture" for exploring specific issues and aspirations important for national and regional development. Is there an imminent resource scarcity looming? What are the prospects for oil imports, and at what price? Are there other global concerns that could constrain regional possibilities or create regional opportunities?

Because of their different objectives, methods, time horizons, and levels of detail, both "top-down" and "bottom-up" analyses are important. They provide the complementary information needed to piece together the puzzle of future energy development possibilities and constraints. However, the differences between the two also mean that their results will not agree perfectly in every detail. One example is that the sum of national development aspirations is generally higher than what global analyses estimate to be feasible (Schrattenholzer and Marchant, 1996). This is because global analyses must consider constraints that are not necessarily binding in narrower national or regional analyses. A second example is that of technological change. A short-term national or regional analysis must focus on technologies that already exist and on possible incremental improvements. Long-term analyses can consider more radical technological change, as is done in this study. Gas turbines provide a good example. They are now one of the most attractive technological options for electric power generation, but 50 years ago they were at best a technological curiosity and would have been appropriately excluded from any short-term energy study. But while it is only long-term studies that can properly include radical technological possibilities, technology diffusion remains essentially national. Thus detailed national and regional perspectives are essential for translating the potential for technological change into realistic next steps toward actual innovation and technology diffusion.

To integrate these two perspectives this study adopted a two-phase approach. During the first two years, the six scenarios described in Chapter 5 were developed with emphases on the use of formal models to ensure scenario consistency and on the principal long-term trends reflected in the scenarios. The first phase involved about 10 person-years of effort, including the work of the core modeling team at IIASA and some 50 additional experts serving as either reviewers, expert consultants, or members of the study's steering committee. The second phase, lasting three years, focused on regional implications of the global scenarios. It had two parts. First, more than 100 regional experts organized into 11 groups corresponding to the study's 11 world regions (see Appendix D) developed detailed regional assessments of the scenarios. Second, regional experts were polled to identify the relative importance of various energy-related concerns within regions and across regions. We begin the chapter with a summary of the regional polls.

### 7.1.1.  Regional polls

*Table 7.1* shows the results of the regional polls. Regional groups were asked to score eight issues central to energy policy debates: population growth, demand for commercial energy services, technology, financing, institutional deficiencies, efficiency and conservation, local environmental concerns, and possible climate change. The issue of local environmental concerns was added to the six issues listed in the 1992 WEC poll (WEC, 1993), and one of the 1992 categories, energy need and population growth, was split into two. Issues were scored according to the following scale: 1 = very important; 2 = important; 3 = of concern; and 4 = of no concern.

In many respects, *Table 7.1* confirms conventional wisdom. But it also holds some surprises. Considering all 11 regions together, institutional deficiencies and financing are the top concerns. Both issues have moved ahead of efficiency and conservation, the top-ranked issue in the 1992 poll.[1] Even NAM, often considered an institutional role model, ranked institutional deficiencies as a number-one priority. For all 11 regions as a group, possible climate change is at the bottom of the list, as it was in 1992. In general, it is of more concern to the OECD regions than to either the developing or reforming regions. In all regions, local environmental concerns are ranked higher than or equal to possible climate change. WEU lives up to its green image, being the only region to give local environmental concerns a number-one ranking. NAM is unique in scoring technology as a number-one issue.

While population growth is generally of greater concern in developing regions than in reforming or OECD regions, it was also ranked relatively highly by PAO. However, the reason is Japan's concern about too *little* population growth, exactly the opposite of the developing regions' concerns about too much population growth. PAO's top ranking for efficiency and conservation, and NAM's top ranking for technology reflect the two regions' positions on the cutting edge of these fields. Given the importance of technology, efficiency, and conservation in the scenarios, the continuing high priority given these issues is reassuring.

### 7.1.2. Global and regional perspectives

The key result of the detailed regional reviews was that the overall conclusions from the global study were confirmed: increasing energy needs; improving energy intensities and continuing decarbonization; the absence of binding global resource constraints; increasing consumer preference for clean, convenient, and flexible energy even as primary supply structures diverge across scenarios; the importance of near-term investments and technology strategies in determining long-term options; the slow pace of long-term change; the critical importance of capital requirements and infrastructure needs; the persistence of regional differences; and the precedence of local over global environmental concerns.

---

[1]This may be partly an artifact of changed regional groupings. The 1992 poll had a single South Asian/Pacific group. Here there are three: PAS, SAS, and PAO.

**Table 7.1:** Regional poll results. Areas of regional concerns are ranked as follows: 1 = very important, 2 = important, 3 = of concern, 4 = of no concern.

| Region | Population growth | Demand for commercial energy supplies | Technology | Financing | Institutional deficiencies | Efficiency and conservation | Local environmental concerns | Possible climate change |
|---|---|---|---|---|---|---|---|---|
| NAM | 4 | 3 | 1 | 4 | 1 | 3 | 2 | 2 |
| LAM | 2 | 2 | 2.5 | 1.5 | 2 | 2 | 2 | 2.5 |
| WEU | 3.5 | 3 | 2 | 3 | 2 | 2 | 1 | 2 |
| EEU | 4 | 2.5 | 2.5 | 1.5 | 2 | 1.5 | 2 | 3.5 |
| FSU | 4 | 4 | 3 | 1 | 1 | 2 | 3 | 4 |
| MEA | 1 | 3 | 2 | 1 | 1 | 3 | 3 | 4 |
| AFR | 1.5 | 3 | 2.5 | 1.5 | 1.5 | 3 | 2.5 | 3 |
| CPA | 1 | 2 | 2 | 1 | 1 | 1 | 2 | 2 |
| SAS | 2 | 1.5 | 2 | 1.5 | 1 | 2 | 1.5 | 4 |
| PAS | 3 | 1 | 2.5 | 1.5 | 2.5 | 2 | 2.5 | 3.5 |
| PAO | 2 | 2.5 | 1.5 | 2 | 2.5 | 1 | 1.5 | 1.5 |

Second, the regional reviews drew attention to additional issues of particular importance in a number of regions. Examples are institutional deficiencies that slow energy development, the need to include in the energy market the world's two billion poor who now have no access to modern energy services, the importance of improving regional cooperation, and the importance of developing new energy infrastructures and transport networks, especially in Asia. Finally, in some cases, the regional reviews differed in the emphasis placed on particular scenario results and offered comparisons between short-term regional projections and the scenarios' long-term results.

### World energy needs will increase

The regional analysis confirms the global conclusions that the demand for energy will increase substantially and that there will be a major shift in energy use toward the South and Asia. Timing and growth rates are uncertain, but from the perspective of the regions where energy demand will grow the most, that is, the developing regions of Asia, there was a clear "winner," the high-growth Case A. The ecologically driven Case C did not find widespread appeal, despite its explicit focus on international equity including massive resource transfers from North to South.

In Asia and several other developing regions, such as LAM and MEA, there was a clear preference for the conventional development model, represented by Case A and characterized by high rates of economic growth with developing countries quickly "catching up." There were differences of opinion about how quickly the high-growth path of Case A could unfold, with regional aspirations generally favoring short- to medium-term economic growth rates even above those of Case A. However, the basic scenario logic of coupling productivity growth in the economy with energy intensity improvements was endorsed. Consistent with this logic, reviewers favoring higher short-term economic growth also suggested faster energy intensity improvements (e.g., CPA, SAS, and PAS). In calculating energy demands, faster energy intensity improvements tend to offset faster economic growth, so that the resulting energy demands are quite close to those of Case A, although they represent an even more productive economy and more efficient energy system.

The recent experience of EEU and FSU demonstrates that economic growth can by no means be taken for granted. The recessions in EEU and FSU during the 1990s have been deep and severe. EEU's economy contracted by more than 30% between 1989 and 1993. In FSU economic output fell 50% from 1990 to 1996. Energy use also declined, although generally less than GDP. The overall paths of the recessions were anticipated quite closely in all scenarios. Unfortunately, both the scenarios and recent developments suggest that the process of recovery will be a long one. In EEU, per capita GDP recovers to 1990 levels by 2010. In FSU, 1990 levels of per capita income are reached between 2010 and 2020. However, by that time the same amount of GDP per capita as in 1990 represents a considerable increase in personal income and welfare as a result of the overall economic shift away from smoke-stack industries toward consumer goods and services.

*Energy intensities will improve significantly*

Overall, the regional reviews endorsed the conclusion that energy intensities will improve significantly in the long term as a result of changes in economic structures and technological improvements, but there was a large difference of opinion across the regions on the rate of future improvements. Some regions considered the improvements reflected in the scenarios overly optimistic, while others judged the potential for rapid improvements to be much higher than in the global analysis. The general pattern was that regions with large domestic resources and less auspicious economic growth over the past decade (MEA, LAM, and NAM) tended to be more conservative regarding future energy intensity improvements. In contrast, regions with more limited domestic resources that have recently experienced rapid economic growth and energy intensity improvements (CPA and SAS) considered further rapid improvements to be not only possible but necessary to sustain rapid growth.

This pattern is consistent with the basic historical relationship between economic growth and energy intensity improvements incorporated in the scenarios – namely, other things being equal, economic growth rates depend on the distance to the productivity frontier and on capital turnover rates. Thus, growth is faster in regions with low per capita income and low labor productivity than in regions with high incomes and productivity. Similarly, energy intensity improvements are faster in regions with high energy intensities than in regions closer to the energy productivity frontier.

*Resource availability will not be a major global constraint*

Although *regional* assessments generally are not well positioned to judge the global sufficiency of energy resources, it is noteworthy that no regional review voiced concerns, so dominant even a decade ago, about imminent resource scarcity. Another change in emphasis – in agreement with the global analysis – is that resource availability is no longer seen as geologically preordained. It is viewed more as a function of the incentives and policies put in place for exploring and developing resources, constructing the necessary long-distance transport infrastructures, and, above all, attracting capital to energy investments. All are necessary to translate potentially vast geological resources into economically and technically recoverable reserves. Some regional reviews (e.g., NAM and PAO) raised the possibility of fossil resources even beyond the range used in this study, emphasizing the potential of "exotic" fossils such as methane hydrates, which constitute the largest occurrence of hydrocarbons in the earth's crust.

Import dependence and export possibilities were important concerns in several regions. The MEA regional review voiced concern about medium-term export potentials, particularly should regional energy demands grow much faster than anticipated in the scenarios, thus "crowding out" possible export volumes. Similar trends were projected in the PAS regional review. Concerns about import dependence were particularly strong in WEU, a region traditionally heavily dependent on energy imports. The potential causes of future increases in import dependence are poor *regional* resource development due to (1) too little RD&D; (2) underinvestment in new energy technologies and oil and gas exploration; and (3) insufficient diversification should coal and nuclear power be abandoned. WEU's share of global energy imports will decline with the shift of global energy markets (and imports) to developing regions. Therefore, WEU's concerns about supply security will increasingly be shared with other regions, and the region will likely be less and less influential in shaping international energy markets. This makes it increasingly important that the region maintain diverse energy supplies, particularly from regional resources such as oil, gas, nuclear, and renewables.

The NAM regional review expressed concern about coal-intensive developments in Scenario A2, with its enormous synfuel production and exports. This concern raises the question of whether NAM would be prepared to invest in the massive coal development and synfuel exports associated with Scenario A2. More generally, with the exception of CPA and SAS, coal resources are concentrated in the affluent North, regions that may not necessarily depend on revenue generated from energy exports.

Overall, the regional reviews concurred with the global perspective that revenues for energy exporters will remain solid. Oil export revenues for MEA are high in all scenarios: up to US$300 billion[2] by 2020 (and even higher thereafter) in Case A, some US$200 billion in Case B, and US$140 billion in Case C. The region need not fear revenue losses due to policy measures aimed at reducing $CO_2$ emissions in industrialized countries. Even in the ecologically driven Case C, combined oil and gas export revenues are higher than in 1990. For FSU, gas exports consistently increase in all scenarios through 2020, reaching around 300 billion cubic meters per year with revenues increasing to about US$50 billion, five times the 1990 value. Beyond 2020 exports and revenues increase further in half the scenarios, with revenues exceeding US$100 billion, 10 times the 1990 value. Given FSU's location between the large, mature energy market of WEU and the new, rapidly growing markets of Asia, and given the increasing importance of grid-dependent

---

[2]Unless specified otherwise, all monetary values in this chapter are expressed in constant 1990 US dollars, US(1990)$.

energy in all scenarios, there are strong incentives for FSU to build a network link-ing its oil and particularly its gas resources to the large markets to its east and west. This would enhance trade and revenue possibilities enormously while facilitating cleaner, less coal-intensive development in Asia.

### *Quality of energy services and forms will increasingly shape future energy systems*

The regional reviews strongly endorsed the global conclusion that the historical trend by which final energy forms become cleaner and more grid-dependent as affluence grows can continue for a wide range of possible future primary energy structures. Solid fuels such as traditional biomass and coal in residential and com-mercial applications are gradually phased out in all scenarios. Grid-dependent fu-els such as electricity, natural gas, and eventually also hydrogen are phased in. Although the speed of the transition varies, the trend everywhere – in global and regional perspectives alike – is toward cleaner, more convenient, and more flexible end-use fuels.

Even the LAM regional review's emphasis on developing agriculture rather than industry, thereby slowing migration from rural to urban areas, is consistent with the scenarios' global trends. Specifically, the development of new renewables and decentralized energy options, as in Case C, could create opportunities for rural, agricultural growth consistent with the trend toward cleaner, modern energy forms.

The consistency between the regional reviews and the global analysis in their emphasis on the structure of final energy use is important, as this feature of the study represents a conceptual and methodological advance over traditional scenario studies focused solely on energy supply perspectives. The agreement between re-gional reviewers and the study team reflects the substantial insights to be gained by replacing the traditional dichotomy between supply and demand with a holistic approach that integrates the two.

### *Energy end-use patterns will converge, even as energy supply structures diverge*

As the energy system shifts toward high-quality, cleaner, and more flexible en-ergy carriers, the range of possibilities for providing high-quality energy forms for consumers widens. The energy supply structures in the six scenarios diverge over the long term, driven by different decisions about RD&D, investments, and energy and environmental policies. The broad range of possible futures was generally en-dorsed by the regional reviews, although it is significant that in terms of energy supply structures the reviews did not identify a single clear "winner" from among the six scenarios.

Among regions primary energy supply opportunities vary greatly in the scenar-ios, with some regions having a much wider range of options than others. This global result was echoed in the regional reviews. NAM has the widest range of possibilities – wider even than the world as a whole (compare *Figures 7.2* and *5.4*). While recognizing this range of possibilities, the NAM regional review nonetheless expressed clear preferences for "middle-of-the-road" energy development avoiding extreme reliance on particular resources and technologies, whether focused on coal, nuclear, or renewables. The WEU regional review also preferred a diversity of energy sources, especially to limit import dependence.

At the other end of the spectrum, for MEA all six scenarios are largely dominated by oil and gas, and the regional review identified a clearly preferred scenario: the high-growth, oil- and gas-intensive Scenario A1. Similarly, PAO, as a high-income region with an outstanding record of promoting improved conservation, efficiency, and diversification beyond fossil fuels, is the region best positioned for a high-growth transition beyond fossil fuels (i.e., Scenario A3). This global conclusion was largely shared by the regional review. Given PAO's importance as a source of Asian investment, and the earlier conclusion about the importance of early investments, PAO's choices are likely to exert a strong influence on the future of other Asian regions.

*Technological change will be critical for future energy systems*

Technological progress is key. Improved efficiency in replenishing reserves from the resource base, in energy conversion, and in end-use processes all depend on technological change. In the regional polls technology was ranked a top-priority issue in NAM and PAO, each of which has a particularly strong RD&D and technology base. The PAO regional review emphasized the long-term possibilities of tapping unconventional fossil resources in the form of methane hydrates, an option that will require continued and concerted efforts in technology development. In AFR the regional review stressed the importance of improved, low-cost technologies tailored to the region's specific circumstances and needs. In CPA the review stressed the importance of technological progress in reconciling rapid economic development and environmental protection.

A key theme of this study is that investment choices in the next two decades will lead to technology improvements, and the technologies that benefit will become cheaper, thus making subsequent investments in the same direction more attractive. Given this conclusion, concerns in the WEU regional review that deregulation may curtail incentives for long-term RD&D are particularly important. In both the public and private sectors, recent declines in RD&D budgets need to be carefully checked and a coherent technology strategy developed to respond to long-term challenges. Of all OECD countries, at present only Japan seems prepared to invest in the kind of long-term thinking and technology development that is so vital from the perspective of the global scenarios.

Technological change increasingly crosses national borders. One example is that of PAO as the dominant source for both capital and technology for much of Asia. However, efforts to transfer technology frequently have failed because successful technology diffusion requires high local expertise. Infrastructures need to be in place, labor must be educated, and local RD&D capabilities are required for the necessary local adaptations, modifications, and extensions of new technologies that make the difference between success and failure. For this reason, capacity building is a resilient strategy for preparing for a transition away from fossil fuels to a more diversified energy supply system. If indigenous technological capabilities are not nurtured right from the outset, however, sunk costs in infrastructure and human capital make it progressively more difficult to change course.

*Rates of change in global energy systems will remain slow*

The scenarios illustrate that pervasive changes in economic development, energy end-use patterns, and energy supply systems take considerable time. It may take some three decades before FSU's economic depression is over and at least five decades before FSU catches up to Western Europe's 1990 GDP level. Catching up to WEU standards takes AFR and SAS even longer, extending all the way through the 21st century. And even with the ambitious efforts of Case C, it will take almost a century to restructure the global energy system away from its current dominant reliance on fossil fuels.

This was too slow for some reviewers. The CPA and SAS regional reviews, for example, argued for higher rates of both economic growth and energy intensity improvements than those considered feasible in the scenarios. Faster economic growth concurrent with faster energy intensity improvements is consistent with historical experience and the logic underlying the scenarios. The two together do not dramatically alter the evolution of energy demand as described in the scenarios, although faster economic development would mean a developing region could catch up to developed countries more quickly. As emphasized in many regional reviews, however, high growth requires a skilled population, a stable political climate, functioning institutions and markets, free trade, and access to technology.

The CPA regional review also raised the possibility of switching between alternative development paths. For CPA, a preferred alternative scenario would start on a coal-intensive trajectory (Scenario A2) but, after 2020, switch to less carbon-intensive fuels (Scenario A3). Such a transition might be partially possible if carefully planned. However, from the perspective of the global analysis, changing horses in midstream creates two principal difficulties. First, infrastructural and technological investments in coal ultimately will have to be duplicated by new investments in energy alternatives. The railway infrastructure for coal transport, the coal-fired power plants, and the work force dedicated to coal will be in place and expanding, making it ever more difficult to change course. Second, the required massive technological shift would cost more after 2020 than if less carbon-intensive fuels had been the focus of investments right from the beginning. There will not have been the same learning, experience, and cost reductions for the new technologies as would have occurred had they been the original focus of investments.

These difficulties reflect a major conclusion from the global scenarios – that while the choice of the world's post–2020 energy systems may be wide open now, it will be a lot narrower by 2020. The decisions made over the next 20 years will largely determine the long-term direction of development even though they will not change the structure of the energy system dramatically over the next two decades. This conclusion is also relevant to the regional perspectives on the ecologically driven Case C. The regional reviews and poll results emphasize how ambitious and difficult it will be for the world to choose the route offered by Case C – and Case C requires a deliberate choice. The longer one waits to initiate the necessary steps in the direction of Case C in terms of new institutions, new forms of international co-operation, new incentives, stepped-up technology development and demonstration programs, and further efforts to accelerate energy intensity improvements and decarbonization, the more difficult it will be to actually meet the challenge. Perhaps the Kyoto Protocol and the even more ambitious carbon reduction targets discussed in Western Europe, are indicative of short-term steps that could lead in the direction of Case C in the long run. But addressing global environmental problems is a long-term issue. Short-term emission reduction targets matter, but what matters even more is whether the necessary steps are taken to prepare for meeting longer-term, even more stringent environmental limits.

*Interconnectivity will enhance cooperation, systems flexibility, and resilience*

One advantage of a global analysis is its ability to identify opportunities for trade – and associated long-term infrastructure needs – that may not be apparent in national and regional studies. The possible magnitude of desirable new transcontinental infrastructures frequently comes as a surprise to national energy planners.

In this study, the global importance of improved connections among energy systems and the need for related infrastructures was echoed in a number of regional reviews. The importance of interconnections among electricity grids was stressed in the AFR, LAM, and MEA reviews, and the importance of continental-scale (i.e., Eurasian) gas pipeline grids was raised by commentators from FSU, CPA, and MEA. However, while the long-term benefits of such interconnections in terms of supply diversification, higher-quality energy, and environmental improvement are clear, implementation strategies are less so. We hope this study's long-term global scenarios will help promote the long-term perspective and renewed spirit of cooperation necessary for the large-scale risk sharing and capital mobilization needed to build up continental-scale energy infrastructures. Current geopolitics provides a number of opportunities. Even during the height of Cold War short-sightedness it was possible to construct a huge pipeline system between Western Siberia and Europe. Thus despite current investment myopia, it might well be possible to mobilize the goodwill, the long-term planning, and the capital required for new energy infrastructure connections.

### *Capital requirements will present major challenges for all energy strategies*

Of all the conclusions from the global analysis, the importance of capital requirements was among the most widely endorsed in the regional reviews. In the OECD regions (NAM, WEU, and PAO) regional reviewers were most concerned that energy sector deregulation might erode incentives for long-term RD&D and might discourage needed energy infrastructure investments that are big, "lumpy," and risky. For these cases it is important to find creative ways to spread risks and costs.

In developing regions, regional reviewers emphasized institutional reforms to attract the investments needed for expansion. There must be a shift from subsidies and flat rates to segmented market pricing. Losses and leakage must be reduced, and efficiencies improved. Such shifts must be politically acceptable, and they must not leave the poor behind. Market pricing only makes an investment attractive if there is a market that can pay the price, and the poor in developing countries do not yet constitute such a market. How quickly they become empowered energy consumers will depend more on the success of policies to reduce poverty than on energy policies.

Thus, market deregulation and policies to "get prices right" will by themselves fall short. The fundamental problem is market *exclusion* rather than market *distortion*. As the history of rural electrification in OECD countries demonstrates, market forces need to be complemented by public policies, particularly in financing upfront infrastructure expenditures. New, decentralized energy options can lower investment needs by helping regions "leapfrog" capital-intensive traditional infrastructures, much as cellular phones may make prior networks of telephone wires unnecessary in some regions. Yet, providing adequate energy infrastructures for the urban poor and affordable decentralized energy services to rural communities remains a daunting challenge, not least because RD&D and technology programs have traditionally paid insufficient attention to the issue.

### *Regional differences will persist in global energy systems*

The future trends central to the scenario results emphasize convergence. Although convergence is sometimes slow relative to current aspirations, in all scenarios economic development is substantial, developing regions narrow the gap between themselves and developed regions, all regions shift toward cleaner, more grid-dependent final energy forms, final energy use per capita converges, energy intensities improve, and decarbonization continues. However, as is abundantly clear from both the regional scenario results and reviews presented in this chapter and the poll results in *Table 7.1*, there will continue to be substantial differences among the regions. These arise from differences in populations and natural resource endowments and different economic and technological starting points.

Important regional differences exist, for example, in terms of population, GDP, import dependence, and energy use. Average population growth between 2000 and 2050 varies from a still rapid 2% per year in AFR to even a slight *decline* in PAO. Average medium-term economic growth (1990 to 2020) in the high-growth Case A ranges between 7% per year in CPA and below 2% per year in both PAO, with its mature economy and aging population, and FSU, whose economy continues to suffer from the recession of the 1990s. GDP per capita in PAO is still between 25 times (Case C) and 35 times (Case B) that in SAS in 2050 (down from nearly 70 times higher in 1990). For CPA, one difference between Scenarios A2 and A3 is the difference between being a net exporter, in Scenario A2, or net importer, in Scenario A3. MEA and FSU are always net exporters; WEU, PAS, and SAS are always net importers. In NAM the post–2020 primary energy structure varies tremendously across scenarios, reflecting the region's diverse resources. In MEA, all scenarios are dominated by oil and gas. Final energy use per capita in 2050 still varies from about 0.6 toe in SAS and AFR to about 5 toe in NAM in Cases A and B (it is closer to 2 toe in Case C in NAM).

However, despite continuing diversity, the consistent message is that the gaps between rich and poor can be narrowed across a wide range of possible energy supply structures. The speed with which they are narrowed will depend on investments in technological progress and infrastructures to integrate regions and reach the poor.

### *Local environmental impacts will take precedence over global change*

Agreement with this conclusion was unanimous in both the regional polls and the detailed regional reviews. Regional air pollution is a major issue in several developing regions, while possible global warming is given low priority. Even in regions such as CPA, NAM, PAO, and WEU, where climate change is ranked as an important long-term issue, local environmental issues are ranked at least as high.

It is important, however, not to see the two as separate issues, and to explore strategies that achieve simultaneous progress on both fronts. Improved efficiency and cleaner energy systems (decarbonization) are the key components of the strategy suggested by the global analysis. This emphasis on more efficient and cleaner energy systems as a resilient environmental strategy was invariably echoed in the regional reviews. If policies and investments correspond to the ecologically driven Case C, both regional and local issues are addressed together. But if regional pollution controls are add-ons to fossil-based energy systems such as in Case B, any future reductions in $CO_2$ emissions only become increasingly remote and difficult. Hence, a number of regional reviews emphasized the need for a long-term perspective and anticipatory RD&D and technology policies to best prepare for future contingencies. After all, climate change is a long-term environmental issue and responses must be prepared in time to prevent disruptions and high costs. This also means that care must be taken not to reduce the flexibility of longer-term energy systems through prematurely locking in existing technologies or exclusively using "add-on" environmental fixes.

*Decarbonization will improve the environment at local, regional, and global levels*

Efficiency improvements and structural changes in energy systems, that is, decarbonization, constitute the most resilient strategy in response both to today's local, regional, and global environmental challenges and to those uncertain environmental challenges that may surface in the future. Such a multipurpose strategy was unanimously endorsed in the regional reviews.

The essential message is that *quality matters*. The trend toward higher-quality, more flexible, and cleaner energy – an energy mix increasingly composed of electrons and hydrogen rather than carbon atoms – is strong. As in the past, progress will be gradual. The energy system changes slowly, and big, long-term shifts need to be prepared well in advance. The initiatives, incentives, and policies put into place over the next two decades will determine how quickly these efforts will bear fruit in the form of energy system decarbonization. From the perspective of both the global and regional analyses, a resilient strategy responds to consumer preferences and the need to preserve the environment both locally and globally.

## 7.2.    North America (NAM)

The region consists of two countries, Canada and the USA. Economically and technologically it is one of the most developed of the 11 regions, with a high degree of economic and political integration. Trade integration already embraces Mexico and may eventually extend to Latin America and the Pacific Rim. In 1990, with just 5% of the world's population, NAM accounted for 29% of GWP, 24% of global primary energy use, and 25% of global energy-related net carbon emissions.

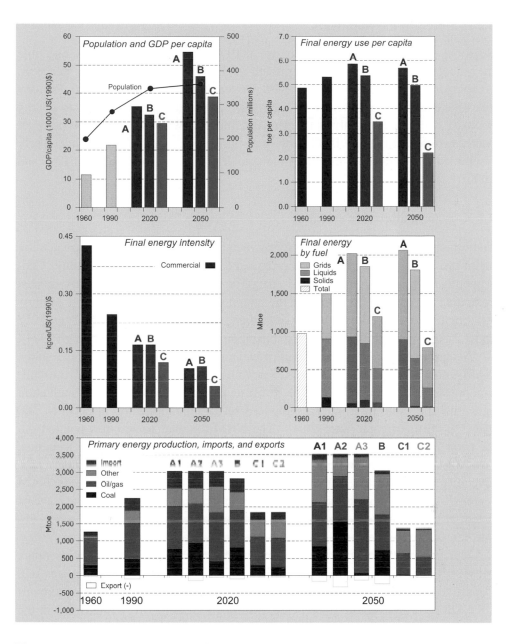

**Figure 7.1**: Scenario summary for **NAM** through 2050. *Top left:* Levels of GDP per capita reached in Cases A, B, and C, plus the population projection used in all three cases. *Top right:* Final energy use per capita. *Middle left:* Decrease in total final energy intensities in all three cases. *Middle right:* Final energy divided into solid fuels, liquid fuels, and energy delivered by grids. *Bottom:* Composition of primary energy production, plus imports and exports, for all six scenarios.

### 7.2.1.  Recent developments

Regional economic developments between 1990 and 1996 most closely resemble the middle-course Case B. GDP growth in the region averaged 2.2% per year, compared with Case B's growth of 2.3% per year between 1990 and 2000. Primary and final energy use grew by 1.8% per year. As a result, regional primary energy intensity decreased by only 0.4% per year, more slowly than the 1.2% per year decrease reflected in Case B. Growth in carbon emissions (1.4% per year) closely resembled that in the coal-intensive Scenario A2 (1.3% per year between 1990 and 2000).

### 7.2.2.  Population and economy

The region is characterized by a slightly increasing but steadily aging population and a mature, high-technology and progressively service-oriented economy. Consequently, long-term economic and energy demand growth are relatively modest compared with world averages and the region's own history over the last century. Although there is substantial diversity among the six scenarios (see *Figures 7.1* and *7.2*), the overall trend is one of stabilization of consumption at very high levels.

According to the 1992 World Bank population projection adopted for this study, NAM's population is expected to increase in line with historical trends to 340 million people in 2020 (see *Figure 7.1*). Subsequently, growth slows as the population ages, and moderate migration assumptions result in stabilization at between 360 and 370 million people after 2050. The NAM regional review has pointed out the need to consider alternative immigration scenarios. Such alternative scenarios, however, are not available, and it would be difficult at this stage to defend a specific alternative reflecting higher immigration to North America. Nor did the NAM regional review attempt to do so. What is important, however, is that the issue of potential immigration increases in North America has been flagged as a concern that should be revisited regularly in future energy assessments.

Medium-term economic growth rates (1990 to 2020) in the scenarios range between 1.7% and 2.3% per year. This is in good agreement with the latest US Department of Energy Annual Energy Outlook (USDOE, 1998), which estimates US GDP growth at 1.3 to 2.4% per year through 2020. Energy intensities in the scenarios improve at between 1.2% per year (Cases A and B) and 2.3% per year (Case C), somewhat higher than improvement rates used in the USDOE projections.

Long-term economic growth rates (2020 to 2050) decline to between 1.1% per year (Case C) and 1.6% per year (Case A), reflecting both demographic trends (aging, retiring "baby boom" cohorts, and lower immigration) and economic developments such as a continuing shift toward services. Energy intensity improvements also slow as the region completes the transition to an economy dominated by the service sector and as capital turnover slows with lower GDP growth.

Unless policies and investments become explicitly focused on environmental protection and international equity, as in Case C, NAM moves toward "stabilization at very high levels." By 2050 per capita income ranges between US$46,000 and US$55,000 in Cases A and B, and per capita final energy demand stabilizes between 5 and 6 toe, quite close to the 1990 value (see *Figure 7.1*, top right chart). By 2100, per capita GDP could reach between US$75,000 and US$100,000, with final energy use remaining stable at between 4 and 6 toe per capita. If international and environmental policies correspond to Case C, however, final energy use in the region decreases substantially in both per capita and absolute terms, as shown in the middle and top right charts in *Figure 7.1*. Economic structural change toward an "information economy" combined with accelerated efficiency and conservation improvements would reduce per capita final energy use to below 4 toe by 2020 and below 2.5 toe by 2050, levels comparable with present-day Europe and Japan. Per capita income would remain high – close to US$40,000 by 2050 and US$60,000 by 2100.

The majority of contributors to the NAM regional review considered the energy intensity improvements in the scenarios too optimistic, especially in Case C. It was argued that energy intensities might even *increase* as electricity and hydrogen gain importance. This position, however, is inconsistent with historical experience, where new energy technologies have improved conversion efficiencies and new energy carriers have improved energy end-use efficiencies. Other reviewers argued that intensity improvements in the scenarios were too low in light of the potential for technology change throughout the energy system. In part, the differences of opinion are definitional. The bulk of intensity improvements in the scenarios occurs outside the traditional energy sector at the level of final and useful energy conversion and economic structural change. The overall results therefore look optimistic compared with expectations based on the more limited boundaries of the traditional energy sector. But part of the differences of opinion are substantial: even in one of the most well-informed, high-technology regions of the world significant disagreement exists about technological potential. The range of possibilities reflected in the scenarios lies within the full range of opinions represented by the NAM regional review – although the scenarios fall more on the optimistic side of NAM's spectrum of opinion. The diversity of opinion in the NAM regional review is unique among the 11 regional reviews. It matches NAM's uniqueness in terms of its diversity of alternative energy futures (see *Figure 7.2*) and its status as the only region to rank technology as a number-one priority in the regional poll.

### 7.2.3.    Energy systems

With high per capita income goes a demand for high-quality energy services. Thus the structure of final energy use is high quality and quite similar across all six scenarios. Coal is essentially eliminated from end-use applications, and the market share of liquid fuels (oil products that are increasingly supplemented by methanol after 2020) declines from 52% in 1990 to between 32% and 43% by 2050. Liquids benefit from their comparative advantage as transportation fuels, which prevents a demise similar to that of coal, at least until around 2050. The preferred end-use fuels in this well-to-do region are clean, grid-dependent energy carriers such as electricity, natural gas, district heat, and, in the very long term, hydrogen. Pollution-emitting energy conversion steps are shifted upstream in the energy system.

If energy investments are focused on coal (Scenario A2) or are relatively unfocused (Case B), additions to oil reserves do not keep pace with oil consumption, and by 2050 oil products are replaced by synfuels (e.g., methanol and synthetic gasoline) or natural gas. The same shift is projected to take place in Case C, although for different reasons, namely, explicit policy decisions that support, in particular, public transportation and the diffusion of new transport technologies such as high-speed trains, fuel cells, and electric vehicles.

The future for fuel cells projected in the scenarios deserves elaboration. It is representative of the critical role of technology dynamics – innovation and diffusion – in any analysis of long-term energy demand and supply. In this study we adopted a full energy systems approach, which means we had to address potential future technological and economic improvements in numerous end-use technologies. The potential for technical surprise is enormous. Our approach was to concentrate on a few key technologies in a largely generic manner, matching their evolution both to historical patterns of innovation and diffusion (see Section 4.4; Grübler, 1998) and current characteristics of related technologies as collected in IIASA's technology databases. Fuel cells are an example of such a "generic" technology, in this case providing for transportation powered directly by hydrogen or indirectly by methanol or natural gas. Depending on the availability of alternative transport fuels and technologies in the different scenarios, a greater or lesser role is projected for fuel cells and their associated fuel supply infrastructures, including storage for hydrogen generated from intermittent renewable sources. In Case C, in particular, fuel cells play a very large role. Concern was expressed in the NAM regional review about the mechanism for providing transportation services in Case C in light of its shrinking oil sector. Fuel cells are the scenarios' answer in this case.

The evolution of the region's primary energy supply ultimately depends on how investments are targeted. Up to 2020, differences across scenarios are small because of the slow capital stock turnover in the energy sector (due to long lifetimes

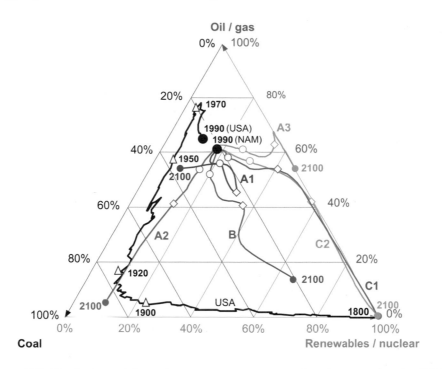

**Figure 7.2**: Evolution of the primary energy structure for six scenarios for **NAM**, shares of oil and gas, coal, and non-fossil sources, in percent. Each corner of the triangle represents a hypothetical case in which 100% of primary energy is supplied by a single source: oil and gas at top, coal left bottom, and non-fossil energy right bottom. The figure shows the historical evolution for the USA since 1800 as well as the six scenarios for NAM to 2100. See Section 5.2 for a further explanation of the figure.

of energy plants, equipment, and infrastructures) and the region's large established energy sector. After 2020, however, significant differences unfold. The range of possibilities is in fact greater in NAM than in any other region, reflecting the fact that in principal NAM faces no serious constraints in terms of know-how, capital, technology, or resources that would preclude particular development paths.

If investment choices target improvements in coal production, conversion, and end-use technologies (Scenario A2) – particularly technologies to convert coal to clean and convenient fuels – then coal production and coal use expand. Production nearly doubles by 2020, and by 2030 the region shifts from being a net energy importer to being a net exporter, with exports expanding steadily. As shown in *Figure 7.2*, by 2100 coal would once again be dominant within the primary energy structure, as it was 200 years earlier at the beginning of the 20th century. The difference is that coal in 2100 would be almost entirely converted to electricity, heat, and chemical fuels such as methanol in large, centralized clean coal conversion facilities. The infrastructure would be fundamentally different from that in the early 1900s, when coal was predominantly an end-use fuel.

If, however, investments are focused on renewables and nuclear power, allowing technological improvements on both the supply and demand sides to mutually reinforce one another, the region moves rapidly away from coal. In Scenario A3, coal's share of primary energy drops from 22% in 1990 to 14% in 2020 and to only 2% in 2050. Oil and natural gas continue to dominate through 2050, when they account for 64% of primary energy (compared with 61% in 1990). By 2100 they still supply 55% of primary energy needs, with gas accounting for nearly all of that share.

If the region chooses the policies and priorities of Case C, the shift to renewables and nuclear power will be much more complete, particularly after 2050. Case C's policy-directed shift away from fossil fuels makes it the only case to see import dependence begin to decline immediately. In Cases A and B, dependence stays relatively constant through 2020 and then begins a long-term steady decline.

A consistent trend across all scenarios is the increasing contribution of natural gas, the preferred transition fuel to a post-fossil-fuel age. Even with investments focused on renewables and nuclear power, Scenario A3 projects natural gas use to double to 1,000 Mtoe by 2020, increasing its primary energy share to 33% from 23% in 1990. In Case C, even with policies and investments emphasizing a shift away from fossil fuels and total primary energy decreasing by 16% between 1990 and 2020, natural gas use still increases by 10%. In Scenarios A1 and A3 natural gas use continues to grow essentially throughout the 21st century.

The NAM regional review's assessment of the global results focused on the alternative technology development paths represented by the scenarios. There was agreement with the conclusion that resource constraints are not a major issue. The range of expectations paralleled that discussed earlier for general technological improvements. A number of reviewers argued that oil would play a greater role even in Scenario A2 and Case B, where investments are directed elsewhere. Others argued for higher shares for natural gas, although insufficient short-term pipeline availability in a deregulated market environment was seen as a potential barrier to the expanded use of gas.

In general, the review considered coal and nuclear technologies to be available as required in the longer run if unconventional gas resources could not be developed. The almost total elimination of coal in the scenarios targeting renewables and nuclear (Scenario A3 and Case C) was labeled unrealistic. Reviewers argued that, even in these cases, technological improvements would allow coal to maintain a larger role and meet strict environmental constraints. The potential for innovation in carbon production (mining), abatement, and disposal certainly exists and may yet yield surprises not anticipated in the scenarios.

It is in connection with renewables that reviewers differed most among themselves and, as a group, from the global analysis. As was the case with technological improvements in general, the scenarios lie in the optimistic half of the spectrum of NAM opinion. But that spectrum is broad, ranging in the NAM regional review from the assessment that "only Case B seems plausible" to the NAM regional review Chairman's conclusion that "the C cases seem to best characterize NAM in [the] 2050 time frame and beyond."

Through at least 2030 nuclear power's importance grows in all scenarios. Its relative importance grows fastest in Case C, as fossil fuels are explicitly discouraged for environmental reasons. In absolute terms there is little difference among Scenarios A1 and A3 and Case B through 2030. Too little of the energy infrastructure turns over before 2030. After 2030, nuclear power sees a renewed growth phase in Case B, where its principal long-term competitor, renewable energy, benefits much less from technology investments. From a regional perspective, nuclear power's growth may also be constrained by electricity deregulation. If deregulation increases the importance of rapid investment returns, capital-intensive technologies like nuclear power (but also clean coal, photovoltaics, and hydropower) are expected to suffer. Another distinctly regional issue is the public attitude toward nuclear power. There is significant public resistance to nuclear power within the region, although some experts believe that export opportunities in Asia, changing perceptions in the wake of the Kyoto Protocol, and an interest in keeping all energy RD&D options open (PCAST, 1997) may lead to improved prospects for nuclear power.

### 7.2.4. Implications

Common to all scenarios is a significant decline in traditional environmental pollutants, for example, sulfur, $NO_x$, particulates, and CO. By 2020, sulfur emissions decline to between 10% (Case C) and 70% (Case B) of 1990 levels as the region's fuel mix changes and new facilities reflect recent environmental policies such as the 1990 amendments to the US Clean Air Act (EPA, 1996). By 2050, sulfur emissions generally decrease to even lower levels, although they subsequently increase slightly in scenarios returning to coal and synfuels in the long term (Scenarios A1 and A2). In no case, however, do they exceed one-third of their 1990 levels (11 MtS).

$CO_2$ emissions vary tremendously across the scenarios.[3] Except for the ecologically driven Case C and Scenario A3, with its focus on renewables and nuclear

---

[3]Unless otherwise noted, $CO_2$ emissions reported in this chapter refer to net emissions (see Section 6.6.3).

power, emissions increase considerably. This is especially true in scenarios with stepped-up exports of unconventional hydrocarbons and coal-based synfuels (Scenarios A1 and A2). Scenario A3's combination of high economic growth and a focus on renewables and nuclear power results in regional carbon emissions essentially holding steady at approximately 1990 levels. Only in Case C are there substantial reductions, with emissions dropping from 1,500 MtC in 1990 to 1,000 MtC in 2020, 400 MtC in 2050, and to negligible amounts by 2100. The bottom line is that carbon emission reductions in NAM will not happen by themselves, but will require the major policy commitments reflected in Case C. Even rapid progress in nuclear and renewable technologies, as in Scenario A3, is not enough by itself to shift energy use far enough away from the region's rich fossil resources to produce significant reductions in carbon emissions.

Investment requirements are large in absolute terms, varying from a total of US$2.0 trillion (Case C) to US$4.0 trillion (Scenario A2) between 1990 and 2020. As a percentage of GDP, the range is from 0.8 to 1.5%, well above the ratio projected for other OECD regions (PAO and WEU) and similar to the aggregate investment fraction for the world as a whole. But while investment requirements in NAM, as a percentage of GDP, are similar to those in some developing regions (e.g., LAM and PAS), NAM has the advantage of well-developed, well-established institutions to facilitate the needed investments. In the near term, the principal question is whether deregulation will discourage capital-intensive energy options and slow the sort of long-term energy RD&D envisaged in all of this study's scenarios – save the middle-course Case B.

### 7.2.5.  Highlights

- Of all the regional reviews, NAM's reflected the greatest range of perspectives within any region. The global scenarios fall generally in the optimistic half of the regional spectrum of opinion, particularly the technological improvements in zero-carbon options reflected in Case C and Scenario A3. But even there, views ranged from the assessment that "only Case B seems plausible" to the NAM regional review coordinator's conclusion that "the C cases seem to best characterize NAM in [the] 2050 time frame and beyond."

- The range of possible developments in regional primary energy structure (as shown in *Figure 7.2*) is greater in NAM than in any other region. In principle, the region faces no serious constraints in terms of know-how, capital, technology, or resources that would preclude any of the six scenarios. At the same time, it remains unclear how the sort of management and coordination policies required to focus development trends as reflected in Cases A and C could emerge given the current emphasis on deregulation, the region's decentralized decision-making institutions, and its powerful and divergent lobbying groups.

- Case C's policy-directed shift away from fossil fuels makes it the only scenario to see import dependence begin to decline immediately. In Cases A and B, dependence stays relatively constant through 2020, and then begins a long-term steady decline. The region has the potential to become a major net exporter of synfuels in the high-growth coal-intensive Scenario A2.

- In the very long term (2100), most of the scenarios lead to future primary energy structures that, in the aggregate, resemble those of the past. In the coal-intensive scenario the primary energy mix in 2100 is dominated by coal, as it was around 1900 (see *Figure 7.2*). In Cases B and C the dominant share of non-fossil energy sources approaches what it was around 1800. The difference is that both per capita and absolute energy needs will be vastly higher, as will the level of technological sophistication. Traditional primary energy sources – whether coal, biomass, or wind power – will all reach the consumer in the form of high-quality "man-made" energy currencies delivered by grids.

- NAM has a highly deregulated economy and energy system, and is often cited as a role model for similar developments in other regions. Therefore, two concerns voiced in the regional review deserve particular attention. The first is the difficulty of balancing short-term and long-term capital risk-taking under a continuously changing policy environment that couples market deregulation with tighter environmental standards. Second, even as an apparent institutional role model, the NAM regional poll ranked institutional deficiencies as a number-one regional concern.

- Given the region's abundant fossil resources, carbon emission reductions will not happen by themselves. They will require major policy commitments, exemplified for instance in Case C. Even rapid progress in nuclear and renewable technologies, as in Scenario A3, is not by itself sufficient to shift energy use far enough away from the region's rich fossil resources to produce absolute reductions in carbon emissions.

## 7.3.  Latin America and the Caribbean (LAM)

The region includes 38 countries that share a common historical and cultural heritage, but also vary tremendously in their size, economies, and natural resources. In 1995 the two largest countries, Brazil and Mexico, accounted for more than 50% of

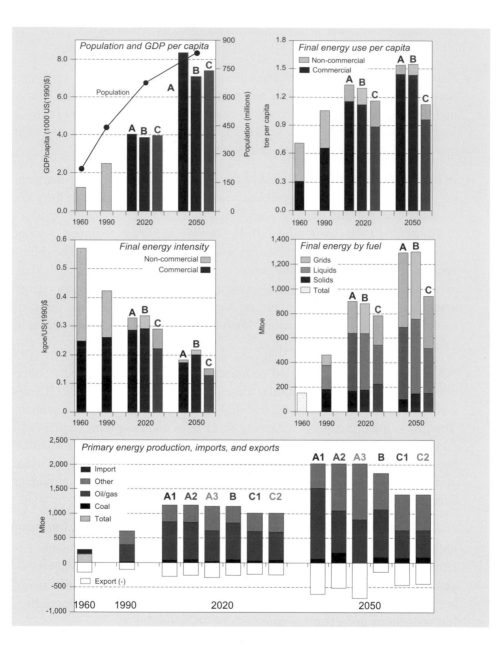

**Figure 7.3**: Scenario summary for **LAM** through 2050. *Top left:* Levels of GDP per capita reached in Cases A, B, and C, plus the population projection used in all three cases. *Top right:* Final energy use per capita. *Middle left:* Decrease in total final energy intensities in all three cases. *Middle right:* Final energy divided into solid fuels, liquid fuels, and energy delivered by grids. *Bottom:* Composition of primary energy production, plus imports and exports, for all six scenarios.

regional GDP. The region is rich in human and natural resources, with ecosystems that represent both much of the world's biodiversity and a large potential $CO_2$ sink.

In 1990, LAM accounted for 8.3% of the global population, 5.2% of GWP, 6.8% of global primary energy consumption, and 4.7% of global energy-related net carbon emissions. Economically it lies between the OECD and the developing countries of Africa and Asia. In 1990, per capita GDP in LAM was US$2,500, almost one-eighth that in the OECD countries and about 40% below the world average. In terms of purchasing power parities, however, per capita income was close to the world average of US$4,900.

### 7.3.1. Recent developments

For two decades, LAM enjoyed rapid economic growth ranging between a very high rate of about 7% per year from 1966 to 1973 and about 5% per year from 1974 to 1980 (OLADE-SIEE, 1997). Growth slowed to a near standstill during the "lost decade" of the 1980s, averaging 1.3% per year from 1981 to 1990 (OLADE-SIEE, 1997). During the early 1990s, economic growth was erratic but generally higher than in the previous decade. Average growth was 3.2% per year from 1991 to 1995. Growth was 0.2% in 1995 and 3.4% in 1996 (World Bank, 1997a).

Primary energy growth averaged 2.9% per year from 1990 to 1996 (OLADE-SIEE, 1997), which compares with average total primary energy growth between 2.1% and 2.5% per year between 1990 and 2000 in the scenarios.

### 7.3.2. Population and economy

The demographic transition is more advanced in LAM than in other developing regions. Population growth is slower, at 1.6% per year from 1990 to 1993, and urbanization is higher, with about 75% of the population living in cities (Rosa and La Rovere, 1997). The 1992 World Bank population projection estimates a near doubling of the population from 430 million people in 1990 to 840 million in 2050, with very little growth thereafter (see *Figure 7.3*). This estimate falls within the range of medium-term projections (through 2020) developed within the region (OLADE-SIEE, 1997) and cited in the LAM regional review.

In all six scenarios LAM eventually reaches high levels of affluence. Per capita GDP increases from US$2,500 in 1990 to surpass the current Western European level of US$16,000 within the next century – around 2070 in Case A and within the next two decades in Cases B and C.

Between 1990 and 2020 economic growth in the scenarios is between 2.9% and 3.1% per year. As would be expected, these long-term development prospects are lower than current short-term World Bank projections of 4.2% per year from 1997 to 2006 (World Bank, 1997a). The LAM regional review is even a bit more optimistic than the World Bank, suggesting growth between 4% and 5% per year between 1990 and 2020. This compares with the latest regional projection of 3.7% per year from 1997 to 2010 (OLADE-SIEE, 1997). As described above, the region has not seen such rapid growth since the 1970s, and of the alternative projections, those in this study come closest to the region's actual experience of 3.2% per year between 1990 and 1996 (World Bank, 1997a).

Continued strong growth assumes that the region can successfully meet a number of challenges, including furthering human and social development, strengthening government finances and financial markets, and improving the regulatory environment for investment. Particular priority must be given to extending financial services to rural areas and improving access to energy, water, and health care as a prerequisite for successful economic and social development. For example, currently 100 million people in the region are without access to formal health care (World Bank, 1998a). The LAM regional review stresses the need for greater regional integration. Given the enormous gains of the past decade in terms of economic stabilization and structural reforms, prospects for greater coordination and cooperation are good (World Bank, 1997a). Greater integration among national energy planning and energy systems, building on the accomplishments of the Latin American Energy Organization (OLADE), will be important in accelerating regional energy intensity improvements.

Where the perspectives of the regional review and the global analysis part ways most notably is on the future of agriculture. The regional review emphasizes an expanded development of agricultural activities beyond what is considered to be consistent with rapid economic development and urbanization in this study. Such a development path would be substantially different from historical experience as reflected in the data and models discussed in Sections 4.2 and 4.3. In the scenarios, traditional agriculture is gradually replaced by modern agricultural systems, including renewable energy production such as biomass.

### 7.3.3.  Energy systems

As in other regions, the scenarios reflect a shift toward more flexible, cleaner energy forms in LAM and an increased share of grid-delivered energy (i.e., electricity and natural gas). By 2050 natural gas surpasses oil as the most important fossil energy source in all but Scenario A1 and Case B, and electricity grows from 9% of final energy in 1990 to between 19% (Scenario A1) and 26% (Cases B and C) of final energy by 2100. Overall efficiency increases with the shift from noncommercial to commercial energy sources, and there is a steady decrease in the direct consumption of solid fuels such as coal and traditional biomass. Eventually, biomass energy reaches the consumer exclusively as motor fuel (ethanol and methanol) and electricity. The share of traditional solid biomass use decreases steadily from 37% in 1990 to negligible amounts in all scenarios after 2050.

There are only a few regional scenarios to compare against this study, and they extend, at most, to 2020. Final energy intensities projected by OLADE decrease at an average annual rate of 0.8% from 1990 to 2010 (OLADE-SIEE, 1997), essentially identical to the decrease projected in the middle-course Case B. Regional short-term projections of economic growth are more ambitious, as cited above, which follows the general pattern discussed in Section 7.1 of more ambitious short-term expectations compared with global long-term analysis. History suggests, however, that for a region to successfully realize these high economic ambitions, a prerequisite is faster technology change leading to more rapid improvements in energy intensity than is suggested in the regional review.

Among the 11 regions, LAM is distinctive for its large renewable resources. Hydropower and noncommercial energy (mostly traditional biomass) already provided a greater share of primary energy in 1990 than oil, and with time, modern renewables replace traditional biomass. Modern biomass and hydropower are the most important renewable energy forms in all scenarios. Hydropower reaches its estimated full potential of 600 gigawatts (GW; OLADE-SIEE, 1997) in Scenario A1 and Case B. Biomass is increasingly converted into electricity, liquids, and gas (ethanol, methanol, and possibly hydrogen in the very long run). In time, liquid fuels from biomass replace oil products as the main motor fuel in all scenarios. In the meantime, the predominantly leaded gasoline in the region is replaced by lead-free gasolines, partially based on alcohol anti-knock additives. In most scenarios synfuels from biomass develop into an important energy export for the region. The scenarios suggest that LAM's unique experience with renewables (at the height of Brazil's fuel-ethanol program,[4] half the cars sold in the country had ethanol engines) has been in fundamentally the right direction, if perhaps a bit ahead of its time.

From a regional perspective, the prospects for renewables (especially hydropower), natural gas, and electricity look even better. For all these energy forms regional reviewers from LAM agree with the basic direction of the scenarios, but expect greater production and consumption. In part the difference reflects the faster economic growth projected in regional studies, as discussed above; it also reflects lower expectations for energy intensity improvements. For natural gas in particular, growth will depend on regional integration and the speed at which the region's natural gas infrastructure can be expanded.

---

[4]The Brazilian fuel-ethanol (Pro-álcool) program was begun in 1975 in response to the first world oil crisis in 1973. At the time Brazil was importing 80% of its oil. The program was designed in part to reduce import dependence by substituting ethanol for gasoline in passenger cars and light vehicles (Goldemberg *et al.*, 1993).

### 7.3.4.   Implications

With increasing affluence in LAM, adverse environmental impacts of energy use diminish. New renewable energy reduces both degradation and pollution. The increasing importance of natural gas, electricity, hydropower, and new renewables all lead to lower urban air pollution. Sulfur emissions, as an indicator of regional environmental damage, are below 1990 levels by 2020 in Case C and Scenario A3, and by 2060 at the latest in all six scenarios.

LAM's forests are potentially important as carbon sinks. The $CO_2$ emissions calculated for each of the scenarios, however, do not assume that they are used explicitly as sinks. New biomass is assumed to be used sustainably and generates no net carbon emissions (see Section 6.6.3). Whatever carbon is released when one tree is burned is effectively absorbed by the next tree as it grows, but no afforestation is assumed solely for the purpose of sequestering carbon. $CO_2$ emissions grow in all scenarios through 2020 as a result of continued reliance on fossil energy, particularly oil. As natural gas and non-fossil fuels gain importance, emissions then peak around 2050 in all but Scenario A1 at between 480 MtC (Case C) and 790 MtC (Scenario A2), that is, between 1.7 and 2.8 times the 1990 level. By 2100 carbon emissions are below their 1990 level in Scenario A2 and Cases B and C, and down to 380 MtC in Scenario A3. Only in the oil- and gas-intensive Scenario A1 do emissions keep climbing after 2050. By 2100 they reach 1,700 MtC, but on a per capita basis are still slightly below WEU's 1990 per capita emissions.

Cumulative investment requirements from 1990 to 2020 range between US$620 billion and US$1.0 trillion. These nearly equal the region's total GDP for 1990 (US$1.1 trillion), but are less daunting as a percentage of regional GDP, ranging between 1.2% and 1.8% of GDP. Investment requirements for NAM and PAS are similar as percentages of GDP. Although needed investments increase in absolute terms, they are relatively constant as a share of GDP through 2050 and then gradually decrease.

### 7.3.5.   Highlights

- Among the 11 regions, LAM is distinctive for its large renewable resources, of which modern biomass and hydropower are the most important in all scenarios. Hydropower reaches its estimated full potential of 600 GW in Scenario A1 and Case B.

- The LAM regional review stressed the need for greater regional integration, and given the enormous gains of the past decade in terms of economic stabilization and structural reforms, prospects for greater coordination and cooperation are good. Greater integration among national systems will be important in speeding energy intensity improvements and in expanding the utilization of the region's natural gas resources.

- Liquid fuels from biomass tend to replace oil products in all scenarios. Equally, synfuels from biomass develop into an important energy export from the region in the long term. The scenarios suggest that LAM's unique experience with renewables (e.g., Brazil's fuel-ethanol program) has been in fundamentally the right direction, if perhaps a bit ahead of its time.

- With increasing affluence in LAM, adverse environmental impacts of energy use diminish despite considerable inertia – not least in the transportation sector. The replacement of traditional biomass by new renewable energy reduces both environmental degradation and pollution, and increasing use of natural gas, electricity, hydropower, and new renewables lowers urban air pollution. Sulfur emissions, as an indicator of regional environmental damage, are below 1990 levels by 2020 in Case C and Scenario A3, and by 2060 in all six scenarios.

- Although it is a rapidly developing region, LAM is expected to be a relatively minor contributor to global $CO_2$ emissions. With development, emissions initially increase, but then peak around 2050 in all but Scenario A1. By 2100 they are below 1990 levels in Scenario A2 and Cases B and C, and are near these levels in Scenario A2. Even in the oil- and gas-intensive Scenario A1 they are below WEU's 1990 per capita emissions by 2100.

## 7.4.  Sub-Saharan Africa (AFR)

The region encompasses the 50 countries of sub-Saharan Africa, including the Republic of South Africa. It reflects extensive linguistic and cultural diversity and a wide range of standards of living and lifestyles. Most of the countries are low income and largely rural, relying on traditional patterns of economic activities and using fuelwood (often unsustainably) as a major energy source. AFR's population growth is among the highest in the world. In 1990, the region accounted for slightly more than 9% of the global population, and its share will rise markedly in the future. In contrast, AFR's share of GWP in 1990 was only slightly above 1% (above 2% if GDP is based on purchasing power parities), and per capita income was thus extremely low. The GDP of the entire region totaled US$265 billion in 1990, a bit less than that of the Netherlands, a country with only 3% of AFR's population and much poorer in natural resources. Per capita final energy use is correspondingly low – less than 0.5 toe per capita. Final and primary energy use in the region are only 3% of the world total, with about two-thirds of final energy use coming from

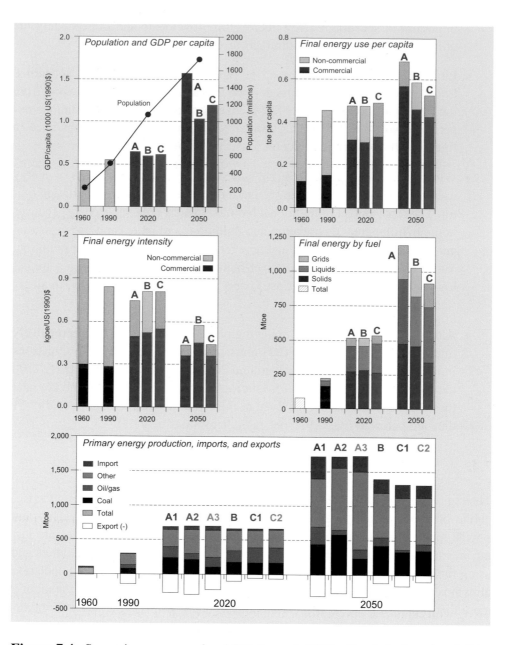

**Figure 7.4**: Scenario summary for **AFR** through 2050. *Top left:* Levels of GDP per capita reached in Cases A, B, and C, plus the population projection used in all three cases. *Top right:* Final energy use per capita. *Middle left:* Decrease in total final energy intensities in all three cases. *Middle right:* Final energy divided into solid fuels, liquid fuels, and energy delivered by grids. *Bottom:* Composition of primary energy production, plus imports and exports, for all six scenarios.

noncommercial traditional fuels, predominantly biomass in the form of fuelwood. Energy-related net carbon emissions in 1990 were 2.3% of the global total.

### 7.4.1. Recent developments

Since 1990, many countries in AFR have experienced far-reaching political changes, including the end of apartheid in the Republic of South Africa and the end of large-scale conflict in southern Africa. Overall, the region's prospects for peace, and thus for sustained economic growth, have notably improved. Economic growth increased substantially to some 4% per year in 1995 and 1996, compared with the dismal 1.6% per year average from 1980 to 1994 (World Bank, 1997a), a "lost decade" for AFR just as it was for LAM. Improved economic prospects are broad based, with 33 countries in the region now achieving positive per capita GDP growth. Private capital flows to the region, which were virtually nonexistent in the early 1990s, reached about US\$12 billion in 1996 (although much of this went to only a few countries – South Africa, Ghana, and the oil exporters Nigeria and Angola; World Bank, 1997a). Statistics available through 1995 (Gürer and Ban, 1997) indicate that commercial energy use grew by some 30 Mtoe from 1990 to 1995 (an average of 4% per year) and noncommercial energy use is likely to have grown by a similar magnitude. This growth is partly due to high population growth. Between 1990 and 1995 the population grew by an estimated 80 million people – a growth rate of 3% per year. Overall, short-term developments since 1990 in AFR have been slightly below trends in the middle-course Case B.

### 7.4.2. Population and economy

Over the medium term (to 2020) all scenarios are dominated by population growth (see *Figure 7.4*). Most of the mothers of 2020 are already alive today, and the inevitable demographic momentum is projected to push AFR's population over 1 billion by 2020. This result is consistent across all population projections discussed in Section 4.1 and takes into account recent mortality increases in the region, especially those from AIDS.[5]

The population issue was also given high priority in the AFR regional review, along with economic development and, to a lesser extent, energy supply. In contrast to lower population projections in more recent studies as discussed in Section 4.1, the regional review suggested a higher baseline projection of 1.88 billion people by 2050, 8% higher than the 1.74 billion used in this study. It also emphasized the uncertainty surrounding long-term projections. That uncertainty is illustrated

---

[5]For an overview see Garenne (1996). Mortality impacts of AIDS in sub-Saharan Africa are estimated to range between 8.5 and 9 million deaths between 1990 and 2005 (Bongaards, 1996), compared with projected total population growth of some 260 million during the same period.

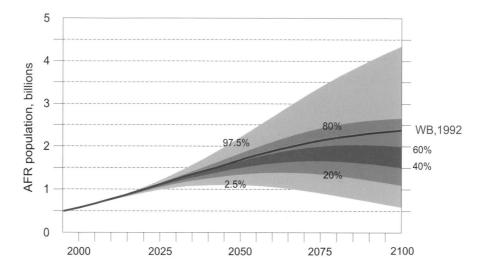

**Figure 7.5**: Comparison of the 1992 World Bank population projection for **AFR** and probabilistic projections developed by IIASA. Source: Lutz, 1996.

in *Figure 7.5*, which compares the 1992 World Bank projection for AFR used in this study with IIASA's probabilistic projections as discussed in Section 4.1 (Lutz, 1996). Based on the judgment of the experts assembled by IIASA, taking into account the recent fertility decline, which was more rapid than was anticipated in 1992, Lutz estimates that there is a 60% chance that by 2050 AFR's population will fall below the 1992 World Bank projection used in this study, and a 40% chance that it will be higher. By 2100, the probability of the actual population being above the 1992 World Bank projection drops to 30%.

Even if the higher population projection suggested by the regional review materializes, the review also suggested regional economic growth below even the modest levels of the middle-course Case B. Lower economic growth is consistent with the view, discussed in Section 4.1, that rapid population growth hinders economic growth more than it helps it: the "more mouths to feed" outweigh the "more hands to work." The lower economic growth suggested in the regional review would thus partially offset the impact of higher population growth on energy use. Overall energy demand would likely remain largely unchanged, albeit at lower levels of energy use per capita. Thus, despite the regional review's different perspectives on population and economic growth, it largely endorsed the scenarios' range of possible energy sector developments as described below.

Despite the slow economic growth suggested by the regional review as the most likely scenario, the review also emphasized prospects for higher growth in individual countries within the region. South Africa's economic growth target is 6% per year by 2000, for example, and the region's low level of economic development means that even isolated and comparatively small investments can trigger large economic growth effects. Growth rates in such cases could approach that of our more optimistic Case A scenarios. Nonetheless, in its judgment that overall economic growth in all scenarios might be too optimistic, the AFR regional review is unique. In most other regions, reviewers projected growth rates exceeding those of the scenarios.

From the global perspective, all six scenarios reflect medium-term overall GDP growth of around 3% per year. This is only slightly ahead of population growth, which is projected to decline to 2% per year by 2020. Despite recent encouraging developments, a definite shift to self-sustained economic growth in the region will take considerable time, even assuming political stability, structural reforms, and a favorable investment climate for foreign capital.

Conversely, the long-term development perspectives provided by the scenarios are brighter than the rather pessimistic short-term outlook. By 2050, the region's large endowment of human capital and natural resources should provide the conditions for sustained economic development. The actual rate of development will depend on both indigenous efforts and the availability of investment capital. These are judged cautiously in Case B and more optimistically in Cases A and C, either because of high growth and a favorable investment climate (Case A), or stimulated by direct resource transfers furthering sustainable development (Case C). By 2050, GDP in all scenarios (expressed in terms of purchasing power parities) surpasses US$2,000 per capita, rising above the threshold for basic food, shelter, and energy services. However, it is not until the very end of the next century that per capita GDP is projected to reach approximately US$10,000 per capita. And even this assumes stable governance, no major regional conflicts, high social capital (education, the empowerment of women, and other conditions conducive to reducing fertility), favorable terms of trade, and foreign investments that help harness the enormous potential of the region's natural and human resources.

### 7.4.3. Energy systems

As shown in the top right and middle left charts in *Figure 7.4*, all three cases reflect a shift from noncommercial to commercial energy. Commercial final energy intensity is therefore projected to *increase* between 1990 and 2020 (middle left chart in *Figure 7.4*), while *total* final energy intensity decreases slightly (at between 0.1% and 0.4% per year). Total energy use increases by a factor of about 2.4 by 2020, while commercial energy use grows by a factor slightly lower than 4. From 2020 to 2050 the decline in total final energy intensity is projected to accelerate to between 1.1% per year (Case B) and 2.0% per year (Scenario C1) as economic growth increases to between 3.5% per year (Case B) and 4.7% per year (Case A). The faster

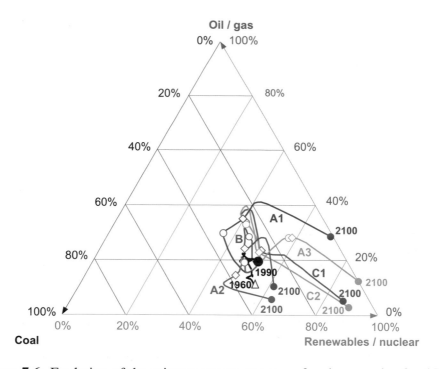

**Figure 7.6**: Evolution of the primary energy structure for six scenarios for **AFR**, shares of oil and gas, coal, and non-fossil sources, in percent. The eventual trend in all scenarios is toward non-fossil fuels, reflecting the region's huge renewable resource potential. Detours are taken by Scenarios A1 and A2 in the directions of relative oil intensity and relative coal intensity, respectively. See Section 5.2 for a further explanation of the figure.

the economic growth, the faster traditional, inefficient energy services can be replaced by more efficient modern ones (a liquefied petroleum gas cooker is at least five times as energy efficient as a traditional wood cooker). By 2050 total final and primary energy use increases by a factor of between 4 and 6 (bottom chart in *Figure 7.4*), while commercial energy use grows even more, by a factor between 8 and 11. Such relative increases are the highest of all 11 regions. Yet, for all these hefty and continued increases, it is only toward the very end of the 21st century that primary energy use in AFR approaches 1.5 to 2.5 toe per capita, levels characteristic of England around 1900 and of Western Europe and Japan in the mid-20th century.

The bottom line across all scenarios is a formidable increase in energy use driven by the twin pressures of population growth (which at least in the near-term is largely unavoidable) and, what is more desirable, income growth. Primary energy use in all scenarios surpasses 1 Gtoe by 2050, and continues to increase substantially.

In all scenarios, oil and natural gas become increasingly important through 2020. After 2020, the dominance of fossil fuels gradually gives way to sustainable, new forms of renewables drawing on the enormous biomass and solar resource potential of the region (*Figure 7.6*). The decrease in the share of noncommercial traditional biomass fuels in Case B is relatively gradual compared with its decrease in Cases A and C where economic growth, and thus structural change, is faster. If either conventional oil and gas or economical new renewable sources are less available, as in Scenario A2, coal readily fills the gap.

The range of possible energy sector developments presented in the scenarios was largely endorsed by the regional review. Energy intensity trends and projected levels of energy demand were judged plausible, and the range and evolution of possible future energy supply structures were also generally accepted. The one exception was the technology-intensive Scenario A3, judged as being overly optimistic given the region's near-term needs for adapted, simple, and especially low-cost energy end-use and supply technologies. Near- and medium-term policy and investment priorities identified in the regional review are consistent with the scenarios. They are, first, to improve the efficiency of traditional biomass use and, second, to mobilize the required investments to harness the region's large resource potential (i.e., oil, natural gas, and hydropower) and to develop infrastructures to use it efficiently (e.g., natural gas infrastructure networks, eliminating natural gas flaring, and urban and rural electrification).

### 7.4.4. Implications

The scenarios emphasize that the challenges ahead for AFR lie less in physical resource availability than in the need to mobilize the necessary capital for resource exploration and development – particularly the infrastructural investments needed to match growing demands for energy services with available resources. This conclusion was echoed in the AFR regional review. Even with the scenarios' rather conservative growth in energy demand, projected cumulative energy sector investments between 1990 and 2020 range between US$320 billion (Case C) and more than US$630 billion (Scenarios A2 and A3), corresponding to 2.5% and 4.7% of regional GDP, respectively. By 2050 energy sector investment requirements reach almost 5% of regional GDP in Scenario A2, and the cumulative investment requirements between 1990 and 2020 are more than twice AFR's 1990 GDP. Energy investments continue to grow beyond 2050, but at a slower pace than GDP. Meeting such investment challenges will depend partly on progress in currently largely embryonic regional cooperation mechanisms. It will also depend in part on the ability to mobilize large sums of foreign capital in a region where, for many decades to come, low regional income per capita will mean low regional capital availability.

If substantial international funding were to flow through mechanisms such as the Global Environmental Facility of the World Bank or, in the longer term, the FCCC, environmental issues might gain importance in AFR. But in the absence of strong incentives from international funding mechanisms, environmental issues are likely to remain low on the regional policy agenda behind more pressing immediate issues of poverty eradication and social and economic development. As a result, both the scenarios and the regional review expect emissions of sulfur and $CO_2$ to increase. In the scenarios, carbon emissions increase between two to three times by 2020, and between four to six times by 2050. But even in Scenario A2, which has the highest emissions, they barely reach 0.5 tC per capita by 2050. This is an order of magnitude lower than current per capita emissions in the industrialized countries.

The overall picture is a long road to development. AFR is likely to require almost a century to reach the level of economic development and access to energy services characteristic of industrialized countries in the second half of the 20th century. This is true even in Cases A and C, which incorporate numerous "leapfrogging" possibilities in energy efficiency (Goldemberg, 1991) due to technology transfer and overall rapid technological progress.

### 7.4.5.  Highlights

- Through 2050 AFR experiences the largest relative increases in energy use of all 11 regions. The increase is driven by the twin pressures of population growth (which at least in the near-term is largely unavoidable) and, what is more desirable, income growth. Yet, it is not until nearly 2100 that primary energy use in AFR is projected to approach 1.5 to 2.5 toe per capita, levels characteristic of England around 1900 and of Western Europe and Japan in the mid-20th century.

- The challenges for AFR lie less in physical resource availability than in the need to mobilize the necessary capital for resource exploration and development, and infrastructural investments, including investment in human capital. Because of the region's low per capita income, successful development will require concerted efforts in regional and international cooperation – efforts that need to move much more quickly than they do today. Too much is at stake in a region where the Malthusian trap of population growth outrunning subsistence, jeopardizing the development prospects for both current and future generations, will remain a persistent danger for decades to come.

- The basic priorities for overall social and economic development in the region fortunately are aligned with the most pressing needs of the energy sector. Urgent attention (and action) to promote social development, education, and the eradication of poverty will decrease fertility and slow population growth while accelerating economic development and increasing the region's attractiveness for international investors. Improvements in institutional frameworks are therefore a high development priority.

- Environmental issues, particularly global warming, are a low priority for AFR. $CO_2$ emissions are projected to increase substantially in all six scenarios, but even in the worst case, per capita emissions barely reach 0.5 tC by 2050. This is an order of magnitude below current emission levels in the industrialized countries.

## 7.5.   Middle East and North Africa (MEA)

The region consists of 19 very diverse countries. Per capita income in the region varies by a factor of more than 40. The 1990 average was US$2,120 per capita, but only 25% of the population had incomes that high. Income disparity reflects the disparity in the region's principal revenue source – oil and natural gas resources. The most populous countries have limited oil and gas resources while several countries with small populations are disproportionately well endowed.

In 1990 MEA accounted for 5% of the global population, 3% of GWP, 4% of global primary energy consumption, and 5% of global energy-related net carbon emissions. Its distinctive feature is its hydrocarbon resources, amounting to almost 60% of the world's proven oil reserves, 43% of conventional oil resources, 37% of proven natural gas reserves, and 25% of conventional gas resources. In 1990, revenues from energy exports equaled US$155 billion, or 26% of GDP; 70% of the region's energy production went to exports.

### 7.5.1.   Recent developments

The World Bank estimates 1990 to 1995 economic growth for most of the region's low- and middle-income countries at 2.3% per year (World Bank, 1997a). These estimates, however, do not include data from Iraq and Iran. Iran accounted for 20% of the region's GDP in 1990. Although high-income countries grew more rapidly, their share of regional GDP is small and overall regional growth is, therefore, likely to have been below the 1990 to 2000 range reflected in the scenarios – from 3.5% per year in Cases B and C to 3.9% per year in Case A. At the same time, the labor force has grown at 3.3% per year (World Bank, 1997a), exacerbating already high regional unemployment of 15%. Rapid population growth, together with investor concerns about sociopolitical tensions and the volatility of export revenues,

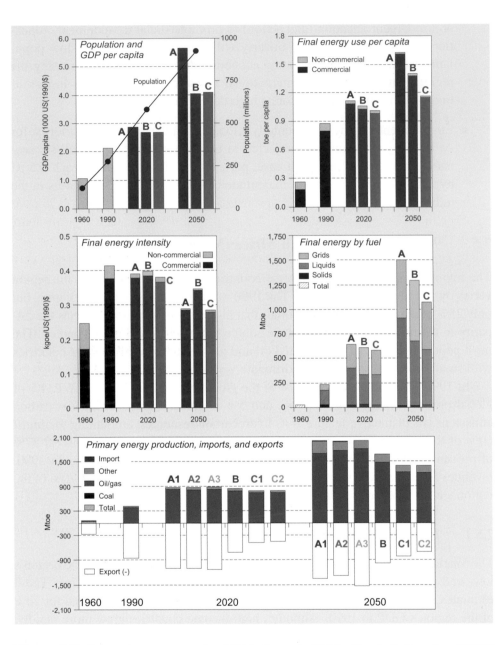

**Figure 7.7**: Scenario summary for **MEA** through 2050. *Top left:* Levels of GDP per capita reached in Cases A, B, and C, plus the population projection used in all three cases. *Top right:* Final energy use per capita. *Middle left:* Decrease in total final energy intensities in all three cases. *Middle right:* Final energy divided into solid fuels, liquid fuels, and energy delivered by grids. *Bottom:* Composition of primary energy production, plus imports and exports, for all six scenarios.

limit prospects for substantial short-term per capita income increases (World Bank, 1997a).

Between 1990 and 1995, total primary energy use grew an average of some 4% per year (BP, 1997; EIA, 1997). This is closest to Case A (3.5% per year). These data imply essentially no improvement in energy intensity, which is consistent with the negligible short-term improvement rates of 0.3% and 0.4% per year between 1990 and 2000 in Cases B and A, respectively.

### 7.5.2. Population and economy

According to the 1992 World Bank projection future population growth in MEA will be high – second only to AFR. Through 2020 the population is projected to more than double, growing at an average rate of 2.6% per year. Growth slows to 1.6% per year from 2020 to 2050. Stabilization is only projected near the end of the century, by which time MEA's population is projected to be 1.3 billion, more than four times its 1990 level of 270 million.

The MEA regional review suggested the 1992 World Bank projection might be too optimistic and proposed an alternative population growth rate of 2% per year after 2020. This contrasts with lower than anticipated fertility during the early 1990s, as discussed in Section 4.1, and a consequent reduction in most population projections. The latest UN projection, for example, anticipates only 1.4% growth in MEA's population between 2020 and 2050 (UN, 1998). The difference between 2% per year (the regional review's suggestion) and 1.4% per year (the latest UN projection) is substantial: by 2050 it would translate into 210 million people. Assuming per capita primary energy use of 1.5 toe in 2050 (a level reached in even the ecologically driven Case C), the difference in energy use is 320 Mtoe, about the same as the region's total primary energy use in 1990.

The population projected in the scenarios therefore represents a middle ground between the higher suggestion of the MEA regional review and the lower, latest projection (1998) of the UN.

Oil exports have been the region's main source of income, and this is not likely to change in the near to medium-term future. Hence, MEA's economic future is tightly linked to future export revenues, which in turn depend on future global oil and gas demands, supplies in other regions, and international oil prices.

*Figure 7.7* shows minor differences among the six scenarios through 2020 in terms of economic growth and the energy sector. The overall trend is one of accelerated industrialization based on the region's natural resources resulting in a gradual growth in per capita income. From 2020 to 2050 economic growth in the scenarios is projected to be between 3% and 4% per year. Development will largely depend on political stability within the region, institutional reforms to attract investment capital, and progress toward economic integration and cooperation. In the energy sector, integration of the North African electricity grid and its interconnections with those of the East Mediterranean countries and Europe is one recent encouraging sign.

Provided there are no major regional political conflicts, the region's extensive energy resources and its proximity to Europe and to growing Asian energy markets should lead to sufficient export revenues for sustained economic development. However, economic growth only gradually raises per capita income and living standards due to continuing high regional population growth. But by 2020, living standards in all scenarios reflect adequate food and shelter, and basic health care and energy services, even in countries currently well below subsistence levels. Per capita income could range between US$4,000 (Case B) and US$5,600 (Case A) by 2050 and could reach 1990 WEU levels near the end of the 21st century.

As shown in the middle left chart of *Figure 7.7*, final energy intensity in the region improves only slightly between 1990 and 2020. Efficiency is generally not a high priority in regions with large, inexpensive energy resources. Moreover, MEA's climate contributes to a relatively high demand for air conditioning and cooling, which will increase with increasing incomes. Water in the region is in short supply, and water demands will also grow with regional wealth. Part of the increased demand is likely to be met through energy-intensive desalination.

After 2020 energy intensities improve a bit more quickly. Between 2020 and 2050, as an industrial base is established and exports gradually shift from raw materials to semi-manufactured and manufactured goods with higher value added, energy intensity improvements in the scenarios increase to between 0.5% per year (Case B) and about 1% per year (Cases A and C). From the regional perspective of the MEA regional review, these rates were judged challenging but realistic.

### 7.5.3.  Energy systems

The interdependence of energy exports and the domestic fuel mix is a central theme across all scenarios. In general, regional oil consumption in MEA complements international export demands. If export demands are relatively high, regional oil consumption is relatively low, preserving oil production for export. Natural gas makes up the difference in meeting regional energy consumption. Thus in Scenario A1, where oil is a high-priority fuel internationally, demand for MEA oil is high and MEA oil exports are higher than in any other scenario, growing from 800 Mtoe in 1990 to 1,200 Mtoe in 2050.

Oil currently covers 63% of final energy demand within the region, compared with a world average of 39%. By 2020 this share decreases in all but Scenario A3. In Scenario A3 the faster global shift toward non-fossil resources limits demand for MEA oil. Lower oil exports correspond to higher regional consumption. But by 2050, oil's share of final energy is still between 46% (Cases B and C) and 62%

(Scenario A3). In the very long run (2100), oil's share of final energy is still about 50% in the middle-course Case B and 40% in the ecologically driven Case C. In the three high-growth Case A scenarios, which leave less oil still in the ground in 2100 and reflect faster technological progress among competing technologies and energy sources, oil's long-term share of final energy is projected to be lower – between 16% (Scenario A3) and 27% (Scenario A2). This compares with long-term global averages of between 5% (Scenario A3) and 23% (Scenario A1).

Electricity use expands in all scenarios at average rates of between 2% and 4% per year. These rates were considered too low by the MEA regional review, and the arguments in support of higher rates are strong. Recent electricity growth in the region has been on the order of 5% per year. Thus, future investment requirements in electrical generating capacity are possibly underestimated in the scenarios. However, the global trend toward clean and convenient end-use fuels still occurs in all scenarios, although at a smaller scale as solids, especially coal, have historically played an insignificant role in MEA. The region is among the first to introduce gas-based methanol, again with the objective of saving oil for export.

Total primary energy in the region grows from 350 Mtoe in 1990 to between 780 and 880 Mtoe by 2020, and between 1,400 and 2,000 Mtoe by 2050. In the high-growth Case A, very long-term (2100) *per capita* primary energy use reaches the 1990 WEU level of 3.4 toe. Ninety percent of primary energy consumption in the region in 1990 was oil and gas. That share remains fairly constant through 2050, although oil's portion decreases while that of gas increases. In 2050 nuclear power's share is small (2% at most), with the share of renewables being considerably higher (between 10% and 14%). After 2050 there is more variation among scenarios. Oil and gas are still much more dominant in MEA than they are globally. In Scenario A1 they maintain their 90% share of MEA's primary energy even through 2100. By then the oil and gas share is lowest in Scenario A2 (50%), in which the world's investments are focused on coal while high economic growth uses up MEA oil and gas more quickly than in Cases B and C. Substantial contributions from renewables and nuclear power occur only after 2050, even in Scenario A3.

The 10 to 14% share for renewables in 2050 appears unrealistic from the regional perspective. In light of the region's inexpensive oil and gas resources, the MEA regional review suggested a maximum contribution from renewables of 100 Mtoe, corresponding to up to 7% of energy supply. The difference between the regional and global estimates is the comprehensive scope of the models used to construct the scenarios. They identify cost-effective global strategies, and these are not necessarily the most cost-effective for each region individually. MEA's supply potentials for solar and wind energy are large, and in scenarios where global demand for oil exports from MEA is high, the region turns to renewables, to make up the difference in regional consumption. It is also true that what might be most cost-effective for MEA as a whole is not necessarily the best for each individual country. The distribution of renewable resources differs from that of oil and gas. In Egypt, for example, solar- and natural-gas-based hybrid systems are a means by which solar thermal energy conversion could leverage natural gas resources.

All things considered, the MEA regional review judged the high-growth, oil- and gas-intensive Scenario A1 (although with fewer renewables and more gas) the most desirable and most realistic scenario for the region. Although the region's potential renewable resources are enormous, the scenarios indicate that they offer a long-term option at best. After 2050 they could be used in combination with nuclear power to generate low-cost electricity, and with chemical fuels to supplement oil and gas exports.

### 7.5.4.  Implications

Export revenues in MEA will depend not only on the international energy market, but increasingly on domestic oil and gas needs. Currently only 30% of all primary energy produced in the region is consumed domestically, but this is expected to change. In Cases B and C, domestic oil and gas demands exceed exports between 2010 and 2020. In Case A this occurs two decades later. Nonetheless, export revenues are high in all scenarios. As shown in *Figure 6.5*, projected revenues are highest in Scenario A1, the MEA regional review's preferred scenario, where global economic growth is high and investments favor oil and gas. Oil exports in Case A range between 18 and 20 million barrels per day (mbd) by 2020, with revenues approaching US$300 billion. In Case B, revenues are US$190 billion in 2020 on an export volume of 13 mbd. In Case C, oil exports drop to below 9 mbd in 2020, but higher prices bring revenues of US$140 billion. Natural gas exports would bring total export revenues to about US$160 billion – that is, above 1990 levels – even in the ecologically driven future of Case C.

Annual revenues in Cases A and B subsequently climb to close to US$250 billion (Case B) and up to US$340 billion (Case A) in 2050, surpassing the historical 1980 revenue peak. In Case C, carbon emission constraints and low innovation in connection with fossil fuels cause oil exports to remain below 10 mbd. But with price increases as shown in *Figure 6.5*, oil export revenues remain slightly ahead of current levels. If natural gas exports are added – the potential of which could be enormous, particularly for Asia – total revenues are yet higher.

The investments needed to increase production to cover both growing domestic needs and exports will be substantial, although as a percentage of regional GDP these decrease over time. Cumulative requirements between 1990 and 2020 range between US$600 and US$700 billion, or between 1.9% and 2.3% of GDP. From 2020 to 2050, cumulative investments are between US$1.1 trillion (Case C) and

US$1.8 trillion (Scenario A3), or between 1.4% and 1.9% of GDP. These investment estimates, and indeed all investment requirements presented in this report, have been revised from earlier estimates (IIASA–WEC, 1995) based on comments and suggestions from regional reviews, particularly the MEA regional review.

As indicated in *Table 7.1*, neither local environmental issues nor possible global warming is seen as a high regional priority. Sulfur emissions, as an indicator of local air pollution problems, change little through 2050. Differences are larger among scenarios after 2050, but still small in absolute terms. To the extent that even minor pollution increases are concentrated in densely populated areas, however, they may still prove large enough to motivate increased local control policies. Net $CO_2$ emissions by 2020 are projected to be nearly double 1990 emissions of 290 MtC. By 2050 there is more diversity among scenarios, with emissions ranging between 750 MtC (Scenario C1) and about 1,200 MtC (Scenario A3). On a per capita basis, however, even the highest emissions in 2050 would still be less than half current OECD levels.

### 7.5.5. Highlights

- Provided there are no major regional political conflicts, the region's extensive energy resources and its proximity to Europe and to growing Asian energy markets should lead to sufficient export revenues for sustained economic development. However, economic growth is projected to raise per capita income and living standards only gradually due to continuing high regional population growth. By 2020, living standards in all scenarios reflect adequate food and shelter, and basic health care and energy services, even in countries currently well below subsistence levels.

- Regional prosperity will largely depend on political stability within the region, institutional reforms to attract investment capital, and progress toward economic integration and cooperation. The historic divisions in the region therefore need to be overcome by a new spirit of cooperation.

- More so than in other regions, the MEA regional review identified a clear winner among the six scenarios. Not surprisingly, the high-growth oil- and gas-intensive Scenario A1 was judged the most desirable, and the most probable – with some adjustment to increase projected gas use and decrease the projected use of renewables.

- Currently only 30% of all primary energy produced in the region is consumed domestically, but this is expected to change. In Cases B and C, domestic oil and gas demands exceed exports between 2010 and 2020. In Case A this

occurs two decades later. For oil, regional consumption generally complements international export demands. If export demands are relatively high,
regional oil consumption is relatively constrained, preserving oil production
for export. Natural gas makes up the difference in meeting regional energy
consumption.

- Export revenues are strong in all scenarios. Annual revenues in Case B climb
  to about US$250 billion in 2050 and in Case A reach US$340 billion, surpassing the historical 1980 revenue peak. In Case C, carbon emission constraints and low innovation in connection with fossil fuels cause oil exports
  to remain below 10 mbd. But with price increases as shown in *Figure 6.5*, oil
  export revenues remain slightly ahead of current levels. If natural gas exports
  are added, total revenues are yet higher.

## 7.6.    Western Europe (WEU)

The region comprises the countries of the EU plus the other European members
of the OECD (Norway, Switzerland, and Turkey). In 1990 the region accounted
for 8% of the world's population, 34% of GWP, 16% of global primary and final
energy use, and 16% of global energy-related net carbon emissions. The region
is characterized by a high degree of economic and political integration that will
continue to deepen with the forthcoming European Monetary Union and the likely
addition to the EU of several Central and Eastern European countries.

### 7.6.1.    Recent developments

Since 1990, GDP growth in the region has averaged 1.5% per year and primary
energy demand has grown at about half that value (0.7% per year). $CO_2$ emissions
remained roughly constant in the first half of the decade, partly due to short-term
reductions in Germany and the UK (Gürer and Ban, 1997). Thus, actual developments have followed a path of modest growth that is intermediate between Cases B
and C. GDP growth has matched that in Case C most closely, while energy demand
growth has been halfway between demand growth reflected in Cases B and C.

### 7.6.2.    Population and economy

The region is characterized by a rather stable, aging population and a mature, increasingly service-dominated economy. As a result, across all scenarios economic
and energy demand growth are comparatively modest (see *Figure 7.8*). The overall
trend is a "drive to maturity," that is, consumption stabilizing at high levels near the
middle of the next century.

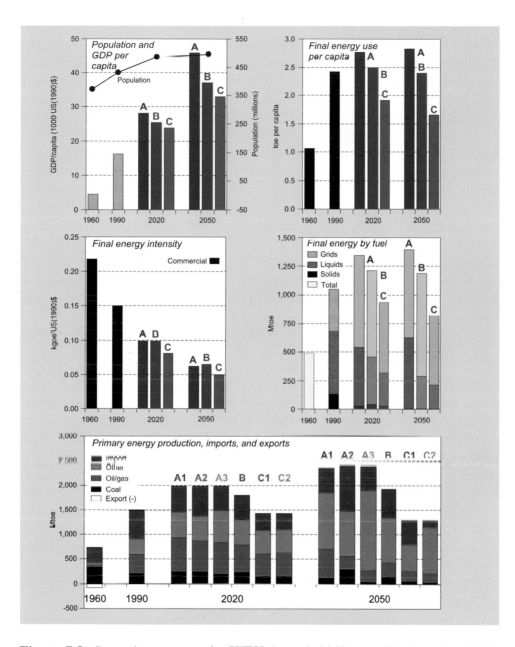

**Figure 7.8**: Scenario summary for **WEU** through 2050. *Top left:* Levels of GDP per capita reached in Cases A, B, and C, plus the population projection used in all three cases. *Top right:* Final energy use per capita. *Middle left:* Decrease in total final energy intensities in all three cases. *Middle right:* Final energy divided into solid fuels, liquid fuels, and energy delivered by grids. *Bottom:* Composition of primary energy production, plus imports and exports, for all six scenarios.

Medium-term economic growth rates (through 2020) across the scenarios range between 1.7% and 2.3% per year and are in good agreement with the range of 1.6% and 2.3% per year projected in the latest regional scenarios of the European Commission (EC, 1996). Final energy intensities continue to decrease at rates between 1.3% and 2% per year, also in line with the 1.2 to 1.7% per year improvements projected in the EC regional scenarios (EC, 1996).

Long-term economic growth rates (the period from 2020 to 2050) decline, reflecting both demographic aging and increasing economic maturity and service orientation. Energy intensity improvement rates also slow down, with slower turnover rates of capital stock and an economic structure dominated by energy extensive services. The result is per capita incomes in 2050 ranging between US$35,000 in Case C and US$45,000 in Case A, and per capita final energy demands stabilizing between 2.5 toe (Case B) and 3 toe (Case A). In the ecologically driven Case C, policies are chosen to continue economic "dematerialization" leading to final energy use levels of 2 toe per capita by 2020, 1.5 toe per capita by 2050, and 1 toe per capita, by 2100 – the level prevailing in the 1960s. Net carbon emissions drop below 0.3 tC per capita, which is about 10% of their level in Scenario A2.

### 7.6.3.   Energy systems

WEU is characterized in all scenarios by high incomes, and this is reflected in high-quality final energy mixes. Direct coal use in end-use applications virtually disappears. Due to their comparative advantage in the transport sector, liquids maintain a stable market share to at least 2020 and, in Scenario A1, to well beyond 2050. Increasingly dominant are clean, grid-dependent energy carriers such as electricity, natural gas, district heat, and, in the very long term, hydrogen. By mid-century, the market share of liquids gradually erodes in most scenarios. In Scenario A2 and Case B, liquids lose market share because of the global depletion of conventional petroleum supplies that are gradually replaced by synfuels (e.g., methanol and ethanol) and fossil fuel alternatives (e.g., natural gas). In Scenario A3 and Case C, the drop in the share of liquids is driven by technological change or environmental policies stimulating the diffusion of new transport technologies such as high-speed trains and fuel-cell and electric vehicles.

The primary energy sources that make up that high-quality final energy mix vary significantly across the scenarios. Until 2020 differences across scenarios are minor (*Figure 7.9*), largely because of existing committed capital investments and the slow turnover of capital stock in the energy sector.

After 2020, however, significant differences unfold. Although all scenarios describe "orderly" transitions away from conventional oil and gas, structural changes are faster in the rapid development "bio-nuc" Scenario A3 and in Case C, with its vigorous energy efficiency and conservation efforts and accelerated diffusion of

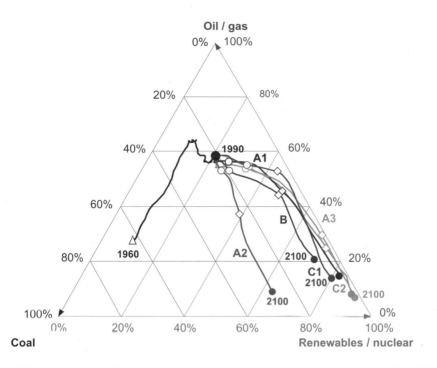

**Figure 7.9**: Evolution of the primary energy structure for six scenarios for **WEU**, shares of oil and gas, coal, and non-fossil sources, in percent. WEU progresses in all scenarios toward a mature, service-oriented economy with small or even negative energy demand growth as a result of environmental policies. In most scenarios the future energy supply is dominated by a transition to indigenous nuclear and renewable energy sources. Scenario A2 is unique in that coal's share of primary energy increases. Case B maintains the second largest role for coal and makes the slowest progress toward non-fossil energy sources and technologies. See Section 5.2 for a further explanation of the figure.

non-fossil energy technologies. In these cases, indigenous regional resources in the form of nuclear and renewable energy sources supply up to two-thirds of primary energy after 2050, reversing the region's historical overdependence on fossil fuels (and imports). Conversely, structural changes are slower when resources are limited (Scenario A2) or technological progress is relatively modest (Case B). In these cases, coal (domestic production, direct imports, and "indirect" imports in the form of synliquids) complements the region's non-fossil technologies and resources. In the very long-term, however, after 2050 even these scenarios undergo a structural transition from an energy system dominated by fossil fuels to one dominated by non-fossil fuels. What emerges as a consistent trend across all scenarios is the increasing role of natural gas as the preferred transition fuel to a post-fossil-fuel future. Natural gas use increases from 230 Mtoe in 1990 to between 450 and 600 Mtoe by 2020, and to between 240 and 730 Mtoe by 2050. Given the consistent importance of natural gas as a transition fuel, a number of contributors to the WEU regional review suggested that future research address the possibility of shorter-term disruptions that might result from low gas availability leading to a gas "price shock" similar to the experience with oil in the 1970s.

The overriding concern identified in the WEU regional review is that of supply security. Import requirements in the scenarios diverge significantly. Toward 2050, energy imports range typically between 0.5 and 1 Gtoe, compared with 0.6 Gtoe in 1990.[6] Import dependence declines from 1990 levels only in the ecologically driven Case C scenarios that combine vigorous energy conservation with accelerated development of non-fossil supplies and technologies.

Import dependence above that in the scenarios could result from more rapid demand growth, a slower expansion of regional energy supplies, or both. The WEU regional review cited the possibility of energy demand rising to 2.3 Gtoe by 2020 and 3 Gtoe by 2050, levels 10 to 25% above those in Case A. Should the increase not be matched by increased domestic supplies, imports would have to rise by an additional 300 to 500 Mtoe. Such an increase would be unlikely to cause major disruptions on international energy markets were it an isolated development. If it were a global development pattern, however, import prices would rise significantly and act as a brake on further demand growth.

The other possible cause of increased import dependence – slow growth in regional supplies – could result from too little RD&D, underinvestment in new energy technologies and oil and gas exploration, and insufficient supply diversification if coal and nuclear power are abandoned or renewables grow more slowly than envisioned in the scenarios. The WEU regional review raised the possibility of regional coal production dropping to 100 Mtoe by 2020, oil and gas production reaching a maximum of 400 to 500 Mtoe, nuclear output limited to less than 200 Mtoe, and a combined maximum contribution of all renewable energy forms of 200 Mtoe. The resulting regional primary energy production of 900 to 1,000 Mtoe in 2020 would be comparable with that in Case C, but this would be a future without the benefits of Case C's early investments in technological progress and diversity. The result would be higher levels of import dependence.

While such developments cannot be ruled out completely, they assume WEU becomes less attentive to energy investments, technological progress, and supply security than has been the case historically. As capital availability is not a major constraint in the region, and technical progress can improve the availability and

---

[6]Caution is required when comparing imports calculated for WEU with those usually published for the European Community. The inclusion of Norway in WEU immediately lowers regional import dependency from an EC value of 50% in 1990 to 40% for WEU.

economics of regional resources, it is unlikely that the low regional supply scenario outlined above, coupled with high demand, would occur. Historically, recurrent fears of shortfalls in regional oil and gas energy production, and imminent price increases have not materialized.[7] North Sea oil production, for example, continues to increase despite repeated forecasts that it would decline by the mid-1990s.

More sobering are concerns about the regional availability of non-fossil energy sources, specifically nuclear and renewables. The WEU regional review raised the question of whether the region might ultimately do without nuclear energy altogether. Judging from our scenarios this is not a realistic possibility in the near to medium term, although Scenario C1 represents a future that is nuclear free in the very long term, after 2050. Given the economic and public-opinion difficulties now confronting nuclear power, the near-term future of nuclear energy will most likely be characterized less by new capacity additions than by extensions of existing reactor lifetimes up to 40 years. With modest capacity additions, this would result in an increasing near-term contribution of nuclear power to WEU's regional energy supply, as is reflected in the scenarios. Such a strategy makes sense in terms of current economics and public opinion. However, it merely postpones the more fundamental strategic issue facing WEU: should the region's energy RD&D portfolio include a new generation of nuclear technologies that address current economic, security, and waste disposal concerns, or should the region choose a future comparable to Scenario C1 and phase out active nuclear power altogether in the long-term while continuing to live with the radioactive legacy of the past?

Turning to renewables, the two principal concerns expressed in the WEU regional review deal with early investments and land availability. The expanded contribution of new renewable energy resources, particularly in Scenario A3, depends on sufficient upfront investments in RD&D and gradually expanding niche market applications that are vital for technological learning. The concern is how to ensure sufficient upfront investments given the increasing preoccupation of WEU industry and government with the deregulation of energy markets. Perhaps the most important concern voiced unanimously in the WEU regional review is that of maintaining a long-term perspective and sufficient investments in RD&D and new energy technologies despite the current emphasis on short-term financial targets and market deregulation. This concern about maintaining incentives that promote long-term technological progress in the midst of deregulated markets applies both to renewables and nuclear power.

Concerns about land availability were unique to the production of renewables – specifically biomass. However, it is unlikely that land availability will become a problem in a region with as highly productive an agricultural sector (and as costly

---

[7] See for example the chapter on Western Europe in *Energy for Tomorrow's World* (WEC, 1993).

surplus agricultural production!) as WEU. Regional increases in agricultural productivity have already resulted in the abandonment of some 15 million ha of agricultural land in Europe since 1950 (Richards, 1990), even with high agricultural subsidies. Given high productivity (see *Box 5.2*), that land alone could have a long-term production potential of 50 to 100 Mtoe of biomass, largely sufficient for biomass production through 2020 in all scenarios except Scenario A3. To put both the issues of import dependence and biomass land requirements into perspective, one must remember that WEU agricultural subsidies, largely for maintaining surplus production (and exports), equaled some US$100 billion in 1990 (OECD, 1991). That is more than the region's entire 1990 energy import costs of US$80 billion for 600 Mtoe of imports. Finally, concerns were expressed about the viability of large-scale biomass on environmental grounds, including the adverse impacts of monocultures and harvesting on a grand scale.

Reviewing regional concerns about supply security, the global perspective provided by this study offers two distinctive insights. First, import dependence should not be seen only as a supply security risk. It is an integral part of the pattern of free trade and international economic cooperation that makes possible the high growth rates reflected in the scenarios. The WEU economies are all dominated by exports. In 1996 exports accounted for 30% of GDP in the European Community, 36% in Switzerland, and over 50% in the Netherlands (OECD, 1998a). Part of these exports go directly to regions such as MEA and FSU, from which WEU buys most of its energy imports. Such regions would not be such good customers for WEU exports without the revenues they earn from their own energy exports. Energy trade therefore should not be considered in isolation from trade in other goods and services. Indeed trade and economic cooperation are important contributors not only to economic growth, but also to international and regional security.

The second distinctive insight is that WEU's share of global energy imports will inevitably decline with the global geopolitical shift of energy markets (and import requirements) to the South. In 1990, WEU energy imports were slightly below 40% of the global total for all energy carriers. The scenarios indicate that by 2020 the WEU share will drop to between 25% and 30%. By 2050 it will drop to 10 to 25%. Thus the region will lose its dominant position as the world's principal energy buyer and increasingly share its supply security concerns with others whose import volumes are likely to grow faster.

### 7.6.4. Implications

Environmentally, WEU is a clean region with ambitions to become even cleaner. In the final December 1997 negotiations on the Kyoto Protocol, the EU was a strong advocate of requiring Annex I countries to cut greenhouse gas emissions by 15% by 2010. Only Case C could match such targets. Indeed, only Case C appears able to match even the limited targets that eventually emerged from Kyoto despite the EU's persistent advocacy for more stringent requirements. The challenge is to translate advocacy into preparedness to act. The environmental emphasis within the region is reflected in very high rankings given to environmental concerns throughout the WEU regional review. The review pointed out that such environmental priorities increase the significance of any uncertainty in the supply prospects for nuclear and renewables. Without the expansions of these non-fossil sources that are estimated in the scenarios, $CO_2$ emissions could turn out to be higher than calculated even in Cases A and B. To meet the Kyoto targets under these circumstances, the region would have to somehow implement emission reductions that are deeper, more difficult, and more costly than any envisioned in the scenarios, even Case C.

The WEU regional review also raised the question of whether WEU emissions at Case C levels might be made compatible with economic growth at Case A levels. The answer based on further detailed analysis (Schrattenholzer and Schäfer, 1996) is, "almost." However, to do so would be inconsistent with Case C at a global level and would not significantly reduce global carbon emissions. The regional energy efficiency gains and emission reductions of Case C could be made compatible (within 10%) with the economic growth in Case A, but only at the expense of the resource transfers to developing countries otherwise reflected in Case C. Without such additional investments and transfers, there would be no direct *economic* benefit to developing countries, and they would have less of an incentive than in Case C to be part of necessary cooperative international agreements restricting $CO_2$ emissions. In the absence of such investments, developing countries are also less likely to experience the sort of dynamic, vibrant growth assumed in Case C that rapidly replaces old inefficient technologies and infrastructures with low-emission, high-efficiency improvements. Without political incentives to reduce their $CO_2$ emissions, and without the economic growth needed to make such reductions possible, developing countries are unlikely to follow low emission trajectories. And because a ton of $CO_2$ makes the same contribution to global warming whether it comes from the South or the North, the fact that WEU adheres to the $CO_2$ emission trajectories in Case C makes little difference if it does so in isolation. It also makes little economic sense, as it is generally cheaper to avoid a ton of $CO_2$ emissions in the developing countries than in WEU. The effort and money that would be spent on further stringent WEU emission reductions could buy much greater reductions if wisely spent in developing regions.

Were $CO_2$ emissions not the major concern that they are for WEU, there would be less to choose among the scenarios on environmental grounds. Traditional pollutants, such as sulfur emissions, decline significantly in all scenarios. By 2020 emissions decrease by 70% (Case B) to 80% (Case C) compared with 1990. The decrease occurs as the region's fuel mix changes over time and as a result of policy measures driven by existing international environmental agreements such as the Second Sulphur Protocol to the Convention on Long-Range Transboundary Air Pollution. By 2050 sulfur emissions stabilize at even lower levels, although they subsequently increase slightly in Scenario A2 and Case B as coal use rebounds for synfuel and electricity production.

### 7.6.5.  Highlights

- The overall trend in WEU is a "drive to maturity," that is, consumption stabilizing at high levels near the middle of the next century.

- The region gives high priority to supply security. Insufficient *regional* resources could result from too little RD&D, underinvestment in new energy technologies and oil and gas exploration, and insufficient supply diversification if coal and nuclear power are abandoned.

- Substantial reductions in WEU energy imports may not be compatible with high global economic growth. The region has high energy imports but also substantial exports of manufactured goods. Both reflect current patterns of free trade and international cooperation that make strong growth possible. Many of WEU's exports are imported by energy exporters. Substantially reducing WEU energy imports would likely upset trade patterns and reduce growth rates below those reflected in the scenarios.

- WEU's share of global energy imports will decline with the shift of global energy markets (and imports) to the South. Thus the region's concerns about supply security will increasingly be shared with other regions and will likely be less and less influential in shaping international energy markets.

- There is unanimous agreement in the WEU regional review about the importance of maintaining a long-term perspective and substantial investments in RD&D and new energy technologies to ensure a desirable long-term energy portfolio despite the current emphasis on short-term financial targets and market deregulation.

- Land availability is unlikely to be a constraint on biomass and renewable energy production. The land taken out of production since 1950 due to increasing agricultural productivity alone could supply all the biomass requirements of the scenarios through at least 2020, except for Scenario A3. Reducing agricultural subsidies would free additional agricultural land and save money. As with all large-scale energy activities, however, there would also be risks of adverse environmental impacts.

## 7.7.    Central and Eastern Europe (EEU)

The region includes Albania, Bosnia and Herzegovina, Bulgaria, Croatia, the Czech Republic, Hungary, Poland, Romania, Slovakia, Slovenia, Yugoslavia, and the former Yugoslav Republic of Macedonia. In 1990 EEU represented 2.4% of global population and 1.4% of GWP, 2.8% of GWP when expressed in terms of purchasing power parities (World Bank, 1993; UN, 1997). GDP per capita in 1990 was US$2,400, or 60% of the global average when calculated at market exchange rates. In terms of purchasing power parities, GDP per capita was higher: US$5,700, 17% *above* the world average (Sinyak and Yamaji, 1990). The region is an energy importer, producing 2.6% of the world's energy in 1990 while consuming 3.8% (IEA, 1993). Final energy use per capita in 1990 was 1.8 toe, 50% above the world average. Final and primary energy intensities of some 0.8 and 1.1 kgoe/US(1990)$ were about 2.5 times higher than the world average. Energy-related net carbon emissions in 1990 were 4.8% of the global total.

### 7.7.1.    Recent developments

All the countries in the region are currently experiencing severe economic crises, with those in the North already much closer to emerging successfully than those in the South. Economic downturns after the collapse of the European planned economies caused overall regional real GDP to drop 30% between 1989 and 1993 (UN, 1997; OECD, 1997). Economic recovery is now under way. Regional GDP in 1996 was estimated to have grown to within 10% of its 1990 value and is expected to pass the 1990 level around or shortly after 2000 (World Bank, 1997a). Primary energy use also declined, falling about 15% between 1990 and 1995 according to British Petroleum (BP, 1997) statistics. With the onset of economic recovery, primary energy intensities in the region have been falling (Gürer and Ban, 1997; European Commission, 1997).

Despite problems, most of the EEU countries have made tremendous progress and are on the way to catching up with Western European standards of living. As part of this progress, all EEU countries are currently restructuring their energy sectors (EC, 1997). In particular, privatization and third-party access to energy grids aim at higher competition and rationalization of energy prices (WEC, 1995d). As a result, energy prices will rise well above the levels experienced during earlier periods of heavy energy subsidies. The alternative economic and energy futures for EEU described by the scenarios all reflect continued convergence toward WEU, a trend that will be further reinforced as several EEU countries join the EU starting in 2002 (World Bank, 1996).

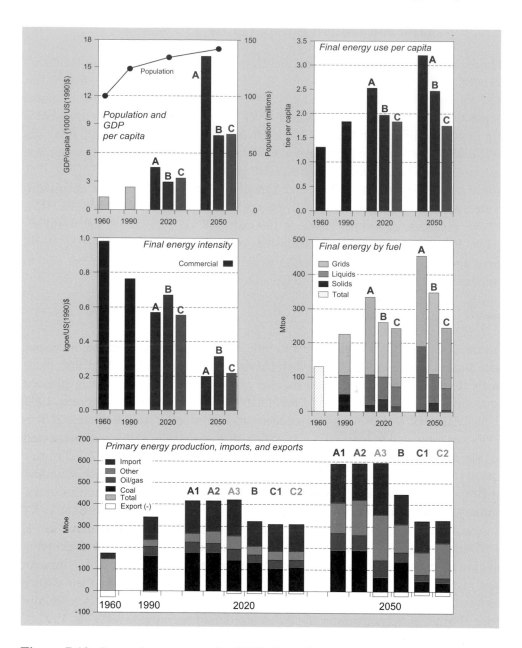

**Figure 7.10**: Scenario summary for **EEU** through 2050. *Top left:* Levels of GDP per capita reached in Cases A, B, and C, plus the population projection used in all three cases. *Top right:* Final energy use per capita. *Middle left:* Decrease in total final energy intensities in all three cases. *Middle right:* Final energy divided into solid fuels, liquid fuels, and energy delivered by grids. *Bottom:* Composition of primary energy production, plus imports and exports, for all six scenarios.

### 7.7.2. Population and economy

With the economic collapse in EEU after 1989, population growth halted and initially reversed. The population has since begun to grow again, and the 1992 World Bank projection for the region assumes growth to increase from essentially zero during the 1990s to 0.4% per year from 2000 to 2010 before gradually dropping again as the region grows wealthier. Through 2100, the only regions with lower population growth than EEU are WEU and PAO.

Following the drop in GDP in the 1990s, economic recovery to 1990 levels is assumed to occur shortly after 2000. From then on, the economic recovery begun in the mid-1990s accelerates. Growth rates for the period 2000 to 2010 range between 2.2% per year (Case B) and 3.4% per year (Case A). In Case A, growth rates peak at 5.5% per year during the 2010s before declining gradually with increasing affluence. Growth peaks in the 2020s for Case B (3.5% per year) and Case C (3.3% per year). Consequently, by 2050 GDP is between four times the 1990 GDP levels (Cases B and C) and eight times the 1990 levels (Case A).

In all three cases the service sector of the economy expands substantially and the share of the industrial sector declines. The shift is most pronounced in Case A. Industry's share drops from 55% in 1990 to 40% in 2050, and services increase from 37% in 1990 to 57% in 2050. The trends in Cases B and C are similar but less pronounced.

GDP per capita recovers by 2010 at the latest, surpassing 1990 levels in all scenarios except in Case B. But by 2010 the same amount of GDP per capita as in 1990 represents a considerable increase in personal income and a marked improvement in economic welfare resulting from the overall economic shift away from traditional, heavy industry toward consumer goods and services. By 2020 the most important aspects of restructuring should be complete, and subsequent growth is strong. Income reaches 1990 WEU levels of some US$16,000 per capita by 2050 in Case A, and within the next three decades in Cases B and C. This may seem a long wait to countries impatient to catch up, but it is comparable with the time it took Western Europe to progress from a level equivalent to today's EEU to where it now stands.

### 7.7.3. Energy systems

Cases A, B, and C differ substantially in terms of final energy consumption levels. In Case A, which appeared to the EEU regional reviewers as the most likely and desirable case, final energy consumption grows steadily from 227 Mtoe in 1990 to roughly 330 Mtoe in 2020 and to 450 Mtoe in 2050, leveling off thereafter. In Case B growth is also steady but more modest, increasing to 260 Mtoe in 2020 and to 350 Mtoe in 2050 before leveling off. In Case C, final energy consumption grows modestly to 240 Mtoe in 2020, remains essentially flat through 2050, and then drops gradually to 20% below 1990 levels by 2100. The fact that final energy consumption levels off, or even drops, in all scenarios reflects the consistent move toward more service-oriented and efficient economies. In Case C the high priority given to the environment accelerates economic structural change and the introduction of highly efficient technologies.

Changes in economic structure lead to changes in the structure of end-use fuels (see *Figure 7.10*, middle right chart). The EEU regional reviewers confirmed that developments in this region would increasingly approach Western European trends and patterns. There would be a continued relative and absolute decline of solid fuels, mainly as a result of environmental concerns, but also because consumer preferences for more convenient flexible end-use fuels increase with rising incomes. The growing share of final energy in *Figure 7.10* that is supplied by energy grids reflects the increasing roles of electricity, district head, and gas and the decreasing role of coal. The share of liquids remains fairly stable at around 25% through 2020 in all scenarios. In Cases B and C it remains at 25% through 2050; it rises to 33% in Scenarios A2 and A3 and to 41% in the oil-intensive Scenario A1.

In all cases the growth of final energy consumption is well below economic growth (see *Figure 7.10*, top two charts). This translates into steadily declining final energy intensities (*Figure 7.10*, middle left chart), although EEU reaches WEU's 1990 values by the mid-21st century at the earliest. The energy intensity improvement rates are thus a perfect mirror image of the region's economic catch-up described in the scenarios.

In terms of primary energy, the six scenarios fall into two broad categories. Through 2050 in Scenarios A1 and A2 and Case B, coal use stays relatively flat in absolute terms at the 1990 level. Meanwhile the share of nuclear and renewable energy grows to about 25% in 2050. By 2100 these $CO_2$-free energy sources could supply about 50% of primary energy. The second category (Scenario A3 and Case C) reflects faster progress in nuclear and renewable energy technologies and, in Case C, the high priority given to the environment. By 2050 $CO_2$-free sources already provide over 35% of primary energy, and after 2060 coal and gas diminish in importance. In all scenarios natural gas plays an initially important role in either covering rapidly growing demand or allowing a faster reduction of coal and oil shares.

Imports remain very important in all scenarios. In 1990 imports provided 29% of primary energy, and this level rises to about 35% by 2020. This rise did not particularly concern the EEU regional reviewers as imports would be spread across more suppliers. Import dependency ranges between 26% and 40% in 2050, with lower values in scenarios where demand is lower and where more use is made of renewables and nuclear. Import sources will be very important, especially for the South. While oil imports are already reasonably diversified, with 40% coming from FSU, all gas imports still come from EEU's eastern neighbors. Increased diversification of oil imports should be reasonably straightforward, but diversifying natural gas imports is more difficult because it entails large infrastructure outlays. With the anticipated decline in the domestic consumption of coal, a modest export capacity (of less than 22 Mtoe by 2050) may become available, not least because of pressure to control unemployment in coal mining and because of proportionally low labor costs.

### 7.7.4. Implications

Investment requirements vary widely across the scenarios. In absolute terms, investments are lowest in Case C. Because of lower energy demand and more sophisticated energy supply systems, Case C's investment requirements in 2050 are only slightly higher than 1990 values. In Case B, where technological progress is slower and the energy system changes the least, annual investments more than double by 2050. Scenario A2 needs the highest investments, projected to be three times as high in 2050 as in 1990. Relative to GDP, investments increase slightly through 2020 in all cases except Case C, where the ratio stays constant. After 2020 investment requirements in all scenarios decrease relative to GDP, largely due to steady improvements in energy intensity.

Cumulative investment requirements through 2020 range between 3.6% (Case C) and 6.0% (Scenario A3) of GDP. Of the 11 regions, only FSU has higher values. The reasons for such a high percentage are, first, only gradual economic recovery and, second, substantial restructuring requirements in the energy sector (e.g., upgrading district heat and pipeline grids, building new infrastructures for import diversification, and retrofitting and reconstructing coal and nuclear electric plants; IEA, 1996). The importance to the region of such substantial investments is echoed in the EEU regional poll, in which both financing and institutional deficiencies received top rankings. The two are closely connected as the desired institutional reforms that will bring countries of the region closer to EU membership will also make the region increasingly attractive to investors (UNEP, 1997; WEC, 1995e). The EEU regional reviewers felt that intraregional investment and technology flows would certainly be less than those from outside the region, but should not be underestimated.

$CO_2$ emissions also vary substantially across scenarios. After dropping with the initial economic contraction in all scenarios, $CO_2$ emissions gradually increase, rising steadily through 2050 in Scenario A1 (up 29% from 1990) and Scenario A2 (up 25%). Emissions stay relatively flat in Scenario A3 and Case B, mainly because of structural changes in energy supply (decline of coal and increase of natural gas and zero-carbon options). Emissions are much lower for the ecologically driven

Case C: by 2020 emissions are 25% below 1990 levels, and they drop further to between 40% and 50% of 1990 levels by 2050 (110 to 150 MtC in 2050 compared with 284 MtC in 1990).

Within the short-term context of the Kyoto Protocol, EEU emissions are below the specified limits in Cases B and C and in Scenario A3 by 2010. EEU is therefore positioned, along with FSU, to be a seller of "emission reduction units" under the provisions of the Kyoto Protocol to regions seeking to supplement domestic actions toward meeting commitments. In Case C it has no buyers, because the $CO_2$ emissions of other Annex I regions are within their Kyoto targets. But particularly in Case B, and to a lesser extent in Scenario A3, the transfer of emission reduction units from EEU and FSU to other Annex I regions (i.e., the OECD) could make the difference in meeting Kyoto targets.

$SO_2$ emissions were also calculated for each scenario and are relevant in connection with the Second Sulphur Protocol to the Convention on Long-Range Transboundary Air Pollution and the high priority given to local environmental concerns by the EEU regional review. Since 1990, $SO_2$ emissions have dropped by a greater percentage than has the EEU economy. In all scenarios they continue to drop (unlike the economy) and never return to 1990 levels. The drop is fastest for Case C, but $SO_2$ emissions are less than half 1990 levels (4.6 MtS) in all scenarios by 2020 and decline further thereafter.

The EEU regional poll ranked financing, institutional deficiencies, local environmental concerns, and efficiency and conservation as the highest regional concerns. All relate to the basic challenge facing the region – overcoming the legacy of a period when efficiency and the environment had low priority. As a result, the complete energy chain, from mining coal to heating households, has to be restructured to meet modern standards for comfort, convenience, and pollution control. The process will take money, time, and patience and will remain initially uneven. But objectives are relatively clear and uncontroversial, largely set with a view to Western European experience and standards. And the technology exists to allow immediate progress. Poland, for example, has already demonstrated how the capital intensity of restructuring electricity and district heat generation can be considerably reduced by the introduction of gas-based combined cycle power plants. Growth in investment requirements relative to GDP in the scenarios is small, and regional reforms can be expected to continue to make the region more attractive to investors. The progress of the strongest economies in the region toward EU membership indicates the future of the whole region.

### 7.7.5.   Highlights

- The recession in the 1990s has been deep and severe. The region's economy contracted by more than 30% between 1989 and 1993 (UN, 1997) and energy use fell even more. Both recent evidence (World Bank, 1997a) and the scenarios, which in all cases anticipated the overall path of the recession, suggest that the regional economy should recover to 1990 levels around the year 2000.

- The financing of investments – a priority issue – appears manageable if market-oriented reforms continue.

- The high priority given to local pollution in the EEU regional review is one reflection of the current and continuing shift in consumer demand toward cleaner, more convenient, and more flexible end-use fuels as incomes gradually catch up to Western European levels. This shift results in reduced local air pollution in EEU in general, and reductions in sulfur emissions in particular, in all scenarios.

- EEU is on a track toward convergence with WEU. The process will no doubt be slower than regional aspirations, but progress is already well under way and will profoundly change the political, economic, social, and environmental dimensions of a new "common house" Europe.

## 7.8.  Newly Independent States of the Former Soviet Union (FSU)[8]

In December 1991 the Soviet Union dissolved into 15 newly independent states, each with distinct, and sometimes conflicting, national interests, and each embarking on its own separate pattern of economic and political development.[9] In 1990 FSU accounted for 5% of the world's population, 4% of GWP calculated at market exchange rates (7% in terms of purchasing power parities), and 17% of global energy-related net carbon emissions.

The region has up to 11% of the world's proven oil reserves, some 40% of proven gas reserves, and about 25% of proven coal reserves (Rogner, 1997; Safranck, 1996; Vyakhirev, 1996; BP, 1997; IIASA–WEC, 1995; WEC, 1993). They are, however, concentrated in only a few countries. Azerbaijan, Kazakhstan, the Russian Federation, Turkmenistan, and perhaps Uzbekistan have the potential

---

[8]Because the base year of this study is 1990, the term "former Soviet Union," has been used to refer to the 15 independent states that, prior to 1992, made up the Union of Soviet Socialist Republics (USSR).

[9]The newly independent states of the FSU are usually grouped geographically. The European republics include Belarus, Russian Federation, Republic of Moldova, Ukraine, and the Baltic states – Estonia, Latvia, and Lithuania. The Central Asian republics are Kazakhstan, Kyrgyzstan, Tajikistan, Turkmenistan, and Uzbekistan. The Transcaucasus includes Armenia, Azerbaijan, and Georgia. The 12 republics of FSU (i.e., excluding the Baltic states) are members of the Commonwealth of Independent States (CIS).

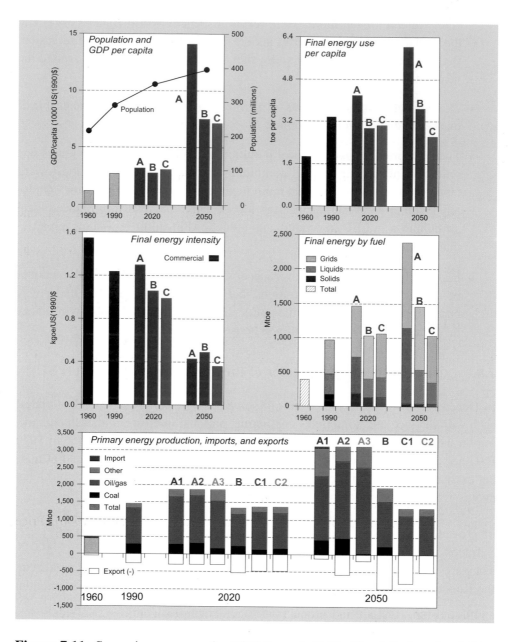

**Figure 7.11**: Scenario summary for **FSU** through 2050. *Top left:* Levels of GDP per capita reached in Cases A, B, and C, plus the population projection used in all three cases. *Top right:* Final energy use per capita. *Middle left:* Decrease in total final energy intensities in all three cases. *Middle right:* Final energy divided into solid fuels, liquid fuels, and energy delivered by grids. *Bottom:* Composition of primary energy production, plus imports and exports, for all six scenarios.

to be self-sufficient in energy and to be major exporters. All other FSU countries continue to be dependent on mostly Russian oil and gas.

### 7.8.1. Recent developments

Political and economic reforms in the FSU since 1990 have shifted all countries away from central planning and toward market economies. Though the pace and extent of change vary among countries, from "shock therapy" to more hesitant incrementalism, the changes are all in the same direction – toward price and trade liberalization, privatization, competition, and the reform of financial and tax systems.

These changes have had tremendous impacts. Economic output fell 50%, and industrial output by more than 50% from 1990 to 1996 (UN, 1997; World Bank, 1996, 1997a). This severe recession was accompanied by hyperinflation in many countries, with inflation rates in the early 1990s sometimes exceeding 1,000% per year (UN, 1997; Russian Energy Pictures, 1997). Living standards dropped dramatically and poverty increased. In Russia the total fertility rate fell from 2.1 in 1985 to 1990 to 1.5 in 1990 to 1995 (UN, 1997).

Energy consumption and production in FSU decreased by about 30% between 1990 and 1996, and energy exports dropped by half (IEA, 1993; BP, 1997; EC, 1997; ESCIS, 1996). Although the drop in energy use was huge, it was still less than the fall in GDP, and the region's already high energy intensity became even higher (Schipper and Cooper, 1991). It is now estimated to be two to four times the energy intensities in OECD countries (UN/ECE, 1996).[10]

At the same time the share of gas in total primary energy consumption increased significantly, from 40% in 1990 to 50% in 1996 (BP, 1997; Russian Energy Pictures, 1997). Although prices have begun to rise from the artificially low levels set by central planning and are now twice the 1990 levels, energy production costs and transit payments are still 20 to 30% lower than in Europe (Makarov, 1996). Price increases have also led to significant nonpayment problems that limit profitability and further investment (Aslanyan and Volfberg, 1996).

Recent statistics (UN, 1996, 1997) indicate that over the past two years the recession may have bottomed out in all but two countries of the region, Ukraine and Tajikistan. Output and productivity have begun to increase, and inflation is down (from 48% in 1996 in Russia, for example, to 11% in 1997 and a target of 5 to 7% in 1998). Poverty rates have stabilized. If these promising trends continue, by the year 2000 regional GDP will be surprisingly close (considering the economic turmoil the region has been through) to the values given in the scenarios. But

---

[10]Some argue that such estimates should be reduced by a factor of two to three to reflect the huge shadow economy in the FSU, which is not included in official statistics. The FSU regional review cited one estimate of Russia's shadow economy as accounting for 40% of economic activity.

that would still leave FSU well behind where it was in 1990. The recession was extremely deep and recovery will take time. As outlined in the scenarios, per capita GDP (at market exchange rates) recovers to 1990 levels between 2010 (Case A) and 2020 (Cases B and C). But by that time, the same per capita GDP as in 1990 represents a considerable increase in personal income and a marked improvement in economic welfare because of the overall economic shift away from FSU's heavy smokestack industry toward consumer goods and services.

### 7.8.2.    Population and economy

According to the 1992 World Bank population projection used in this study (see *Figure 7.11*, upper left), population growth in FSU is slow and getting slower. Projected growth averages 0.7% per year between 1990 and 2020. It drops to 0.4% per year between 2020 and 2050, and to 0.1% between 2050 and 2100. The FSU regional review noted that, for the near term, even the modest growth rates in the scenarios may turn out to be an overestimate.

Near-term economic dynamics in all scenarios reflect the recession and initial recovery that characterize the end of the 1990s. Actual developments have been closest to the middle-course Case B. This is true for both economic developments and, as will be discussed below, energy developments.

As the region begins its recovery, economic growth increases in the scenarios to between 2.2% and 2.5% per year from 2000 to 2010, and to between 3.3% and 3.9% per year from 2010 to 2020. These values are in good agreement with the 3.0 to 3.5% annual growth for the period 2005 to 2010 projected for the Russian Federation in the regional scenarios of the Institute of Energy Research (Russian Academy of Sciences, 1995; Makarov *et al.*, 1995). In the longer term (2020 to 2050) growth accelerates to 5.3% per year in Case A and to 3.7% per year in Case B, and holds essentially steady at 3.2% per year in the ecologically driven Case C. In the very long term (2050 to 2100), economic growth rates in the scenarios decline to between 2.2% and 2.7% per year with increasing levels of affluence. GDP per capita reaches 1990 WEU levels (of US$16,000) around 2050 in Case A and within three decades later for Cases B and C.

The FSU regional review emphasized that a better indicator of economic activity for the region is GDP in terms of purchasing power parities, although even that measure is incomplete in FSU because of the difficulty of incorporating any sort of realistic measure of the shadow economy. With this caveat in mind, measuring GDP in terms of purchasing power parities results in FSU catching up to WEU standards before 2050 in Case A and two decades later in the other two cases. Were it possible to include the contribution from the shadow economy, WEU standards would be reached sooner.

In connection with the 1990s recession in FSU, energy intensity increased, a trend accurately reflected in the scenarios (see *Figure 4.6*). With the onset of economic recovery, energy consumption in all scenarios grows more slowly than GDP. Primary energy intensity thus starts to improve and drops to the 1990 level of 1.8 kgoe/US(1990)$ by 2010 in Cases B and C and around 2015 in Case A. Between 2000 and 2010 it improves at a rate of 1.1 to 2.1% per year, and between 2010 and 2020 at a rate of 2.0 to 3.2% per year. These values are similar to those suggested by the IEA for the region: between 2.0% and 2.7% per year from 1996 to 2010 (IEA, 1996). From 2020 to 2050 energy intensity improvements vary from 2.5% per year in Case B to 3.3% per year in Case C and 3.6% per year in Case A, as economic growth and investments pick up. Beyond 2050 the region's improvement rates drop to a long-term average (2050 to 2100) of 2.0 to 2.4% per year. Although energy intensity improvements are consistently faster than in WEU, FSU still only reaches WEU's 1990 energy intensity level of 0.2 kgoe/US(1990)$ almost a century later – in 2080, in Case C and even later in Cases A and B. NAM's 1990 level of 0.4 kgoe/US(1990)$ is reached earlier – 2060 in Case C, and within the following two decades in Cases A and B. Thus, the energy intensity improvement rates adopted here are hardly overly optimistic, a view shared by the FSU regional review.

### 7.8.3. Energy systems

In the medium term (2000 to 2020) total primary energy consumption grows at rates between 0.2% per year (Case C) and 1.6% per year (Case A). From 2020 to 2050 growth picks up very slightly in Cases A and B, while in Case C it begins an equally slight downward trend. Total primary energy use in 2050 ranges between 1,340 Mtoe (Case C) and 3,140 Mtoe (Case A), compared with 1,400 Mtoe in 1990. These values correspond to per capita final energy consumption levels in 2050 ranging between 6.2 toe per capita (Scenario A2), that is, NAM levels at that time, and 2.6 toe per capita (Case C), compared with FSU's 1990 level of 3.4 toe per capita. From 2050 to 2100, final energy consumption per capita declines by 10% in Cases A and B. In the ecologically driven Case C it declines to between 1.6 and 1.9 toe per capita.

Already in 1990 the share of final energy reaching consumers through grids was higher in FSU than in any of the other 10 regions. Fifty-one percent of final energy was delivered in the form of high-quality energy carriers: electricity, district heat, or gas. This is ahead of 40% and 36% in NAM and WEU, respectively, the two OECD regions with the highest shares of grid-delivered final energy. The quality of the final energy mix improves further across all scenarios. By 2050 between 52% (Scenario A1) and 66% (Case C) of final energy is delivered by grids. Between 30% (Scenario C1) and 46% (Scenario A1) is in the form of liquids, with solids (direct use of biomass and coal) making up the remainder – less than 5%. Solids are entirely phased out as end-use fuels in the second half of the 21st century.

Through 2020, in all scenarios and also in the view of the FSU regional reviewers, there are no significant changes in the structure of the primary energy supply relative to 1990 (*Figure 7.12*). Gas and oil account for between 70% and 80% of the primary energy supply (with gas' share increasing steadily), coal accounts for between 9% and 18%, and nuclear and renewables for the remaining 9 to 18%. Only after 2020, as fewer and fewer of today's energy facilities and less of its infrastructure remain in service and new facilities are installed, do differences in the primary energy structure take shape. All scenarios (except Scenario A2) basically extend the historical global trend of a shift from coal to oil, gas, and non-fossil energy sources. Gas and oil maintain their dominance in all scenarios but two – Scenario A2 and Case B. Even in Scenario A2, however, where coal supplies about half the primary energy toward the end of the 21st century, coal is essentially eliminated from final energy consumption. Coal is converted to electricity, heat, and synthetic fuels before it reaches the final user.

Natural gas will continue to dominate the energy supply. From 41% in 1990 (572 Mtoe), its share of primary energy use rises to between 44% and 56% in 2020 (610 to 840 Mtoe). Production in Scenario A3 reaches 1,700 Mtoe by 2050 and could reach a peak as high as 2,850 Mtoe thereafter. Production volumes in the other scenarios are lower, but even in the lowest scenario they do not fall below 770 Mtoe by 2050. Oil in 1990 provided 29% of primary energy, a share that increases gradually in Case A through 2050 but otherwise declines slowly everywhere. In absolute terms, production in 2050 varies from 600 Mtoe (Scenario C2) to as high as 1,200 Mtoe (Scenario A1), compared with 570 Mtoe in 1990. Coal's importance declines in all but Scenario A2, where it could reach production levels of 480 Mtoe by 2050 and could continue to grow (to 2,370 Mtoe) toward 2100. Conversely, nuclear power and renewables gain importance across the board. In 1990 they provided 10% of primary energy (140 Mtoe). By 2050 they provide as much as 27% of primary energy in Scenario A1 (850 Mtoe). Their role is smaller where technological progress in slower, such as in Case B (18% of primary energy, or close to 340 Mtoe).

Investment requirements are substantial, as discussed below, and energy trade and exports are therefore especially important to the future of FSU. A striking result of this study is the continued growth in import demand outside FSU. Western Europe and particularly Japan have limited oil and gas resources, and energy demand is growing rapidly in the developing economies of Asia. The largest potential exporters of primary energy are MEA for oil and FSU for gas. As a result, FSU gas exports increase in all scenarios to around 2020, and also thereafter, except in Scenarios A1 and A3, where regional demand and relative prices make the gas more

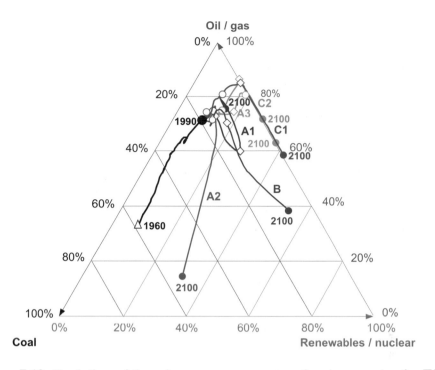

**Figure 7.12**: Evolution of the primary energy structure for six scenarios for **FSU**, shares of oil and gas, coal, and non-fossil sources, in percent. The figure shows the dominance of gas and oil in all scenarios except Scenario A2 and Case B, and the general shift away from coal toward gas, oil, nuclear, and renewables. See Section 5.2 for a further explanation of the figure.

valuable within the region. Gas exports grow consistently across all scenarios from 116 Mtoe in 1996 to between 250 and 270 Mtoe by 2020. This development is consistent with the estimates and expectations of the regional reviewers. By 2050 there is more diversity among scenarios, from 65 Mtoe (Scenario A1) to 610 Mtoe (Case B) with export revenues approaching US$150 billion. A robust finding across all scenarios is that gas exports will be a true bonanza for FSU. By 2020, export revenues approach US$50 billion per year and in half the scenarios exceed US$100 billion by 2050, that is, 5 and 10 times the 1990 revenues, respectively (see *Figure 6.6*).

In choosing its development strategy and seeking investment capital, FSU has enormous opportunities to substantially shape international energy relations (Makarov, 1996). In the past few years, major Western oil companies have concluded a series of multi-billion dollar contracts with Azerbaijan, Kazakhstan, and Russia to explore and develop the oil and gas deposits of the Caspian Sea, one of the biggest oil fields in the world. But implementation of these contracts still depends on the construction of a network of pipelines to export gas and oil to Western and Eastern markets, similar to the pipelines running from Yamal to Europe. Azerbaijan, Kazakhstan, Turkmenistan, and Uzbekistan in particular are all landlocked and isolated from potential markets. The relative attractiveness of alternative pipeline routes depends on both economics and politics. A number of proposed routes exist that are technically feasible and financially viable but face political uncertainties. For its new gas export pipeline to Turkey, for example, Gazprom has chosen the expensive and technically demanding, but politically secure, trans–Black Sea underwater route after rejecting alternative routes running west through Ukraine, Republic of Moldova, and Romania, or east through Georgia and Armenia due to "instability and high political risk."

In some cases it may be more attractive to build to the East rather than to the West, as the large growth markets of the future are in the East, not the West. By 2010, or shortly thereafter, energy demand in the developing countries of Asia (3 to 3.7 Gtoe across all six scenarios) surpasses that of WEU, EEU, and FSU combined. One initiative in this direction is that of the China National Oil Corporation, which will develop the Uzen and Aktyubinsk oil fields in Kazakhstan and will build a 3,000-km pipeline to China's western border and a 250-km pipeline to the border of Turkmenistan.

Taking into account Russia's Siberian gas resources in particular, total FSU gas resources are plentiful and within reach of the fastest-growing markets of the future. Gas is a key transition fuel in every scenario because it is so well matched to the pervasive trend in consumer preferences for high-quality, clean, flexible, and convenient final energy delivered by grids. Should FSU build a network linking its oil and particularly its gas resources, both in central Asia and Siberia, to the large markets to its east and west, it would enhance its trade capacities enormously, alter the political economy of energy supply, enhance the region's export revenues, and position itself for the post-fossil-fuel era at the center of trade patterns and facilities that might move synfuels derived from alternative energy sources as easily as gas and oil. Such a strategy would be resilient given the role of gas in all scenarios. FSU's interest may be mainly economic in the near to medium term, but in the long term it may also be environmental: gas exports may alleviate the enormous environmental impacts that could arise from coal-intensive development in the rapidly growing economies of Asia.

### 7.8.4.    Implications

The FSU regional poll ranked financing and institutional deficiencies as the region's most pressing concerns. The two are related, as more rapid institutional reforms could make the region's energy system significantly more attractive to foreign investors. The investment needs implied by the scenarios challenge the capacity of the region's governments and private investors. From 1990 to 2020, energy sector investments in the scenarios vary from 5.8% of GDP in Case C to 9.0% of GDP in Scenario A3. In absolute amounts, cumulative 1990 to 2020 investments range between US$1.3 and $2.1 trillion. As a percentage of GDP, these are the highest of any region in the world as a result of the considerable investment backlog and the substantial investment needs for energy sector restructuring and construction of production facilities and infrastructure development for both domestic consumption and exports. From 2020 to 2050 investment requirements drop to between 2.3% and 4.5% of GDP. In absolute amounts they range between US$1.3 and US$3.9 trillion.

The near-term investment requirements in the scenarios were higher than expected by the regional reviewers, who noted that investment costs in FSU may be lower than costs based on Western projects due to cheap labor and cheap energy. If energy in FSU continues to be priced below international levels, however, it will be correspondingly more difficult to attract investment capital to energy projects in the region. It must also be remembered that the high investment calculated in the scenarios results in the successful restructuring and reconstruction of the energy sector. Lower investment levels are also possible, but would result in both a less efficient energy sector and, most likely, lower export volumes and revenues.

Particularly in the medium term (prior to 2020), the investment levels in the scenarios represent a major challenge for the region. In 1990, energy investments in Russia were estimated at about 4% of GDP (UN, 1993b), compared with the range of 5.8 to 9.0% calculated above. A comparison can also be made between investments in the electric power industry in the scenarios and regional estimates. Dyakov (1996) estimates investment requirements for electric power in CIS for the period 1995 to 2000 at above US$45 billion. This is consistent with FSU investments for the same period in the scenarios for electric power totaling between US$61 billion (Case D) and US$77 billion (Case A).

Although possible climate change is currently a low priority according to the FSU regional poll, the poll anticipates greater priority being given to the issue in 2020 and 2050. Concerns about possible climate change are currently ranked low for two reasons. First, the recession of the 1990s has left the region with other huge pressing economic and social priorities. Second, largely because of the recession, FSU greenhouse gas emissions decreased substantially from 1990 to 1997. In all six scenarios, net $CO_2$ emissions remain below 1990 levels until at least 2010. In some scenarios this situation lasts even longer, such as in the middle-course Case B, where 1990 emission levels are not reached until 2040, and in Case C, where they never exceed the 1990 levels.

The drop in greenhouse gas emissions may become an important asset for FSU (and for EEU) as a result of the Kyoto Protocol (UN/FCCC, 1997). The Protocol sets emission limits for the period 2008 to 2012 for nearly all Annex I countries, which include Russia, Ukraine, and the Baltic states, as well as OECD countries as of 1992 and most of Central and Eastern Europe. The Protocol also permits Annex I countries to trade "emission reduction units" to other Annex I countries seeking to supplement domestic actions toward meeting commitments. For Russia and Ukraine the Kyoto limit is 100% of 1990 levels. Keeping in mind all the caveats discussed in Section 6.6.3, the difference between this level and the much lower emissions since 1990 plus those anticipated through 2010 in Cases B and C and Scenario A3 is an asset potentially worth money if and when the Kyoto Protocol enters into force. In Case C FSU would have no buyers, because the $CO_2$ emissions of other Annex I regions are within their Kyoto targets. But particularly in Case B, and to a lesser extent in Scenario A3, the transfer of emission reduction units from FSU to other Annex I regions could make the difference in meeting Kyoto targets.

Looking beyond the short-term targets of the Kyoto Protocol to the long-term developments of the scenarios, by design Case C shows FSU carbon emissions declining steadily from 1,030 MtC in 1990 to 800 MtC in 2020, to 700 MtC in 2050, and, in Scenario C1, to less than 290 MtC in 2100, more than a 70% decrease from 1990 levels. At the other end of the spectrum emissions in Scenario A2 rise to 1,210 MtC in 2020, 1,990 MtC in 2050, and 2,790 MtC in 2100, a 170% increase from 1990.

### 7.8.5.  Highlights

- The distinctive characteristic of FSU is its emergence from the deep and far-reaching recession of the 1990s, during which GDP dropped by half. The recovery will take time. In the scenarios, per capita GDP only recovers to 1990 levels between 2010 and 2020.

- Principally gas and secondarily oil dominate the region's energy supply in all scenarios. Gas covered 41% of primary energy in 1990; its share rises to between 44% and 56% in 2020 in the scenarios, and could rise even more according to the FSU regional review. Gas exports consistently increase in all scenarios. Revenues approach US$50 billion by 2020, and in half the scenarios exceed US$100 billion by 2050, that is, 5 and 10 times the 1990 values, respectively. The "gas bonanza" could turn out to be even greater if a transport infrastructure covering all of Eurasia to the East and West were to be put in place.

- FSU is situated between two major energy-consuming regions. Although growth is slowing in WEU, consumption will stabilize at high levels and gas imports will rise. To the East are the fastest-growing markets of the future.

By 2000, or shortly thereafter, energy demand in the developing countries of Asia will reach 3 Gtoe, surpassing that of WEU, EEU, and FSU combined. Should FSU build a network linking its oil and particularly its gas resources, both in central Asia and Siberia, to the large markets to its East and West, it would enhance its trade capacities enormously, alter the political economy of energy supply, enhance the region's export revenues, and position itself for the post-fossil-fuel era at the center of trade patterns and facilities that might move synfuels derived from alternative energy sources as easily as gas and oil.

- The regional poll gave a very low ranking to concerns about possible climate change. Emissions have dropped with the recession and will not return to 1990 levels until 2010 at the earliest (in the coal-intensive Scenario A2) and at the latest by 2040 (in the middle-course Case B). Russia and Ukraine (and the Baltic states) are Annex I countries under the Kyoto Protocol, which, if and when it enters into force, allows Annex I countries to trade "emission reduction units" among themselves. The difference between current emissions and FSU's higher 1990 levels (the so-called "Russian bubble") is thus an asset that FSU might be able to trade for cash or investments from other Annex I countries (in NAM, PAO, and WEU). However, such trading could be blocked if it were seen not to be supplemental to domestic action to curb emissions.

## 7.9. Centrally Planned Asia and China (CPA)

The region includes Cambodia, China, Korea Democratic People's Republic (DPR), Lao People's Democratic Republic (PDR), Mongolia, and Viet Nam. In 1990, CPA accounted for 24% of the world's population, 2.3% of GWP, 11% of global primary energy use, and 12% of global energy-related net carbon emissions. In terms of population, economy, and energy use, the region is dominated by China. China accounts for 91% of the region's population, 78% of its GDP, and 92% of its commercial energy consumption. Consequently, any discussion of the region and its future is largely a discussion of prospects and options for China.

### 7.9.1. Recent developments

Throughout the 1990s, CPA has enjoyed vigorous economic growth, substantial energy intensity improvements, and rapid growth in commercial energy use. GDP growth since 1990 has averaged almost 10% per year (World Bank, 1997a), closest

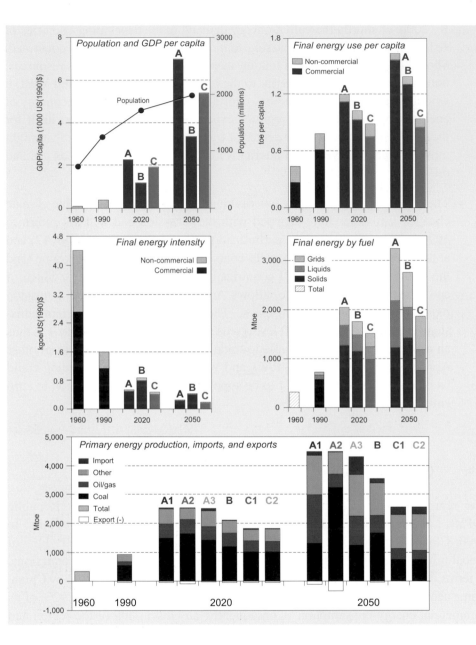

**Figure 7.13**: Scenario summary for **CPA** through 2050. *Top left:* Levels of GDP per capita reached in Cases A, B, and C, plus the population projection used in all three cases. *Top right:* Final energy use per capita. *Middle left:* Decrease in total final energy intensities in all three cases. *Middle right:* Final energy divided into solid fuels, liquid fuels, and energy delivered by grids. *Bottom:* Composition of primary energy production, plus imports and exports, for all six scenarios.

to the 9.5% per year projected in Case A between 1990 and 2000. Commercial energy use has grown at an estimated 4% per year (BP, 1997) and, as a result, commercial energy intensities have decreased substantially, by 6% per year. Nevertheless intensities remain high compared with other regions (see *Figure 3.4*), indicating a large potential for continued improvement.

### 7.9.2. Population and economy

China's population is the largest in the world. Between 1950 and 1990 it grew at an average annual rate of 1.8%, reaching 1.1 billion in 1990 (UN, 1992). Future growth is expected to be significantly slower, corresponding to an average annual growth rate of only 0.5% between 1990 and 2050. For the CPA region as a whole population is projected to increase in the scenarios from 1.2 billion people in 1990 to 1.7 billion in 2020 and 2 billion in 2050 before stabilizing near 2.1 billion around 2100 (Bos *et al.*, 1992). The scenarios' population projections match quite well the Chinese population projections cited in the CPA regional review of 1.4 to 1.6 billion people in 2020 and 1.3 to 1.7 billion in 2050. According to the regional review the most likely demographic scenario for China suggests a peak of 1.5 billion people by 2050 with a subsequent decline (Fengqi, 1998). All population projections for the region assume declining fertility rates, leading to decreasing population growth rates and allowing the region to complete the demographic transition within the next century.

Since the late 1970s, China's economic development has been among the fastest ever achieved (World Bank, 1997a). Average annual economic growth between 1978 and 1990, measured in constant yuan, was close to 9%. Through the 1980s it was closer to 10%. Macroeconomic imbalances and high inflation rates reduced growth to about 4.5% in 1989 and to 5.0% in 1990 (UN, 1993b). Rapid growth exceeding 11% per year resumed again in the early 1990s and has continued at about 10% annually. Such high growth rates have been generally attributed to economic reforms in China and high investment rates, especially in infrastructures. Other countries in the CPA region have largely shared in China's rapid growth. Their growth has averaged 7% in real terms since the mid-1970s, accelerating to 9% in the 1990s. Even the less developed Indo-Chinese economies have grown rapidly in recent years, with Viet Nam expanding at 8% per year in the first half of the 1990s, and Cambodia and Lao PDR growing at 6% per year (World Bank, 1997a). Korea DPR, where economic decline has reached catastrophic dimensions including acute food shortages, is a dire counterexample to the impressive performance of the other CPA economies.

Not all aspects of the region's rapid economic development have been positive. Development has widened income disparities and caused adverse environmental impacts and pollution-related health problems that disproportionately affect the poor. Some 350 million people, almost a third of CPA's population, are considered to live in conditions of deprivation and poverty (World Bank, 1997a). Average per capita income expressed at market exchange rates remains low at US$380 in 1990. Incomes expressed in terms of purchasing power parities are about five times as high. Income disparities, particularly between urban and rural populations and among regions have increased with rapid economic growth (Siddiqi *et al.*, 1994). Reducing such geographic inequalities will be important for widespread continued growth. China attaches great importance to both environmental protection and equitable income growth (Gan, 1998). In fact, China was the first developing country to establish a national Agenda 21 (China's Agenda 21, 1994; and Progress on China's Agenda 21, 1996) after the United Nations Conference on Environment and Development (UNCED) held in Rio in 1992.

In the wake of the Asian economic crisis that began in 1996, growth has slowed in CPA (e.g., to 7.2% per year in 1997), but the region's potential remains high. China has already achieved two decades of sustained rapid economic growth and a major reduction in population growth. Overall, the region's natural resources and labor force are richer than those of the "Asian Tigers" and even Japan; equally important, there appears to be a strong and widespread will to achieve ambitious development goals. Thus, China and the whole CPA region have reasonable expectations to reach standards of living now prevalent in southern parts of Europe by 2050 with per capita incomes greater than US$10,000 when measured at purchasing power parities. In the past 30 to 40 years, the "Asian Tigers" and to a much greater extent Japan have all surpassed this level of per capita income (Siddiqi *et al.*, 1994).

This potential translates into strong continued economic growth in all six of the study's scenarios. Through 2020, present high growth rates of 10% per year converge to levels sustainable over more extended time periods. Growth rates average 7% per year in Case A, 5% in Case B, and 6.5% in Case C from 1990 to 2020. As the region develops, growth rates across the six scenarios converge and gradually decline to levels comparable with current rates in countries with comparable per capita income. Between 2020 and 2050, growth averages about 4% per year. Total regional GDP surpasses 1990 levels for WEU between 2030 and 2060 in all six scenarios. In the high-growth Case A, it reaches US$20 trillion – exceeding the entire 1990 global GDP – shortly before 2070. Even in the most cautious scenario (Case B), CPA's regional GDP surpasses 1990 global GDP by 2100. Parity comes more slowly on a per capita basis. By 2050 CPA reaches per capita GDP levels similar to those of current middle-income developed countries – between US$3,400 and US$7,000 at market exchange rates (expressed in 1990 dollars), and from US$7,200 to US$11,100 by 2050 measured at purchasing power parities. Only by 2100, and only in the high-growth Case A, does per capita income exceed the 1990 OECD average of US$19,000.

### 7.9.3.    Energy systems

Along with CPA's impressive economic growth has come an enormous increase in energy consumption. In China alone commercial primary energy increased from about 200 Mtoe in 1960 to some 900 Mtoe by 1996 (BP, 1997). Noncommercial energy is estimated to provide another 200 Mtoe (Sinton *et al.*, 1992; Gürer and Ban, 1997). The result is that China is now the second largest energy consumer in the world, having surpassed Russia in the early 1990s. China has also become a net energy importer, no longer able to cover all its oil requirements from domestic production.

Coal is by far the dominant energy source, accounting for almost 60% of total primary energy in CPA. Biomass and other renewable energy sources have the second largest share, providing slightly more than 25% of total primary energy. This is because noncommercial renewable energy sources such as wood, dung, agricultural wastes, and other forms of biomass are the main source of energy for most of the rural population. Oil is next with an almost 15% share. Natural gas is still embryonic, providing a very modest share.

Energy use in the region is still rather inefficient despite large past reductions in energy intensities. Since the 1960s, final energy intensity has declined at an impressive 2.7% per year. In 1990 final energy intensity was about 1.6 kgoe/US(1990)$ while primary energy intensity was much higher at about 2.0 kgoe/US(1990)$. Comparable 1990 figures for PAO were 0.1 kgoe/US(1990)$ for final energy intensity and less than 0.2 kgoe/US(1990)$ for primary energy intensity – more than a factor 10 lower. Thus, the potential for improvements in CPA remains large.

A main reason for high energy intensities in CPA is the current structure of the energy system. The dominance of solid energy forms both in the primary energy mix and at the point of final energy consumption implies relatively low energy conversion and end-use efficiencies. Commercial energy grids do not reach the estimated 350 million people in the region who live in poverty, and only a very tiny share of all final energy is delivered by grids in the form of electricity and gas, or as liquids (i.e., oil products).

As shown in *Figures 7.13* and *7.14*, a prominent feature across all six scenarios in CPA is the future shift from noncommercial energy and rather inefficient energy use toward higher-quality and efficient fuels, that is, to liquids and grid-dependent forms of energy. This transition leads to large efficiency improvements and greatly reduced indoor and urban air pollution. Per capita final energy consumption exceeds 1.6 toe by 2050 and reaches 2 toe by 2100 in the high-growth Case A, levels comparable with 1990 values in affluent Western European countries and Japan.

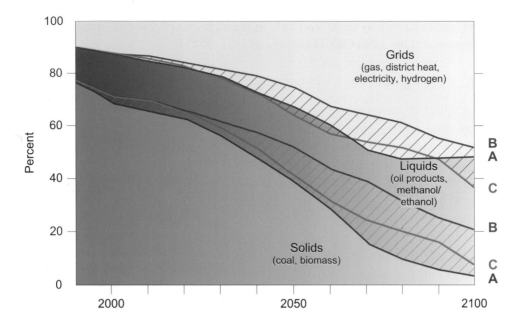

**Figure 7.14**: Final energy by form in **CPA**, in percent, as solids, liquids, and grids. Overlapping shaded areas indicate variations across Cases A, B, and C.

The share of solids decreases from 77% in 1990 to between 50% and 33% of all final energy by 2050. One result is that both final and primary energy intensities decline. Between 1990 and 2020 primary energy intensity declines at an average rate of between 2.2% and 4.2% per year. As CPA completes the transition from noncommercial to commercial energy, improvement rates decrease. Between 2020 and 2050, primary energy intensity declines at rates between 2.2% and 2.8% per year. These improvements reflect a virtual elimination of direct uses of solid fuels by the end of the century. All coal-based and biomass energy reaches consumers in the form of electricity and liquid (methanol) and gaseous fuels.

The energy intensity improvements in the scenarios were considered too modest by the CPA regional review, which suggested both faster energy intensity improvements and faster economic growth. These two modifications are consistent with each other and do not radically change the energy demand levels of the scenarios. The faster an economy grows, the higher its level of investment and the faster old physical capital is turned over in favor of new, less energy-intensive alternatives. As a result, energy intensities improve more quickly. This is what actually happens in the scenarios. The CPA regional review, while agreeing with this basic logic, suggested pushing it further. The review proposed energy intensity improvements of 4.6% per year for the period 1990 to 2020, somewhat higher than the range of 3.7% in Case A to 4.2% in Case C. As stated above, in terms of energy demand, the regional review and the scenarios agree, with the regional review's suggestion being almost identical to Case B.

As is discussed in other sections of this chapter, a key feature of a global study is that energy intensity improvement rates around the world are tied to a common set of historical data and a consistent interpretation of that data. The data and methods tying energy intensity improvements in each scenario to economic development patterns in that scenario are described in Section 4.3. Having said that, it is clearly possible that regions that are now catching up technologically might discover opportunities to leapfrog development stages that were essential in the development paths originally taken by today's technological leaders. We can imagine that technology may develop more quickly in CPA than it did in other regions when they were at CPA's stage of development. Cellular phones, for example, may make telephone wires less necessary. But two cautions are in order. First, opportunities for technological leapfrogging are not new. They have existed in the past, some have been exploited, and these are reflected in the data described in Section 4.3. Second, while it may be possible to leapfrog old infrastructures, new infrastructures may now be essential. There may be less need today for infrastructural networks of copper wire, but there is certainly more need for networks of fiber-optic cable. Thus, while the ambitions of the CPA regional review are the right ones – accelerating technological progress in the direction of yet faster economic growth, and faster energy intensity improvements, is usually a good strategy – the global analysis cautions against ignoring limits to rates of change, especially in the long run. Technological progress now has a long and increasingly well-documented history (see, e.g., Nakićenović and Grübler, 1991; Grübler, 1998), and scenarios for the future should incorporate the lessons that can be learned from that history: technology diffusion takes time.

For the energy sector, the development toward higher-quality and more efficient energy carriers has different consequences across scenarios (*Figure 7.15*). Scenario A2 and Case B rely heavily on CPA's ample coal resources. In Scenario A2, coal's share of primary energy consumption grows even beyond today's level, exceeding 70% by 2050. In the other scenarios the primary energy mix shifts toward oil and gas, which are cleaner and more flexible, plus renewable sources and nuclear energy. Not surprisingly, this shift is especially prominent in Case C, which assumes activist policies explicitly emphasizing non-fossil energy sources. Except for Scenario A2, which is dominated by coal, the most abundant of the region's resources, the region's net energy imports increase.

In light of regional coal resources, the CPA regional review describes a preferable alternative scenario that begins on a coal-intensive trajectory as in Scenario A2 but after 2020 gradually shifts toward less carbon-intensive fossils fuels, nuclear power, and renewables as in Scenario A3. Such an alternative scenario reflects the

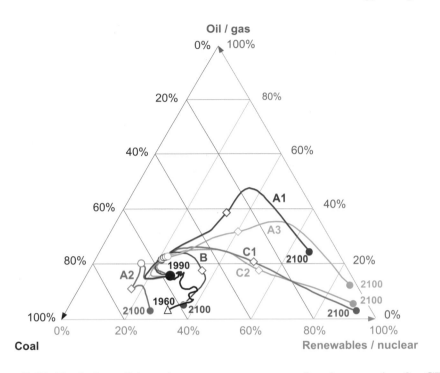

**Figure 7.15**: Evolution of the primary energy structure for six scenarios for **CPA**, shares of oil and gas, coal, and non-fossil sources, in percent. CPA's progression in all scenarios, except for Scenario A2 and Case B, is toward a future dominated by nuclear and renewables. The shift is reasonably direct for Scenarios C1 and C2. Scenarios A1 and A3 make greater use of oil and gas. See Section 5.2 for a further explanation of the figure.

current trends in the region and a recognition that with current momentum, it might take until 2020 to switch to lower reliance on domestic coal. However, such a scenario incurs two penalties relative to the six used in this study. Depending on how quickly it is done, changing horses in midstream will create what are referred to in utility liberalization debates as "stranded assets." Infrastructural and technological investments designed to move large amounts of coal will cease to generate a return for their investors. (Currently 40% of all railway operations in China are dedicated to coal transport; Siddiqi *et al.*, 1994.) Such investments will become at least partially stranded. Moreover, they will have to be duplicated by new infrastructural investments to transport natural gas and renewable energy and convert them to the forms in which they are demanded (i.e., electricity, liquids, and gases). The second penalty is that it would be more expensive to put the new system into place because there will not have been the same degree of learning and experience with new energy technologies that accumulates in, say, Scenario A3. Fewer learning opportunities mean less progress along the learning curves discussed in Section 4.4 and thus higher costs. It is precisely because such premiums must be paid in order to change direction that cost-effective scenarios are characterized by lock-in effects. Experience with the technologies that are chosen first creates customized infrastructures and cost reductions that make them increasingly cost-effective relative to their competitors. Changing direction requires progressively more and more money and political will. Costs can change if there are new technological breakthroughs favoring alternative paths, but the barrier that any breakthrough must overcome becomes progressively higher. In the case of CPA, the more miners there are digging coal, the more railroads that are built to move it, and the more power plants that are built to burn it, the harder it will become for the region to change course after 2020. Some shifts (e.g., from coal-fired to nuclear electricity generation) would be easier to implement than others, such as those that require substantial anticipatory infrastructural buildup (e.g., gas pipelines and distribution networks).

### 7.9.4. Implications

Given the importance of oil for transportation and of gas as the preferred transition fuel in all scenarios, a rapid expansion of oil and gas pipeline capacity is needed. We believe that the major political, institutional, infrastructural, technological, and financial challenges that CPA will face in the near term, at least in the energy sector, will arise from this need to expand pipeline capacity and interconnections.

Looking further into the future, such pipeline networks could facilitate the development of the vast Caspian and Siberian hydrocarbon resources, and possibly the unconventional gas resources locked in hydrate formations in Asia. Networks might be extended to the rest of Asia with the long-term strategic objective of a truly *Eurasian* energy infrastructure. China might become an important hub for energy trade in Asia, with a role similar to that of Germany in Europe's natural gas trade. Although there could be reluctance in CPA to increase import dependence to the extent implied by such a focus first on natural gas and later on electricity, the region's economic growth will ultimately depend on free trade, and this includes the energy trade.

The large increases in energy use anticipated for CPA require large investments. Across the six scenarios cumulative investments between 1990 and 2020 vary from US$1 to US$2 trillion. About two-thirds of these investments go to the power sector. These estimates are consistent with other recent studies. Malhotra (1997), for example, estimates East Asian power sector investments from 1994 to 2003 at US$550 billion, and the World Bank estimates *total* infrastructure requirements (not just energy) from 1998 to 2007 at between US$1.2 and US$1.5 trillion. At these levels, energy sector investments, like other infrastructure investments, will have to come increasingly from private sources. In China reforms carried out in the 1980s have resulted in a gradual shift in public sector investments from the central government to provincial authorities. Increasingly, energy investments are also coming from banks in the form of enterprise loans. In view of the huge investment needs, more private sector participation will be needed, and that will require continuing reforms.

Rapid economic growth and increasing energy needs in CPA have taken a heavy toll on the environment (World Bank, 1997b), and the scenarios indicate that the problems could get worse. Section 6.6.2 estimates that if sulfur emissions were to remain unabated in a high-growth, coal-intensive scenario similar to Scenario A2, by 2020 up to one-third of China would suffer annual sulfur deposition levels exceeding 5 grams per square meter. This is comparable with deposition levels in Europe's "black triangle," the border area of the former German Democratic Republic, Poland, and the Czech Republic, which has been classified by the UN as an "ecological disaster zone" (Amann *et al.*, 1995). In the high-growth coal-intensive Scenario A2, unabated sulfur emissions in CPA would approach 50 MtS by 2050, comparable with the 1990 sulfur emissions of the entire globe. Coal consumption would be well over 2 billion tons coal equivalent (tce) by 2015. For comparison, coal consumption in England at the time of the infamous 1952 "killer smog" was only about 200 million tce.

In light of the damage potential from unabated sulfur emissions, stepped-up levels of abatement were assumed in the investment requirements estimated above. For Scenario A2, in particular, 25% of sulfur emissions are assumed to be controlled by 2020, rising to close to 50% by 2050. As a result, sulfur emissions peak before 2050 at about 30 MtS and gradually decline to the 1990 level of 11 MtS (Foell *et al.*, 1995) by 2100, even in the coal-intensive Scenario A2. Emissions in Scenarios A1 and A3 and Case B peak about the same time and at the same level. The subsequent decline is faster, however, in the less coal-intensive Scenarios A1 and A3. It is slower in Case B, where technological progress moves more slowly. In the ecologically driven Case C, emissions peak in 2020 at 17 MtS and by 2050 have already dropped below their 1990 level.

Not surprisingly, Case C also has the lowest $CO_2$ emissions, rising from 690 MtC in 1990 to peak at about 1,350 MtC around 2030. More important, however, is the result that $CO_2$ emissions steadily increase in Scenario A2 and Case B. The sulfur controls that eventually decrease sulfur emissions in those scenarios do not stop the steady rise in carbon emissions. (In Scenarios A1 and A3, CPA's $CO_2$ emissions eventually do decrease as those scenarios shift to cleaner fuels and post-fossil technologies.) Thus CPA has several alternative approaches to controlling sulfur emissions, but not all of them also control $CO_2$ emissions. To limit both sulfur and $CO_2$ emissions, the best strategies focus on rapid technological progress and the use of energy resources other than coal.

### 7.9.5. Highlights

- The CPA region is changing rapidly into an economic and energy "giant" of global significance. It is currently the most populous of our 11 regions (although SAS is anticipated to overtake it by 2010) and has the fastest-growing economy. In terms of purchasing power parities, it could become the largest regional economy as early as 2020. In Case A, energy use reaches North America's 1990 levels as early as 2020 (more than 2 Gtoe) and total OECD 1990 levels by 2050 (more than 4 Gtoe).

- Projections in the CPA regional review were even more optimistic than in the global analysis. For the next several decades, the scenarios indicate that energy intensity in CPA could improve faster than in any other region, decreasing by between 2% and 5% per year. While other regional reviews generally considered the energy intensity decreases in the scenarios to be too optimistic, the CPA regional review found even its high improvement rates to be too low. The regional review also expected more rapid economic growth than that reflected in the scenarios. Both factors combined result in projected energy growth very close to that in the scenarios, at least up to 2020.

- The most significant qualitative change expected from growing affluence in the region is a transition away from solid fuels as the main energy currency to higher-quality fuels in the form of liquids and grid-dependent natural gas and electricity. This shift will contribute to further efficiency gains and decreases in final energy intensity. It will also help reduce indoor air pollution and improve urban air quality.

- We believe that the major near-term political, institutional, infrastructural, technological, and financial challenges that CPA will face in the energy sector will arise from the need to expand pipeline capacity and connections. Looking further into the future, such pipeline networks could facilitate the development of the vast Caspian and Siberian hydrocarbon resources, and might be extended to the rest of Asia with the long-term strategic objective of a truly Eurasian energy infrastructure.

- The most important environmental issues for the region are local – in particular, poor urban air quality resulting from particulate and sulfur emissions. Particulate and sulfur emissions have immediate human health impacts and longer-term acidification impacts that seriously threaten agricultural production and ecosystems. CPA has several alternative approaches to controlling sulfur emissions and particulates, but not all of them also control $CO_2$ emissions. To limit both sulfur and $CO_2$ emissions, the best strategies focus on rapid technological progress and the use of energy resources other than coal.

- The CPA regional review proposed an alternative scenario that would begin on a coal-intensive trajectory but shift to less carbon-intensive fuels after 2020. Such a scenario incurs two penalties relative to the six used in this study. First, infrastructural and technological investments designed to move large amounts of coal would have to be duplicated by new infrastructural investments to transport and convert alternative fuels. Second, it would be more expensive to put the new system in place because there would not have been the same degree of learning and experience with new energy technologies that accumulates in say Scenario A3. Such lock-in effects are an important characteristic of this study's scenarios and are the reason that while energy options may be wide open today, they will be much narrower by 2020.

## 7.10.    South Asia (SAS)

The region includes Afghanistan, Bhutan, India, Maldives, Nepal, Pakistan, and Sri Lanka. In 1990 India accounted for 80% of the region's GDP and 75% of its population. With Pakistan and Bangladesh, the region also includes two more of the world's 10 most populous countries.

In 1990, the region accounted for 20% of the world's population, less than 2% of GWP, 4.9% of global primary energy consumption, and 3.2% of global energy-related net carbon emissions. Per capita income in 1990 was the lowest of all 11 regions at US\$334, less than 10% of the world average. In terms of purchasing power parities, 1990 per capita income was US\$1,200, or about 25% of the world average. Recent economic growth, however, has been rapid, making South Asia the second fastest-growing region in the world, after East Asia, with an average annual rate of 5.3% over the past decade (World Bank, 1998b). The challenge for the region will be to achieve development for the poor parts of the population. Poverty is widespread, and if economic development is to bring today's poor up to adequate levels of affluence, a prerequisite for development is the level of energy consumption envisioned in our scenarios. To attract energy investments, the region will have to be politically stable and will have to decrease or eliminate energy subsidies, which means that consumers have to be well-off enough to pay their own energy bills. To the extent that growth can be directed toward less energy-intensive and less material-intensive sectors of the economy, the investment challenge will be less severe (India's success as a computer software producer is an example). Scenarios C1 and C2, for example, have the lowest investment requirements of all six scenarios.

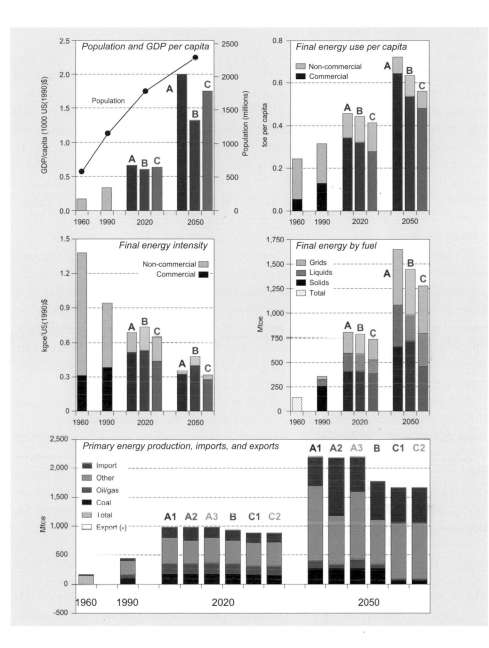

**Figure 7.16**: Scenario summary for **SAS** through 2050. *Top left:* Levels of GDP per capita reached in Cases A, B, and C, plus the population projection used in all three cases. *Top right:* Final energy use per capita. *Middle left:* Decrease in total final energy intensities in all three cases. *Middle right:* Final energy divided into solid fuels, liquid fuels, and energy delivered by grids. *Bottom:* Composition of primary energy, plus imports and exports, for all six scenarios.

### 7.10.1. Recent developments

As noted above, short-term economic growth in SAS has averaged 5.3% per year over the past decade (World Bank, 1998b). Between 1990 and 1996, commercial primary energy use in the region grew at 4.5% per year, from 223 Mtoe to 291 Mtoe according to British Petroleum (BP, 1997) statistics. These numbers do not include noncommercial energy sources such as fuelwood, agricultural residues, or dung-cake, which are important traditional energy forms used in the region (TERI, 1994). Short-term commercial energy growth is thus higher than the 1990 to 2020 growth rates in the six scenarios in this study (3.4 to 4.0%), reflecting the higher short-term economic growth rates in the region, as cited above, and possibly also faster substitution of traditional energy carriers in urban areas. Taken together, the World Bank (1998b) economic data and the British Petroleum (BP, 1997) energy data imply an average short-term decrease in commercial energy intensity of 0.9% per year between 1990 and 1996. This is a high improvement rate in comparison with the long-term improvements in the scenarios. It comes closest to the 0.3% per year decrease in Case C between 1990 and 2020, the only case in which commercial primary energy intensity decreases prior to 2020. In Cases A and B, commercial primary energy intensity *increases* as the shift from traditional noncommercial fuels to commercial fuels in the region outweighs any actual engineering or structural improvements.

In contrast to their Western European and North American counterparts, the SAS regional reviewers viewed the energy intensity improvements in the six scenarios as being too modest. Both the regional optimism concerning technological progress and the recent, relatively rapid short-term energy intensity decreases in connection with high economic growth are consistent with the fundamental model underlying the scenarios. The more rapid the economic growth, the faster the energy intensity will improve as old equipment is replaced by new equipment and the economic structure shifts toward less energy-intensive activities.

### 7.10.2. Population and economy

SAS' population is projected to double between 1990 and 2050, from 1.1 to 2.3 billion (*Figure 7.16*). By 2010 it will pass CPA as the most populous region in the world. After 2050 growth slows, and the population is projected to stabilize at around 2.6 billion sometime after 2100. As was the case in several other regions, the SAS regional review noted that fertility rates have dropped more quickly in the 1990s (Pachauri, 1997) than anticipated in the 1992 World Bank (Bos *et al.*, 1992) projection used in this study. Thus, as discussed in Section 4.1, actual population growth may prove to be below the 1992 World Bank projection.

Economic growth is vigorous in all scenarios through 2050. For 1990 to 2020 annual GDP growth is close to 4% per year, averaging 3.8% in Case A, 3.7% in Case C, and 3.5% in Case B. For 2020 to 2050 it rises to 4.5% in Case A, 4.2% in Case C, and 3.5% in Case B. Economic growth moderates after 2050 as the region becomes more affluent. But even with relatively high growth through the first half of the next century, per capita income in 2050 ranges only between US$2,000 in Case A (about where MEA was in 1990) and US$1,300 in Case B (15% below where PAS was in 1990). Only in the very beginning of the 22nd century is the region likely to reach the per capita GDP levels of WEU in 1990, although in terms of purchasing power parities SAS reaches 1990 WEU standards within the 21st century.

Economic projections within the region exist mainly for India, but because India accounts for 80% of the region's economic output, it is reasonable to compare these with the scenarios' developments for the whole region. The projections for India are mostly short term, to which the caveats mentioned above apply. Estimates for India reported by the 1997 International Energy Workshop (Manne and Schrattenholzer, 1997) project average annual growth between 1990 and 2020 on the order of 5%. Projections, also for India, cited in the SAS regional review (and in Bhattacharya, 1997; INFRAS, 1996) for 1990 to 2020 are between 5% and 6% per year. Based on the projections for India, the SAS regional review estimated 1990 to 2020 growth for the region as a whole at 5.8% per year. These numbers reflect both the recent high growth in the region and the pattern, seen both in this study and in others (Schrattenholzer and Marchant, 1996), of growth estimates from within a country or region being generally higher than globally comprehensive estimates. The regional projections compare with medium-term (1990 to 2020) economic growth rates of some 4% per year (Case A), which subsequently accelerate to 4.5% per year. Both the regional review and recent research, as discussed in Section 4.2, emphasize the importance of intangible factors on economic growth, for example, appropriate institutional arrangements, the market orientation of an economy, trade, and education. Also the SAS regional poll ranked institutional deficiencies as a high-priority concern, and the region's prospects for strong economic growth in the future will largely depend on how quickly and successfully the region is able to address that concern.

### 7.10.3. Energy systems

From seventh place among the 11 regions in terms of total energy consumption in 1990, SAS will grow to first or second place, along with CPA, by the end of the next century. This happens fastest in Case C, where SAS reaches second place by 2050 due to international resource transfers and slower energy growth in the North (thus making Case C particularly interesting to regional experts; Parikh, 1998).

This change in the geopolitics of energy demand is slowest in Case B, where SAS is still in sixth place in 2050. In all cases, the rise of SAS and CPA as the world's largest energy consumers shifts the center of the global energy market from the Atlantic to Asia. To date, SAS and CPA have had a limited history of cooperation. The efficiency and speed with which they can expand energy services and attract the necessary investment to lead the global energy market will partially depend on how quickly they can develop new mechanisms for cooperation. Given the growing importance of grid-dependent energy forms in all scenarios, the construction of integrated Eurasian energy grids, as discussed in Sections 7.8 and 7.9, is an important near-term opportunity in which SAS should want to be an integral player.

In the scenarios, while GDP grows by a factor of 8 (Case B) to 12 (Case A) between 1990 and 2050, final energy consumption grows only by a factor of 3.6 (Case C) to 5 (Scenario A2). Total final energy intensities fall fairly modestly at first. Between 1990 and 2020 they improve at rates between 0.8% per year (Case B) and 1.3% per year (Case C). Improvements then accelerate to between 1.4% and 2.4% per year for 2020 to 2050, and to between 2.4% and 3.3% per year for 2050 to 2100. Note that while total final energy intensities fall in all scenarios, commercial final energy intensities continue their historic increase through 2020 before finally beginning to decline (see *Figure 7.16*, middle left chart). This reflects the shift from noncommercial to commercial energy sources that accompanies economic development. In absolute terms, per capita final energy consumption rises from 316 kgoe in 1990 to around 450 kgoe in 2020 and to between 550 kgoe and 800 kgoe in 2050. The 1990 level of per capita final energy consumption in PAS was 800 kgoe. By 2100, the region's per capita final energy consumption reaches 1.5 toe in Cases A and B, still well below Western Europe's 1990 value of 2.3 toe.

Projections reported by the International Energy Workshop for India for commercial primary energy growth from 1990 to 2020 range between 4.1% and 5.6% per year (Manne and Schrattenholzer, 1997), compared with values in this study between 3.4% per year (Case C) and 4.0% per year (Case A). Total primary energy growth averages 3.5% per year through 2020 in the regional review's alternative projection, and primary energy intensity improves at 2.3% per year. These values compare with primary energy growth rates between 2.3% and 2.7% in this study, and primary energy intensity improvements between 1.0% per year (Case B) and 1.4% per year (Case C). As noted above, the short-term projections of the regional review are consistent with the scenarios' long-term pattern of more rapid economic growth leading to faster energy intensity improvements.

Through 2020, the principal change in the primary energy mix will be from traditional renewables to commercial fossil fuels. Initially, few differences arise among the scenarios because of the long lifetimes of energy infrastructures and the resulting limited turnover before 2020. Partly because of the shift from noncommercial to commercial fuels, fossil fuels grow to between 54% and 60% of total primary energy by 2020, up from 47% in 1990. Their share only begins to decline after 2020, falling most quickly in Case C to 6% in 2100 as new renewables (e.g., biogas and solar) penetrate the energy market. The fossil fuel share falls most slowly in Case B and Scenario A2, declining to 45%. But while the share of fossil fuels increases only eight percentage points from 1990 to 2020 in Scenario A3, for example, that would still amount to an increase in absolute terms by a factor of 2.6, from 210 mtoe to 540 mtoe. In 1990, nuclear power in SAS provided only 0.3% of primary energy, compared with its 5% share of world primary energy. According to the scenarios, this gap will close largely because of the twin pressures of demand growth due to a growing population and increases in the currently low electrification rates.

### 7.10.4. Implications

As SAS grows to be the world's first or second largest energy market by the end of next century, its share of global energy investments also grows. In 1990 SAS ranked among the lowest of the 11 regions in terms of energy investments. By 2100 it will rank first or second, depending on the scenario. Again, the region's rise is fastest in Case C, in which resource transfers explicitly target investment in the South, and SAS moves to second out of 11 by 2050. Progress is slowest in Case B, which has neither targeted investments nor high overall growth, with the result that SAS rises to only eighth place by 2050. The amounts of money are large, between US$540 and US$710 billion from 1990 to 2020, and between US$1.5 and US$2.5 trillion from 2020 to 2050. As a share of regional GDP, however, they rise initially only slightly above recent energy investment levels before beginning a gradual continuing decline. Thus, the required funds are available on international capital markets to cover the investment requirements of SAS – the question is whether those investments can be made attractive enough to bring in the needed funds. To reach the necessary high investment levels, the region will have to attract considerable private funds. To do so it must rectify the region's institutional deficiencies, which were a top-ranked concern of the regional reviewers. There must be a shift from subsidies and flat rates to segmented market pricing, losses and leakage must be reduced, and efficiencies must be improved. Moreover, such shifts must be politically acceptable, and they must not leave the poor behind. Market pricing only makes an investment attractive if there is a market that can pay the price, and the region's poor do not yet constitute such a market. In short, we deal here with a problem of market *exclusion* rather than one of market *distortion*. As emphasized repeatedly by experts in the region, how quickly the poor become empowered energy consumers will depend more on the success of policies to reduce poverty than on energy policies proper.

The SAS regional poll was one of only two regional reviews to emphasize concerns not on the standard list. For SAS, the greatest additional concern was hydro-carbon imports. The scenarios echo this concern. Import dependence increases in all scenarios at least through 2050. Dependence is generally lowest in Scenario A1 (high growth, ample oil and gas), with Scenario A3 (high growth, renewables and nuclear) a reasonably close second. Dependence is highest in Scenario A2 (high growth, coal) and the middle-course Case B. Although in the very long run import dependence is low in both Case C scenarios, it is as high as in Case B through 2050, growing from 9% today to 37% in 2050. Thus import dependence in SAS is generally greater in those scenarios (Scenario A2 and Case B) where coal is more important. To the extent that import dependence is a major concern in SAS, the region's interests are quite different from those of its neighbor CPA, for whom import dependence is lowest in the high-coal Scenario A2.

The SAS regional poll ranked potential climate change as a very low priority for the region. Not only does South Asia have more immediate pressing development concerns on its agenda, it currently also has the lowest per capita emissions of all 11 regions in this study. In 1990, at 0.17 tC per capita, emissions totaled 190 MtC. Future carbon emissions vary substantially among the scenarios, ranging up to a 13-fold increase between 1990 and 2100 in the coal-intensive Scenario A2. In all cases, however, SAS remains among the regions with the lowest per capita emissions. Even in Scenario A2, after a century of coal-intensive development, per capita carbon emissions in SAS in 2100 are less than one-fifth of what they were in NAM in 1990.

In contrast, local environmental impacts were a very high priority for the regional reviewers. $SO_2$ emissions are the best indicator available from the scenarios, and it is only Case C and Scenario A3 that offer any eventual decrease in $SO_2$ emissions relative to 1990 levels. All other scenarios show a steady, persistent increase. Case C is best, with emissions rising to at most 30% above 1990 levels in 2050 before dropping below 1990 levels by 2080. In Scenario A3 the peak is higher at about 2.5 times the 1990 levels around 2060. Emissions grow fastest in Case B and Scenario A2 and more slowly in Scenario A1.

### 7.10.5. Highlights

- From seventh place among the 11 regions in terms of energy consumption in 1990, SAS will grow to first or second place, along with CPA, by the end of the next century. This happens fastest in Case C, due to international resource transfers, and slowest in Case B. The rise of SAS and CPA as the world's largest energy consumers shifts the center of the global energy market from the Atlantic to Asia. Given the growing importance of grid-dependent energy forms in all scenarios, the construction of integrated Eurasian energy grids, as discussed in Sections 7.8 and 7.9, is an important near-term opportunity in which SAS should want to be an integral player.

- Hydrocarbon imports and institutional deficiencies are the two concerns ranked highest in the SAS regional review. Import dependence proves to be lowest in Scenario A1 (high growth, ample oil and gas), with Scenario A3 (high growth, renewables and nuclear) a reasonably close second. The worst cases in this respect are Scenario A2 (high growth, coal) and the middle-course Case B.

- The other top-ranked concern was institutional deficiencies. Progress in this regard is essential if SAS is to attract the investments it needs to grow as projected in the scenarios. There must be a shift from subsidies and flat rates to segmented market pricing, losses and leakage must be reduced, and efficiencies must be improved. Such shifts must be politically acceptable and must not leave the poor behind. Market pricing only makes an investment attractive if there is a market that can pay the price, and the region's poor do not yet constitute such a market. In short, we deal here with a problem of market *exclusion* rather than one of market *distortion*. How quickly the poor become empowered energy consumers will depend more on the success of policies to reduce poverty than on energy policies proper.

- Local environmental impacts were also a high-priority concern for the regional reviewers. $SO_2$ emissions are the best indicator available from the scenarios. Only Case C and Scenario A3 offer any eventual decrease in $SO_2$ emissions relative to 1990 levels. Case C is the best, with emissions rising to at most 30% above 1990 levels in 2050, before dropping below 1990 levels by 2080. In Scenario A3 the peak is higher at about 2.5 times 1990 levels around 2060. Emissions grow fastest in Case B and Scenario A2 and more slowly in Scenario A1.

## 7.11. Other Pacific Asia (PAS)

The region consists of a diverse group of countries. It includes most of the "Asian Tigers" – Taiwan, the Republic of Korea, and Singapore. In 1990 GDP per capita was as high as US$13,000 in Singapore, but it was only US$600 in Indonesia, and the regional average was US$1,500, less than 40% of the world average (World Bank, 1993). National resource endowments vary from negligible amounts in several small countries to the large oil and gas resources of Indonesia. In 1990 PAS' population was 430 million, 8% of the world's total, and regional GDP was 3% of GWP. Over 40% of the region's population lives in Indonesia. The Philippines, Thailand, and Republic of Korea each have population shares larger than 10%. In

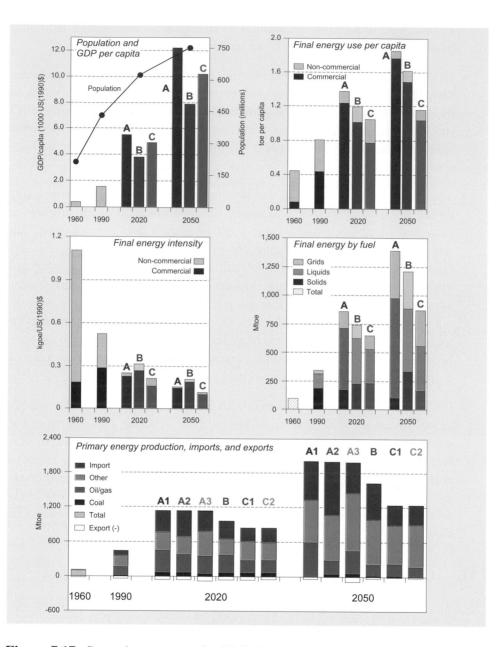

**Figure 7.17**: Scenario summary for **PAS** through 2050. *Top left:* Levels of GDP per capita reached in Cases A, B, and C, plus the population projection used in all three cases. *Top right:* Final energy use per capita. *Middle left:* Decrease in total final energy intensities in all three cases. *Middle right:* Final energy divided into solid fuels, liquid fuels, and energy delivered by grids. *Bottom:* Composition of primary energy production, plus imports and exports, for all six scenarios.

1990, PAS accounted for 4.7% of global primary energy consumption and 3.5% of global energy-related net carbon emissions.

### 7.11.1.  Recent developments

Through the mid-1990s economic growth in the region was strong.  Prior to 1990 growth averaged nearly 10% per year.  As the region became more affluent, growth moderated in accordance with the historical patterns discussed in Section 4.2.  Growth rates were high throughout the region at about 7% per year until 1996.  The past two years, however, saw major reductions in exports and lower economic performance.  Despite a partial recovery in early 1997, by the end of the year much of the region was in a major economic crisis.

With the crisis, short-term expectations for growth were reduced substantially (IMF, 1998).  Coupled with the higher rates in the first half of the decade, the resulting average for 1990 to 2000 may well be close to the 6% per year reflected in all scenarios in this study except Case B.  (There, 1990 to 2000 growth would be distinctly lower at 4.7% per year.)  This agrees well with the PAS regional review, which anticipates growth at 6.2% per year through 2010.

From the long-term perspective of this study, the very recent crisis is a short-term effect that does not change the region's basic potential for relatively stable long-term growth.  To the extent the crisis speeds reforms that improve financial regulation and macroeconomic policy mechanisms, its ultimate effect will be to improve the region's long-term prospects.

### 7.11.2.  Population and economy

Population growth in PAS is projected to decline steadily with continued economic development (see *Figure 7.17*).  Nonetheless the end result is essentially a doubling of the region's population over the next century to 820 million.  World population also approximately doubles, leaving PAS with essentially the same share of the global population in 2100 as in 1990.

After initially high growth rates in the scenarios of 6% per year, growth moderates so that average annual growth between 1990 and 2020 is projected to be 5.7% in Case A, 5.3% in Case C, and 4.4% in Case B.  Economic development as described in the scenarios would lead to high levels of affluence by 2100, ranging between US$19,000 and US$30,000, compared with 1990 PAO levels of US$23,000.  At that stage, regional GDP would total up to US$25 trillion – exceeding the 1990 total global GDP.  The difference would be a regional PAS population of only 820 million, compared with today's global population of 6 billion.

### 7.11.3.  Energy systems

Total commercial and noncommercial primary energy consumption per capita was 1.0 toe, about 60% of the world average in 1990. However, total primary energy intensity was 50% *above* the world average at 0.65 kgoe/US(1990)$. But PAS's energy intensity is the third lowest among the six developing regions. This is an encouraging indication that PAS is well on its way to a modern, energy-efficient economy.

With increasing affluence, energy intensity declines in all scenarios (*Figure 7.17*). Between 1990 and 2020, final energy intensity in PAS improves at rates of between 1.7% per year (Case B) and 3.0% per year (Case C). By 2050 it could reach 0.1 kgoe/US(1990)$, which is the 1990 value for PAO. The PAS regional review suggested that in the short term (through 2010) commercial primary energy intensity would improve at approximately 1% per year. This is consistent with values reflected in the scenarios, which range between 0.2% per year (Case B) and 2.0% per year (Case C). Relatively slow improvements in commercial energy intensity are due to the fact that noncommercial energy is steadily being replaced by commercial energy. Such replacements add to the region's commercial energy consumption without immediately increasing the GDP. This accounting effect partially offsets the actual improvements that take place due to engineering and structural changes. Improvements in *total* commercial and noncommercial primary energy intensity are faster, averaging between 1.6% per year (Case B) and 3.0% per year (Case C) from 1990 to 2010.

Key features of the energy system in 1990 were a large share of noncommercial energy (37% of total primary energy) and a high dependence on crude oil (38% of total primary energy and 60% of commercial primary energy). In the scenarios, overall final energy use grows steadily from 340 Mtoe in 1990 to between 860 Mtoe (Scenario C2) and 1,450 Mtoe (Scenario A2) by 2050. Although noncommercial energy use in PAS continues to be important, it declines in all scenarios in both absolute and relative terms from 160 Mtoe in 1990 to between 60 and 95 Mtoe in 2050.

In 1990 *regional* primary energy dependence on oil imports was only around 15% because of Indonesia's oil supply, but in other countries in the region *national* dependence on oil imports was often very high. PAS was a net exporter of natural gas, with more than half the natural gas produced in the region exported as LNG. Oil consumption expands considerably in all scenarios. By 2020 it ranges between 320 Mtoe (Case C) and 590 Mtoe (Scenario A1). Only in Case C does its share of primary energy consumption remain unchanged at 38% through 2020 (which still means almost a doubling relative to 1990 levels in absolute terms). Otherwise oil's share rises to between 44% (Scenario A3) and 52% (Scenario A1). From a near-term regional perspective (through 2010), the PAS regional review considered such increases feasible. Beyond 2020, however, oil's continued dominance will depend on whether technological progress and exploration make it possible to exploit most of the estimated available oil resources, including unconventional sources (Rogner, 1997). If near-term investments lead in this direction, as in Scenario A1, by 2050 oil consumption could increase by more than a factor of six to 1 Gtoe and would account for half of all primary energy consumed in PAS. However, 70% of that oil would have to be imported to the region, compared with 40% today. Oil use is distinctly lower in Scenario A3 (460 Mtoe) and Case C (360 Mtoe) although oil import dependence is higher. After 2050 oil use decreases in all scenarios, and by 2100 it is less than the 160 Mtoe consumed in 1990 (except in Scenario A1). By then PAS is a very prosperous region, with sufficient know-how and capital to exploit alternative energy sources, which eases the transition away from oil.

For natural gas, short-term expectations are more moderate. By 2020, market shares above 9% of primary energy are reached in all scenarios, up from 7% in 1990. In absolute terms, gas use approaches 100 Mtoe in 2010 in Cases A and B, which corresponds well with the regional perspective reported in the PAS regional review. Consumption continues to grow in all scenarios through 2050, as was the case with oil. And while oil imports grow in all scenarios, gas exports start to decline by 2030 and nearly cease after 2050. Total gas production peaks in the second half of the next century (soonest in Case C and latest in Scenario A1) and in most scenarios drops back to near or below 1990 levels by 2100. From the long-term perspective, oil and gas again play a transitional role in the eventual shift in the energy system to a mix of alternative sources including nuclear energy, renewables, and synfuels.

In all scenarios electricity generation grows between 7 and 10 times to levels of up to 3,400 TWh by 2050. Its expansion is thus distinctly faster than overall energy growth. Electricity generation shifts from a mix of fossil fuels toward gas, nuclear, and renewables. From 60% in 1990, the share of fossil fuels declines steadily to between 5% (Case B) and 26% (Scenario A3) in 2050. Oil-based power generation is effectively phased out, and except for Scenario A2, the same is largely true for coal. Gas is left as the principal fossil fuel for electricity generation. Growing shares come from nuclear energy and especially renewable electricity sources. With the region's large hydropower, wind, and solar potentials, electricity from these sources could reach almost 2,000 TWh by 2050 and more than 3,000 TWh in the very long term (toward 2100). If renewable technologies develop more slowly and oil and gas resources are more limited (Case B), nuclear energy becomes the largest source of electricity. In Case B, 150 GW of nuclear power are on-line in 2050 supplying over one-third of PAS's electricity. Assuming an economic plant life of 30 years, Case B therefore requires steady additions of more than 5 GW of nuclear power annually. Such an expansion is well within the region's capabilities according to the PAS regional review, which indicated that by 2010 nuclear power could already contribute 27 to 30 Mtoe electricity, four times the 1990 levels. This is well above the values in all six scenarios. In the scenarios, nuclear power grows more rapidly in the long run than in the near term. Thus there appears less uncertainty about the eventual important role of nuclear power in the region than about its timing – whether substantial increases come sooner (as suggested in the regional review) or later (as suggested by the scenarios).

Despite heavy investments in nuclear and renewable electricity generation, import dependence grows initially in all scenarios. In 1990, 22% of all energy consumed was imported into the region. By 2020, this value grows to between 28% (Scenario C1) and 40% (Scenario A2). In 2050 it approaches 50% in Scenario A2 as a result of the region's limited coal resources and the slow progress in other energy technologies. In absolute terms, PAS's import requirements are between 340 Mtoe (Scenario C2) and 930 Mtoe (Scenario A2), comparable with the import requirements of WEU at the same time. These figures illustrate the growing importance of Asian energy importers on the world oil market. But in all scenarios, even Scenario A2, import dependence declines after 2050 as renewable-based synliquids and electricity generation become increasingly competitive. The conclusion for PAS is that, in light of its relatively scarce fossil resources, technology is especially important. An appropriate model might be that of Japan. Japan has invested heavily in energy-related RD&D and conservation to avoid extreme import dependence, reduce energy requirements, and enhance domestic energy production that must come largely from non-fossil sources.

### 7.11.4. Implications

Investment requirements for the energy sector between 1990 and 2020 run between 1.1% (Scenario C2) and 1.6% (Scenario A3) of GDP. These levels are less than recent investment levels of closer to 2.3%. (Malhotra, 1997, estimates that East Asian countries have recently devoted 4.7% of GDP to infrastructure investments, nearly 50% of which has gone to the power sector.) Although investment requirements drop in all scenarios as a percentage of GDP, they grow in absolute terms. From 1990 to 2020, cumulative investments total between US$550 and US$860 billion. Between 2020 and 2050 they run from US$1.0 to US$2.0 trillion. In the near term they are highest in Scenario A3; in the long term they are highest in Scenario A2. Investments are generally lowest in Case C. Although investments decline as a share of GDP, they are still likely to exceed what can be accomplished with public financing. Regional reforms to make these investments more attractive to private capital are therefore necessary. We expect that time and perseverance will be needed before substantial private money begins to flow into the region, and in the meantime governments and international financial organizations will necessarily continue to be involved in the energy sector. Otherwise, development options including electrification, especially in more remote areas and for the region's poor, will remain elusive.

Growing affluence and urbanization in a number of countries are already causing environmental problems. Due to increasing populations and centralized energy and transport requirements, major cities like Bangkok and Jakarta face severe and worsening air pollution. Sulfur deposition already significantly exceeds "critical loads" in much of the Republic of Korea.[11] The bad news is that sulfur emissions increase initially in all scenarios. The good news is that in all cases emissions eventually decline. In the very long term (2100) they are below 1990 levels in all scenarios, despite the 2.5- to 6-fold increase in final energy use. The decline begins soonest in Case C, before 2010, with $SO_2$ emissions below 1990 levels by 2020. It begins latest (around 2050) in the high-growth, high-coal Scenario A2 and the middle-course Case B. In Scenario A2 emissions reach 6.4 MtS – 2.5 times the 1990 levels – before beginning their decline.

$CO_2$ emissions from PAS were 210 MtC in 1990, or 3.5% of global emissions, which is in line with the region's share of GWP. By 2020, or shortly thereafter, $CO_2$ emissions at least double in all scenarios. The increase is highest in the coal-intensive Scenario A2, where emissions more than triple. In Case C, with its explicit emphasis on limiting greenhouse gas emissions, $CO_2$ emissions begin to decline after 2030, and after 2050 they decrease in all cases. In the very long term (2100), they fall below 1990 levels in all but Scenarios A1 and A2. That translates into per capita emissions of less than 0.2 tC, half the 1990 levels.

Climate change and $CO_2$ emissions, however, were not ranked as high priorities in the PAS regional poll. Local pollution problems are more important. These will become easier for the region to handle if international agreements are reached to make environmental protection and international equity explicit shared objectives, as assumed in Case C. Otherwise, regional strategies emphasizing oil and gas (Scenario A1) or nuclear and renewables (Scenario A3) are the more promising route.

## 7.11.5. Highlights

- In its "Asian Tigers" – Singapore, Taiwan, and the Republic of Korea – PAS already contains probably the best models for regional development, and through 2020 economic growth in PAS in the scenarios is second only to growth in CPA. Regional interests would be well served in Case C, where environmental protection and international equity are explicit shared global

---

[11]Critical loads are defined as the maximum deposition levels at which ecosystems can function sustainably (see also Section 6.6.2).

objectives. Otherwise the best strategies are those emphasizing high growth and technological progress related to the region's resource endowments – primarily renewables (Scenario A3) and secondarily oil and gas (Scenario A1). An appropriate model might be that of Japan, which has invested heavily in energy-related RD&D and conservation to avoid extreme import dependence, reduce energy requirements, and enhance domestic energy production that must come largely from non-fossil sources.

- For the range of development strategies examined in this report, investment requirements for PAS drop as a percentage of GDP. Nonetheless, they grow in absolute terms and are likely to exceed the region's capacity for public financing. Regional reforms are needed to attract private investment. In the meantime, governments and international financial organizations must remain involved in the energy sector; otherwise development options, especially in remote areas and for the region's poor, will remain elusive.

- PAS is highly dependent on oil (which accounted for 38% of total primary energy and 60% of commercial primary energy in 1990). *Regional* dependence on oil imports is low because of Indonesian oil. *National* dependence is much higher in many countries. Oil consumption expands considerably in all scenarios, and import dependence increases. At the same time the region's gas exports decrease. Except for the very near term, import dependence is lowest if the region emphasizes conservation and development of its renewable and nuclear energy potentials (Case C or Scenario A3), rather than its fossil resources.

- Local air pollution, particularly in large cities, is an important immediate concern. It will become easier for the region to handle if international agreements are reached to make environmental protection and international equity explicit globally shared objectives as assumed in Case C. Otherwise, regional strategies of particulate and sulfur control together with an emphasis on oil and gas (Scenario A1) or nuclear and renewables (Scenario A3) are the more promising route.

## 7.12.  Pacific OECD (PAO)

The region consists of three Pacific OECD countries – Japan, Australia, and New Zealand. With 2.7% of the global population in 1990, it accounted for 16% of GWP, 6.0% of total primary energy consumption, and 6.3% of global energy-related carbon emissions. These shares indicate that the region as a whole is the most energy efficient regional economy. Primary energy intensity was the lowest of all study regions at 0.17 kgoe/US(1990)$, below 40% of the world average of

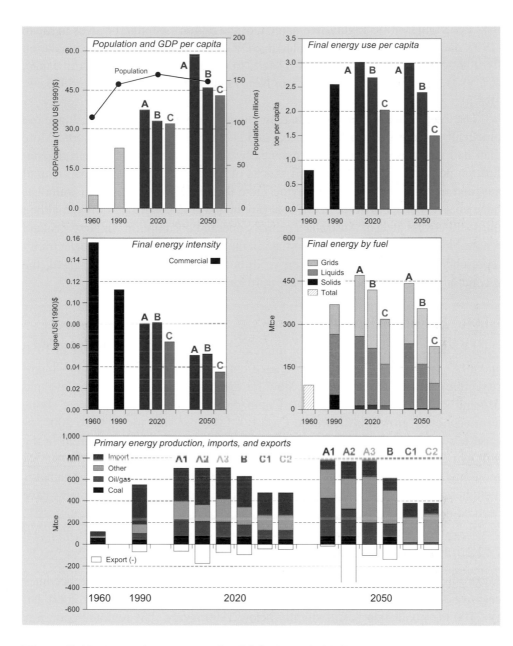

**Figure 7.18**: Scenario summary for **PAO** through 2050. *Top left:* Levels of GDP per capita reached in Cases A, B, and C, plus the population projection used in all three cases. *Top right:* Final energy use per capita. *Middle left:* Decrease in total final energy intensities in all three cases. *Middle right:* Final energy divided into solid fuels, liquid fuels, and energy delivered by grids. *Bottom:* Composition of primary energy production, plus imports and exports, for all six scenarios.

0.43 kgoe/US(1990)$. Eighty-five percent of the region's population lives in Japan (1995 data) producing 89% of regional GDP. Twelve percent of the population lives in Australia and 2% in New Zealand. In 1995 income was highest in Japan, at US$25,100 per capita and lowest in New Zealand at US$13,700 (World Bank, 1997a).

The three countries also differ in their energy systems and resources. Japan has practically no domestic fossil resources and is highly dependent on imports. More than half its energy supply is oil, essentially all of it imported, although diversification policies triggered by the oil shocks of the 1970s have led to coal, natural gas, and nuclear energy each supplying more than 10%. In contrast, Australia has huge coal resources. Two-thirds of coal production is exported, and the remaining third still covers 40% of Australia's primary energy consumption. In New Zealand renewable energy is more important than in the rest of the region. Hydroelectricity represents one-third of primary energy. In both Australia and New Zealand, the second largest energy source is oil, the third largest is natural gas.

### 7.12.1.  Recent developments

Until recently, PAO was a region of impressive economic growth. Between 1990 and 1995, however, GDP growth averaged only 1.4% per year, and energy intensity improvements have been very slow for the past decade (mostly due to rapid demand growth for transportation). Compared with the scenarios, recent short-term economic growth is closest to Case B's 1.8% per year for 1990 to 2000, or Case C's 1.7% per year. The causes of the decline in growth are different in the different countries. Japan's problems were initiated by the collapse of the bubble economy at the end of the 1980s, which left a shaky financial structure and investment houses and banks holding large amounts of suspect assets. The problems have been compounded by the wider Asian economic crisis that hit in 1997.

### 7.12.2.  Population and economy

According to the 1992 World Bank projection, PAO's population growth becomes negative after 2020 (see *Figure 7.18*). Very modest growth returns after 2060, but by 2100 the population is still slightly below 1990 levels. The regional pattern is driven mainly by Japan, with its combination of affluence, high life expectancy, and declining fertility. As a result, the Japanese contribution to the PAO regional poll ranked population growth as a top priority for exactly the opposite reasons that it was ranked high in some developing regions. In particular, an increasing proportion of retirees in Japan might decrease traditionally high saving rates, with negative repercussions for the availability and cost of capital, and consequently economic growth (Ito, 1997).

In accordance with the patterns discussed in Section 4.2, economic growth gradually slows in all scenarios to between 1.4% and 1.9% per year from 1990 to 2020, and to between 0.8% and 1.3% per year from 2020 to 2050. Nonetheless, per capita GDP remains the highest of all regions in all scenarios throughout the time period. The PAO regional review was more optimistic about the short term (through 2010), projecting growth at 2.7% per year. Short-term projections in the scenarios (1990 to 2010) range between 1.6% and 2.0% per year. Thus they lie between the actual developments from 1990 to 1995 (1.4% per year) and the more optimistic 2.7% per year projected in the regional review.

### 7.12.3. Energy systems

As stated earlier, PAO is the most energy efficient of all 11 regions and its energy intensity continues to improve in all scenarios. These improvements are consistent with the region's historical experience. Of all OECD countries, Japan has always had among the lowest energy intensities and has achieved among the highest energy intensity improvements. Between 1990 and 2020, PAO's energy intensity decreases at an average rate of between 1% per year (Cases A and B) and 1.8% per year (Case C). The rate of improvement gradually slows along with GDP growth so that between 2020 and 2050 improvements average between 1% and 1.5% per year. The relatively rapid improvements of Case C, coupled with a population that starts to decline slightly after 2020 and modest economic growth reflecting the region's affluence and resource transfers to developing regions, as assumed in Case C, all add up to a *decrease* in primary energy consumption in Case C. By 2050 primary energy consumption is 30% below its 1990 value. The same pattern holds for final energy, as shown in the middle right chart in *Figure 7.18*. Between 2020 and 2050, final energy consumption drops for all cases in absolute terms, partly because of the region's declining population. The decrease continues after 2050 in Case C, which combines modest economic growth with relatively rapid energy intensity improvements, and more slowly in Case B, where economic growth is also modest, as are energy intensity improvements. By 2050 final energy consumption in Case C is around 220 Mtoe, 40% lower than in 1990. By 2100 it drops to approximately 130 Mtoe. Final energy use is 350 Mtoe for Case B and about 450 Mtoe for Case A by 2050. Even for Case A, very long-term final energy consumption growth (to 2100) would be less than 20% compared with 1990.

The energy intensity improvements reflected in the scenarios are consistent with projections in the PAO regional review, which cited regional short-term estimates of 1.5% per year through 2010. Given the higher economic growth projections in the regional review and similar projections for energy intensity improvements, the regional review results in higher short-term energy consumption projections than seen in the scenarios. Again, the key for the short term will be how quickly Japan, in particular, recovers from the end of its bubble economy and the subsequent Asian economic crisis.

Oil continues to be the principle energy source in the scenarios through 2020, although its share declines (from 53% in 1990), as does the percentage of oil that is imported (91% in 1990). The PAO regional review projected higher dependence on oil imports than in the scenarios in the short term, through 2010. In part this reflects the higher energy consumption projections in the regional review; it also reflects lower expectations about near-term contributions from alternative energy sources. Japan has been a leader in diversification, and in extending that pattern into the future, the scenarios are evidently somewhat more optimistic than the regional reviewers. In 1995 Japan spent 1% of GDP on energy-related RD&D (EIA, 1995), and efficiency and conservation were ranked as continuing top priority issues in the regional review.

If early investments are focused on oil and gas (Scenario A1), PAO's unconventional oil and gas resources become competitive, oil and gas production increases to 370 Mtoe in 2050 (more than six times the 1990 levels), and oil import dependence drops to 30%. Even in Scenario A1, however, oil's share of primary energy in 2050 drops. The reason is the region's technological leadership in alternative energy sources. After 2020, as existing energy facilities are retired and replaced, non-fossil sources grow most quickly.

If near-term investments focus instead on coal (Scenario A2), coal production in 2050 could amount to 430 Mtoe, nearly four times 1990 levels. Exports soar to 350 Mtoe, five times 1990 levels. However, within the region coal's share of primary energy continues to decline as the region takes advantage of alternative energy technologies while exporting 80% of its coal production.

In the event that the environment and equity become agreed international priorities (Case C) and investments follow suit, the shift toward non-fossil resources is accelerated. Due to enhanced conservation, total primary energy use is only 380 Mtoe in 2050. That is 30% lower than primary energy consumption in 1990. All of PAO's coal production is exported in 2050, and renewables approach 50% of primary energy consumption.

If investment priorities correspond to the middle-course Case B, coal production increases by 85% by 2050, second only to the high-coal Scenario A2. Oil consumption declines in both absolute and relative terms – to 120 Mtoe by 2050. Gas use amounts to 120 Mtoe by the same year. In contrast, renewables grow substantially, from 7% of primary energy in 1990 to 31% in 2050 (190 Mtoe). Nuclear energy grows from 8% of primary energy in 1990 to 19% in 2050 (120 Mtoe).

Electricity generation in 1990 was dominated by coal and renewables. Each had a 23% share of generation, with hydropower (at 130 TWh) accounting for 52% of the share of renewables. Unless investments are either focused on coal

(Scenario A2) or are relatively unfocused (Case B), coal's share in electricity generation drops after 2020 as new power plants replace old ones. Growth is greatest for new renewables, especially wind generators and, to a lesser extent, solar and geothermal power. Hydropower grows only very slightly. The regional review's perspective on renewables is somewhat more conservative even though a doubling of electricity from renewables by 2030, as in Scenario A3, is consistent with a straightforward extension of Japan's recent record in applying and accelerating advanced technologies. In the end, the future of renewables will depend on the success and future expansion of current policies and RD&D investments, such as those in Japan focusing on renewable and energy-efficient technologies, or the restructuring of electricity tariffs in Australia as suggested by the International Energy Agency (IEA, 1997a).

Through 2020, nuclear power's share increases in all scenarios except where the focus is on coal (Scenario A2). In the case of high growth with investments focused on nuclear power and renewables (Scenario A3), nuclear's share of electricity grows from 20% in 1990 to 29% in 2020, or to almost 500 TWh. Such growth is consistent with the regional development goals assessed in the PAO regional review. By 2050 nuclear power in Scenario A3 reaches 800 TWh, 40% of electricity generation.

### 7.12.4. Implications

As a percentage of regional GDP, the range of energy sector investments implied by the scenarios is modest. Cumulative investments from 1990 to 2020 range between US$440 and US$930 billion and average between 0.4% and 0.7% of regional GDP. Investments in PAO are the lowest of all 11 regions as a percentage of GDP, not least because energy is used sparingly in PAO's economy. In both relative and absolute terms, they are highest in the high-growth coal-intensive Scenario A2 and lowest when the focus is on conservation and environmental protection (Case C).

Air pollution from power plant emissions has been a major concern in Japan and has been significantly reduced through both pollution control technology and an increased share of nuclear power plants. Air quality and water acidification remain important issues due to increased road traffic and imported pollution from continental Asia. Reflecting this concern, $SO_2$ emissions from the energy system decline in all scenarios. By 2020 emissions are between 0.3 MtS (Case C) and 0.5 MtS (Case B) – that is, 70 to 80% less than the 1990 emissions of 1.5 MtS – and long-term emissions decline even further.

In 1990 PAO generated 6% of global energy-related $CO_2$ emissions, or 370 MtC. If policies are explicitly focused on limiting carbon emissions (Case C), these emissions begin to decline in the short term, dropping by 30% by 2020 and by 70% by 2050. Cases A and B project initial growth in $CO_2$ emissions, but by 2050 decarbonization of the energy system will have driven emissions below 1990 levels in all cases. The decarbonization rates of primary energy reflected in the scenarios range between 0.6% per year in Scenario A1, where fossil energy remains attractive, to almost 2% per year if investments are focused on either renewables and nuclear power (Scenario A3), or on limiting $CO_2$ emissions (Case C). These decarbonization rates are higher than the short-term regional perspectives reported in the PAO regional review, reflecting lower regional expectations about how quickly Japan's leadership in diversification can and will be extended in the future.

### 7.12.5. Highlights

- Of all 11 regions, PAO has the best record and the most aggressive policies promoting energy conservation, efficiency, and diversification beyond fossil fuels. PAO's combination of natural, human, and financial resources put it in the best position for a high-growth transition beyond fossil fuels – that is, Scenario A3.

- Even if energy investments are not focused on non-fossil sources, the region's renewable resources and nuclear power are likely to play an increasing role in its energy mix. Through 2020 the absolute contribution of nuclear and renewables grows at least 60% in all scenarios.

- It is a repeated theme of this study that, while there is little divergence among scenarios over the next 20 years because of long-lived energy infrastructures, the investment decisions of those 20 years will determine the direction energy systems take after 2020. As the most likely Asian source of investment capital, PAO's choices over the next 20 years will have a critical influence on the future of the world's fastest-growing energy markets. This influence will be particularly important in connection with Eurasian energy grids (as discussed in Sections 7.8 and 7.9), the pace of future efficiency improvements, the future of renewables and nuclear power, the future of unconventional oil and gas, and, more speculatively, the future of the world's immense resources of methane hydrates.

# Chapter 8

# Conclusion

## 8.1.    Why This Study

We believe that adequate energy services are a prerequisite for human development. Therefore, it is essential to understand long-term energy possibilities if humanity is to build a more prosperous and equitable future. For this reason, we have considered the future of energy systems to be interwoven with broader issues of human development and have analyzed jointly the main driving forces of change and their implications. The main conclusion is that future energy services need to be more efficient, cleaner, and less obtrusive.

The study builds on the long history of studies analyzing energy futures, but it departs from conventional wisdom in both its perception of key issues and the details of the analysis. After nearly 30 years of energy studies, we sought in this study to introduce new perceptions (detailed in the next section) about long-term energy developments, to apply a unified methodological framework integrating near-term strategies with long-term opportunities, to incorporate a dynamic treatment of technological change, to harmonize regional aspirations with global possibilities, and to treat development and environmental concerns as joint forces of change. Conventional studies have focused either on short-term possibilities or long-term opportunities. Most have treated both technologies and resources as largely static and have analyzed the energy system from only the supply or the demand side. They have either considered global development as just the sum of regional aspirations or have aggregated all world regions into one global analysis, and they have analyzed environmental implications as an outcome of energy development rather than as an inherent, endogenous driving force of change.

This study developed a new approach to analyzing energy futures that is methodologically sound and meets the study objectives.

## 8.2.    New Features

First, this study integrates changing perceptions about resources and the environment. It incorporates environmental effects at all scales – local, regional, and global – as important driving forces of change. It reflects the new perception that resources and the environment are not merely external constraints set independently of energy system developments. Rather, they are an integral part of the overall patterns of change characterizing alternative futures. Evidence is increasing that energy resources are more plentiful globally than has usually been assumed and that they are dynamic. Resource estimates change with improvements in technology and scientific knowledge and with new economic conditions. They are certainly plentiful enough that extensive use could create critical environmental threats and possibly cause irreversible global change.

Second, technology is treated dynamically. In each scenario, technological change reflects the scenario's distinctive choices, leading to an increasing divergence of development paths among the six scenarios. Both the high-growth Case A and the ecologically driven Case C contain multiple scenario branches within a single pattern of overall development. In each case, the difference between branches leading to different directions of technological change is path dependent; early investments and initial steps in one direction reduce the costs and obstacles of continuing in that direction. The performance and competitiveness of future technologies, and indeed the path that the global and regional energy systems take, is thus shaped by RD&D choices and early investments in new technologies and infrastructures. Future development depends on the path of technological learning, experimentation, and cumulative experience taken in each scenario. In each, the future becomes increasingly locked into a particular technological development paradigm – some are resource intensive, some are environmentally benign.

Third, the study takes a new methodological approach. Different state-of-the-art models describing individual dimensions of overall development are integrated into one modeling framework. The framework encompasses models of energy–economy interactions, energy systems and technologies, regional and global environmental impacts, and related databases. It adds new quantitative measures such as noncommercial energy and purchasing power parities to account better for the informal energy and economic activities that are so important especially in developing regions. Both the models and databases are incorporated into a global assessment integrating the study's 11 world regions. Another new feature is the study's interactive scenario formulation within the modeling framework. The models are used interactively to allow adjustments to reflect numerous feedbacks within and between models and to ensure overall consistency (see Appendix A).

Fourth, the study integrates different temporal and spatial dimensions, unifying near-term considerations with longer-term time horizons, and global perspectives with regional and local spatial resolutions. Both innovations required new data and solutions to major methodological challenges. The data and the quantitative study findings are now available to the next generation of energy studies through an interactive database on the Internet (see Appendix E), in addition to the descriptions of the study and its findings in this book. Features addressing spatial integration include modeling trade flows among regions; matching the high spatial detail of sulfur depositions and other environmental impacts to the increasingly global nature of energy development; distinguishing between rural and urban developments within regions; and reproducing technological diffusion patterns progressing from the source of innovation outward and from developed to developing countries. Features addressing temporal integration include the combination of the six long-term scenarios and the short-term regional assessments of global results, and, most important, the path dependence of technological developments in each scenario tying long-term opportunities and constraints to near-term RD&D and investment decisions.

Fifth, among the wide range of alternative futures, the study developed two scenarios that are environmentally benign at all scales, from local and regional to global. Unprecedented progressive international cooperation focused explicitly on environmental protection and international equity is required to achieve this challenging future in the two ecologically driven Case C scenarios. The future described by Case C includes a broad portfolio of environmental control technologies and policies, including incentives to encourage energy producers and consumers to utilize energy more efficiently and carefully, "green" taxes, international environmental and economic agreements, and technology transfer. This results in substantial environmental protection at all levels while also being consistent with RD&D and technological investment choices over time, with economic and energy developments, and with international trade and resource transfers. This is in stark contrast to other studies that focus on only local, regional, or global environmental issues.

This study is also distinctive in that it takes a positive view of the future. All three cases – all six scenarios – provide for substantial social and economic development throughout the world, particularly in the South. They provide for improved energy efficiencies and environmental compatibility, and thus for associated growth in both the quantity and quality of energy services. We make the strong assumption of a future free of major discontinuities and catastrophes. These are not only inherently difficult to anticipate, but also offer little policy guidance on managing an orderly transition from today's energy system, which relies largely on fossil fuels, toward a more sustainable system with more equitable access to energy services.

There remains, of course, considerable room for improvement. This study both lays out the possibilities and indicates the way forward.

## 8.3.    The Message

*World energy needs will increase*

World population is expected to double by the middle of the 21st century, and economic development needs to continue, particularly in the South. According to the scenarios of this study, this results in a 3- to 5-fold increase in world economic output by 2050 and a 10- to 15-fold increase by 2100. By 2100, per capita income in most of the currently developing countries will have reached and surpassed the levels of today's developed countries. Disparities are likely to persist, and despite rapid economic development, adequate energy services may not be available to everyone, even in 100 years. Nonetheless the distinction between "developed" and "developing" countries in today's sense will no longer be appropriate. Primary and final energy use will grow much less than the demand for energy services due to improvements in energy intensities. We expect a 1.5- to 3-fold increase in primary energy requirements by 2050, and a 2- to 5-fold increase by 2100.

*Energy intensities will improve significantly*

As individual technologies improve, conversion processes and end-use devices progress along their learning curves, and as inefficient technologies are retired in favor of more efficient ones, the amount of primary energy needed per unit of economic output – the energy intensity – decreases. Other things being equal, the faster the economic growth, the shorter the obsolescence time, the higher the turnover of capital, and the greater the energy intensity improvements. In the six scenarios of this study, improvements in individual technologies are varied across a range derived from historical trends. Combined with the economic growth patterns of the different scenarios, the average global reductions in energy intensity range between 0.8% per year and 1.4% per year. These bracket the historical rate of approximately 1% per year and cumulatively lead to substantial energy intensity decreases. Some regions improve faster, especially where current intensities are high and economic growth and capital turnover are rapid.

*Resource availability will not be a major global constraint*

The resource scarcity perceived in the 1970s did not occur as originally assumed. With technological and economic development, estimates of the ultimately available energy resource base will continue to increase. A variety of assumptions about the timing and extent of new discoveries of fossil energy reserves and resources (conventional and unconventional), and about improvements in the economics of their recoverability, are reflected in the range of scenarios reported here. All, however, indicate that economic development over the next century will not be constrained by geological resources. Regional shortages and price increases can occur, due to the unequal distribution of fossil resources, but globally they are not a constraint. Environmental concerns, financing, and technological needs appear more likely sources of future limits. The short-term volatility of international politics, speculation, and business cycles will periodically upset the long-term expansion of resources.

### *Quality of energy services and forms will increasingly shape future energy systems*

The energy system is service driven, from consumer to producer, while energy flows are resource and conversion-process driven, from producer to consumer. The scenarios demonstrate the need to consider energy end use and energy supply simultaneously, both from an analytical and a policy perspective. In addition to prices and quantities, energy *quality* matters increasingly. Quality considerations include convenience, flexibility, efficiency, and environmental cleanliness. The energy system is end-use driven. Under market liberalization, special competitive advantages will arise for those companies prepared to deliver a full range of energy services beyond just fuels and electricity.

### *Energy end-use patterns will converge, even as energy supply structures diverge*

The six scenarios indicate that the historical drive toward ever more convenient, flexible, and clean fuels reaching the consumer can continue for a wide range of possible future energy supply structures. In all scenarios there is a shift toward electricity and toward higher-quality fuels, such as natural gas, oil products, methanol, and, eventually, hydrogen. In contrast to these converging trends, primary energy supply structures diverge, particularly after 2020. Fossil sources continue to provide most of the world's energy well into the next century but to a varying extent across the scenarios. There is a shift away from noncommercial and mostly unsustainable uses of biomass, and direct uses of coal virtually disappear. Sustainable uses of renewables including modern biomass come to hold a prominent place in all scenarios. They reach the consumer as electricity, liquids, or gases, rather than as solids.

### *Technological change will be critical for future energy systems*

Technological change drives productivity growth and economic development. Across all scenarios the role of technological progress is critical. But progress has a price. RD&D of new energy technologies, and the accumulation of experience in niche markets require upfront expenditures of money and effort. These are increasingly viewed as too high a price to pay in liberalized markets where the maximization of short-term shareholder value generally takes precedence over longer-term socioeconomic development and environmental protection. Yet, it is the RD&D investments of the next few decades that will shape the technology options available after 2020. A robust hedging strategy focuses on generic technologies at the interface between energy supply and end use, including gas turbines, fuel cells, and photovoltaics. These could become as important as today's gasoline engines, electric motors, and microchips.

*Rates of change in global energy systems will remain slow*

Capital turnover rates in end-use applications are comparatively short – one to two decades. Therefore, pervasive changes can be implemented rather quickly, and missed opportunities may be revisited. Conversely, the lifetimes of energy supply technologies, and particularly of infrastructures, are five decades or longer. Thus, at most one or two replacements can occur during the time horizon of this study. Betting on the wrong horse will have serious, possibly irreversible consequences. The RD&D and investment decisions made now and in the immediate future will determine which long-term options become available after 2020 and which are foreclosed. Initiating long-term changes requires action sooner rather than later.

*Interconnectivity will enhance cooperation, systems flexibility, and resilience*

Despite energy globalization, *market exclusion* remains a serious challenge. To date, some two billion people do not have access to modern energy services due to poverty and a lack of energy infrastructures. Many regions are overly dependent on a single, locally available resource, such as traditional fuelwood or coal, and have limited access to the clean flexible energy forms required for economic and social development. Policies to deregulate markets and get "prices right" ignore the poor. Even the best functioning energy markets will not reach those who cannot pay. To include today's poor in energy markets, poverty must be eradicated, and that requires policy action that goes beyond energy policies alone. What energy policies can accomplish is the improvement of old infrastructures – the backbone of the energy system – and the development of new ones. New infrastructures are needed in Eurasia, in particular, to match the large available resources of oil and gas in the Caspian region and Siberia with the newly emerging centers of energy consumption in Asia. Extended interconnections are also needed in Latin America and Africa. Governments and industry need an expanded spirit of regional cooperation and a shared commitment to infrastructure investments now if benefits are to accrue in the long term.

## *Capital requirements will present major challenges for all energy strategies*

For all scenarios the capital requirements of the energy sector are large, but not infeasible. Although investment requirements expand more slowly than overall economic growth, the energy sector will have to raise an increasing fraction of its capital from the private sector. It will face stiffer competition and return-on-investment criteria than it has in the past. Moreover, the greatest investment needs are in the now developing world, where current trends in the availability of both international development capital and private investment capital are not auspicious. How available capital is best mobilized remains a critical issue.

## *Regional differences will persist in global energy systems*

Regional energy supply trends diverge across the scenarios even more than those at the global level due to differences in resource availability, trade possibilities, and national and regional development strategies. Regional development aspirations often exceed even the wide range of possibilities outlined by the six global scenarios analyzed here. Yet, for all their diversity, regional perspectives confirm the essential global conclusion: while a range of possible energy sources can fuel the future, there is a persistent trend toward cleaner, more convenient energy forms reaching the consumer. In all regions, success will depend on improved efficiency, continued technological progress, a favorable investment climate, free trade, and enhanced regional and international cooperation, particularly in shared energy infrastructures.

## *Local environmental impacts will take precedence over global change*

The natural capacity of the environment to absorb higher levels of pollution, particularly in densely populated metropolitan areas, will become the limiting factor for the unconstrained use of fossil fuels. Local environmental problems are of greater concern to local decision makers than global problems and therefore will have a greater near-term impact on policy. In the developing world indoor air pollution is an urgent environmental problem. A shift away from cooking with wood in open fireplaces will reduce indoor pollution levels currently estimated to be 20 times higher than in industrialized countries. A second urgent problem is the high concentration of particulate matter and $SO_2$ in many urban areas. Regional air pollution could also prove problematic, especially in the rapidly growing, densely populated coal-intensive economies of Asia. Without abatement measures, sulfur emissions could cause serious public health problems and subject key agricultural crops to acid deposition 10 times sustainable levels.

*Decarbonization will improve the environment at local, regional, and global levels*

The continuing shift to higher-quality fuels means a continuing reduction in the carbon content of fuels, that is, decarbonization of the energy system. Decarbonization means lower adverse environmental impacts per unit of energy consumed, independent of any active policies specifically designed to protect the environment. And at the global level, it translates directly into lower $CO_2$ emissions. But decarbonization is not enough – additional active policies will be required. In some cases energy and environmental policies are mutually reinforcing. Policies to reduce global $CO_2$ emissions, for example, also reduce acidification risks. In others, energy and environmental policies work at cross purposes. Restrictions on nuclear power, for example, mean a possibly greater dependence on fossil fuels, and vice versa. In all cases, however, more rapid technological improvement means quicker progress toward cleaner fuels, and cleaner fuels mean a cleaner environment.

## 8.4.    Outlook

This study developed six scenarios to explore how different energy futures might unfold in the 21st century. We recognize that no analysis can ever turn an uncertain future into a sure thing, and studies such as this are part of an ongoing process rather than a onetime activity. At all levels of the analysis there are next steps to be taken. In particular, global and regional scenarios cannot generate customized guidance for individual countries, industries, or firms. Continued, complementary, bottom-up analyses by those in the midst of regional and national energy policy are needed, as was demonstrated by the regional reviews of this study.

Therefore, while the two-phase, five-year research effort described in this book is complete, we list here open questions suggested by our analysis that might productively serve as the focus for the next generation of energy analyses.

First, while the six scenarios cover a wide range of possible futures, they purposely exclude the high and low extremes of the literature on energy scenarios and projections. It might be desirable to go beyond the three cases presented here and analyze the conditions, possibilities, and implications of greater economic growth, particularly in the South. The challenge of including in energy markets the two billion poor who are now excluded certainly deserves attention. Second, the analysis could be extended beyond final energy in the direction of energy services. Pursuing the analysis only as far as end use does not allow a detailed examination of approaches labeled variously as demand-side management, integrated investment planning, and least-cost energy strategies. We recognize that at the global level even improved long-term assessments can provide only very broadbrush insights into such approaches. Nonetheless, a major message of this study, with its emphasis on end use, is that future research has much to gain by focusing more on energy services.

Third, more analysis is needed of new potential institutional mechanisms to answer the key question of where the financing for required investments is going to come from. Recent declines in RD&D budgets need to be reversed, and analysis is needed to identify the most effective mechanisms and incentives for enhancing technology diffusion and transfer. The possible mismatch between the short-term priority given to fiscal discipline and profitability and the long-term benefits of greater near-term, but forward-looking, investments needs continued attention. The analysis of the macroeconomic implications of policies, such as in the ecologically driven Case C scenarios, that involve substantial taxation and income redistribution – nationally and internationally – needs to be continually refined and updated. The bottom line is that the evolution of the energy system, its main driving forces, and their implications can only be analyzed by adopting a truly long-term perspective. Analytical skills and data need to be complemented by creative imagination if they are to produce visions of a more prosperous and equitable future.

## 8.5.    Alternative Images

Our current perceptions and beliefs determine how we envision the world in 30, 60, or even 100 years. The Mercator projection of the world[1] originally enabled European sailors to discover the world's oceans and distant continents. Now it represents to many a familiar world view, perhaps unduly "Eurocentric." Another view is the equal area geographical projection of the world shown in *Figure 1.1* in *Box 1.1*. Such a projection gives a more accurate representation of the geographical size of different continents.

*Figures 8.1* and *8.2* offer an alternative and unconventional image of economic and energy development in the world. Economic maps in *Figure 8.1* show the land area of the 11 world regions weighted by their relative shares in global GDP expressed in terms of market exchange rates. Energy maps in *Figure 8.2* show the land area of the 11 world regions weighted by their relative shares in global primary energy. The maps in *Figures 8.1* and *8.2* are normalized to have the same total global area in 2100.[2] The growth in each region's size from 1990 to 2100

---

[1]The Mercator projection uses straight lines to indicate latitude and longitude. It is especially convenient for navigational purposes and its use is widespread, but it has the disadvantage that, for high latitudes, it greatly exaggerates land area.

[2]The maps only compare regional areas. The exceptions are that the relative sizes of Japan, Australia, and New Zealand have been altered to reflect their relative shares in economic and energy activities within their regions.

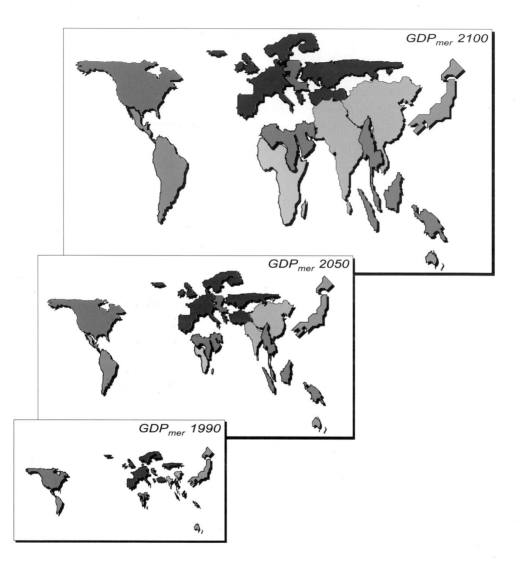

**Figure 8.1**: The changing geography of economic wealth, Case B. Areas of world regions are weighted by their relative shares in 1990, 2050, and 2100.

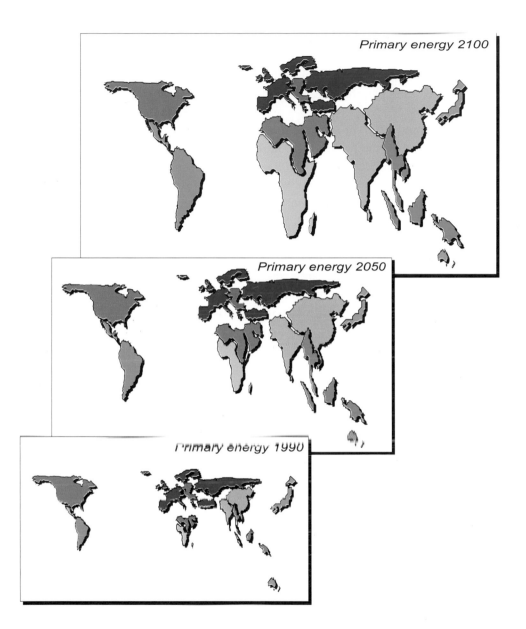

**Figure 8.2**: The changing geography of primary energy, Case B. Areas of world regions are weighted by their relative shares in 1990, 2050, and 2100.

illustrates how much its GDP (*Figure 8.1*) and its energy use (*Figure 8.2*) grow during the period.

In 1990, both the economic map of the world and the energy map look highly distorted as a result of current disparities among regions. Most developing regions are barely discernible compared with Japan, Western Europe, and North America. Compare, for example, the size of Japan in 1990 with that of China or the Indian subcontinent.

For 2050 and 2100, the economic and energy maps shown in the figures correspond to Case B, the middle case presented in this study. In both cases, they look much less distorted by 2100. Not only are regions larger, as their economies and energy use grow, but disparities among the regions are fewer, bringing the economic and energy maps much closer to the geographic map with which we are so familiar.

An essential feature of all six scenarios is the move toward a more balanced global distribution of income and increased access to energy services. This process requires time, wise policy guidance, the resolution of currently divergent policies on economic and social development, and the preservation of the environment for future generations. Energy will play a central role in promoting development, and the energy sector will be a major stakeholder in the process. Expanding and improving access to energy services, reducing present disparities, and improving the quality of energy services will remain high-priority objectives for the energy industry throughout the 21st century. Only with continued concerted efforts will it be possible to shape the global distribution of economic activities and energy use to become more equitable while preserving the habitability of the earth.

# Appendices

**Appendix A: Methodology**

Reproducibility of results adds to their validity. This is especially the case when presenting long-term scenarios for which there are no established, rigorous validity criteria. A necessary – though not sufficient – condition for reproducibility is a good description of the methods that produced the results. This appendix presents a summary description of the methods used in this study and references sources where more complete descriptions can be found. *Figure A.1* illustrates the IIASA modeling framework for scenario formulation and analysis.

We started with two principal exogenous variables: population growth by region, and per capita economic growth by region (see Chapter 3). Levels of energy demand were derived using a model developed at IIASA labeled simply Scenario Generator (SG). It combines extensive historical data on national economies and their energy systems with empirically estimated equations of past economic and energy developments. For each scenario, the SG generated plausible future paths of energy use consistent with historical data and with the specific features that were specified for the scenarios, for example, high or moderate economic growth, rapid or more gradual energy intensity improvements.

Two other models were then used iteratively for testing the consistency of each scenario. A model of energy–economy interactions called 11R (Schrattenholzer and Schäfer, 1996) was used to check for consistency between a region's macroeconomic development and its energy use. 11R is a modified version of Global 2100, originally published in 1992 (Manne and Richels, 1992) and subsequently used widely in energy studies throughout the world. IIASA's energy systems engineering model MESSAGE III (Messner and Strubegger, 1994, 1995) estimated detailed energy demand and supply patterns consistent with the evolution of primary and final energy consumption produced by the SG. MESSAGE III is a dynamic linear programming model, calculating cost-minimal supply structures under the constraints of resource availability, the menu of given technologies, and the demand for useful energy. The two models are used in tandem because they correspond to the two different perspectives from which energy modeling is usually done: top-down (11R) and bottom-up (MESSAGE III). The general model-linking methodology is described by Wene (1995). Its specific application in this study is described by Schrattenholzer and Schäfer (1996). For each scenario, MESSAGE III and 11R were harmonized for a number of key variables including total primary energy, cumulative consumption of primary fuels, carbon emissions, cumulative carbon emissions, and total electricity demand. Harmonization was achieved by iteratively adjusting useful-energy inputs in the case of MESSAGE III and GDP growth and energy intensity reduction rates in the case of 11R.

The acidification estimates presented in Chapter 6 were calculated using IIASA's RAINS model (Alcamo *et al.*, 1990). RAINS is a simulation model of $SO_2$

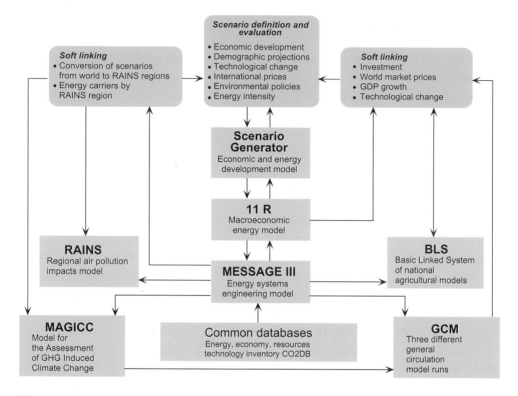

**Figure A.1.** IIASA modeling framework for scenario formulation and analysis. Formal models and databases are shown in light red rectangles; model links and the main scenario assumptions are shown in green rectangles.

and $NO_x$ emissions, and their subsequent atmospheric transport, chemical transformations of those emissions, deposition, and ecological impacts. The last of these are calculated based on a spatial resolution of grid cells of 150 km by 150 km. The atmospheric concentrations and the potential warming that might result from the scenario's net carbon emissions were calculated using MAGICC, a carbon cycle and climate change model developed by Wigley *et al.* (1993, 1994).

The impacts of energy biomass production on land use and potential conflicts with food production were calculated using IIASA's Basic Linked System (BLS) of national agricultural models (Fischer *et al.*, 1988, 1994). BLS is a sectoral macroeconomic model that accounts for all major inputs (land, fertilizer, capital, and labor) required for the production of 11 agricultural commodities. Economic growth rates in BLS are harmonized with those in 11R by adjusting production factors and rates of technological progress in BLS. Acidification impacts on Asian agriculture (Chapter 6) were also based on the BLS model, drawing on the methodology described in Fischer and Rosenzweig (1996).

# Appendix B: Definition of World Regions

## North America (NAM)

Canada

Puerto Rico

Virgin Islands

Guam

United States of America

## Latin America and the Caribbean (LAM)

Antigua and Barbuda

Bahamas

Belize

Bolivia

Chile

Costa Rica

Dominica

Ecuador

French Guyana

Guadeloupe

Guyana

Honduras

Martinique

Netherlands Antilles

Panama

Peru

Santa Lucia

Suriname

Uruguay

Argentina

Barbados

Bermuda

Brazil

Colombia

Cuba

Dominican Republic

El Salvador

Grenada

Guatemala

Haiti

Jamaica

Mexico

Nicaragua

Paraguay

Saint Kitts and Nevis

Saint Vincent and the Grenadines

Trinidad and Tobago

Venezuela

## Sub-Saharan Africa (AFR)

| | |
|---|---|
| Angola | Benin |
| Botswana | British Indian Ocean Territory |
| Burkina Faso | Burundi |
| Cameroon | Cape Verde |
| Central African Republic | Chad |
| Comoros | Côte d'Ivoire |
| Congo | Djibouti |
| Equatorial Guinea | Eritrea |
| Ethiopia | Gabon |
| Gambia | Ghana |
| Guinea | Guinea-Bissau |
| Kenya | Lesotho |
| Liberia | Madagascar |
| Malawi | Mali |
| Mauritania | Mauritius |
| Mozambique | Namibia |
| Niger | Nigeria |
| Reunion | Rwanda |
| Sao Tome and Principe | Senegal |
| Seychelles | Sierra Leone |
| Somalia | South Africa |
| Saint Helena | Swaziland |
| Tanzania | Togo |
| Uganda | Zaire |
| Zambia | Zimbabwe |

## Middle East and North Africa (MEA)

| | |
|---|---|
| Algeria | Bahrain |
| Egypt (Arab Republic) | Iraq |
| Iran (Islamic Republic) | Israel |
| Jordan | Kuwait |
| Lebanon | Libya/SPLAJ |
| Morocco | Oman |
| Qatar | Saudi Arabia |
| Sudan | Syria (Arab Republic) |
| Tunisia | United Arab Emirates |
| Yemen | |

## Western Europe (WEU)

| | |
|---|---|
| Andorra | Austria |
| Azores | Belgium |
| Canary Islands | Channel Islands |
| Cyprus | Denmark |
| Faeroe Islands | Finland |
| France | Germany |
| Gibraltar | Greece |
| Greenland | Iceland |
| Ireland | Isle of Man |
| Italy | Liechtenstein |
| Luxembourg | Madeira |
| Malta | Monaco |
| Netherlands | Norway |
| Portugal | Spain |
| Sweden | Switzerland |
| Turkey | United Kingdom |

## Central and Eastern Europe (EEU)

| | |
|---|---|
| Albania | Bosnia and Herzegovina |
| Bulgaria | Croatia |
| Czech Republic | The former Yugoslav Rep. of Macedonia |
| Hungary | Poland |
| Romania | Slovak Republic |
| Slovenia | Yugoslavia |

## Newly independent states of the former Soviet Union (FSU)

| | |
|---|---|
| Armenia | Azerbaijan |
| Belarus | Estonia |
| Georgia | Kazakhstan |
| Kyrgyzstan | Latvia |
| Lithuania | Republic of Moldova |
| Russian Federation | Tajikistan |
| Turkmenistan | Ukraine |
| Uzbekistan | |

## Centrally planned Asia and China (CPA)

Cambodia                China
Hong Kong               Korea (DPR)
Lao (PDR)               Mongolia
Viet Nam

## South Asia (SAS)

Afghanistan             Bangladesh
Bhutan                  India
Maldives                Nepal
Pakistan                Sri Lanka

## Other Pacific Asia (PAS)

American Samoa          Brunei Darussalam
Fiji                    French Polynesia
Gilbert-Kiribati        Indonesia
Malaysia                Myanmar
New Caledonia           Papua New Guinea
Philippines             Republic of Korea
Singapore               Solomon Islands
Taiwan, China           Thailand
Tonga                   Vanuatu
Western Samoa

## Pacific OECD (PAO)

Australia               Japan
New Zealand

**Appendix C: Summary of Six Scenarios, 1990, 2020, and 2050**

# World

| | 1990 | 2020 | | | | | | 2050 | | | | | |
|---|---|---|---|---|---|---|---|---|---|---|---|---|---|
| | | A1 | A2 | A3 | B | C1 | C2 | A1 | A2 | A3 | B | C1 | C2 |
| Population, $10^9$ | 5.26 | 7.92 | 7.92 | 7.92 | 7.92 | 7.92 | 7.92 | 10.06 | 10.06 | 10.06 | 10.06 | 10.06 | 10.06 |
| GDP$_{ppp}$, US(1990)\$, $10^{12}$ | 25.7 | 56.8 | 56.8 | 56.8 | 49.3 | 50.8 | 50.8 | 115.0 | 115.0 | 115.0 | 87.8 | 91.3 | 91.3 |
| GDP$_{mer}$, US(1990)\$, $10^{12}$ | 20.9 | 46.9 | 46.9 | 46.9 | 40.2 | 40.5 | 40.5 | 101.5 | 101.5 | 101.5 | 72.8 | 75.0 | 75.0 |
| Agriculture | 1.4 | 3.1 | 3.1 | 3.1 | 2.6 | 2.9 | 2.9 | 4.9 | 4.9 | 4.9 | 4.1 | 4.4 | 4.4 |
| Industry | 7.0 | 14.5 | 14.5 | 14.5 | 12.8 | 13.3 | 13.3 | 33.2 | 33.2 | 33.2 | 24.7 | 27.2 | 27.2 |
| Services | 12.4 | 29.3 | 29.3 | 29.3 | 24.8 | 24.3 | 24.3 | 63.4 | 63.4 | 63.4 | 44.0 | 43.4 | 43.4 |
| Final energy, Gtoe | 6.45 | 11.41 | 11.40 | 11.33 | 10.07 | 8.55 | 8.54 | 17.01 | 17.47 | 17.17 | 14.18 | 10.05 | 9.90 |
| Solids | 1.93 | 2.65 | 2.79 | 2.72 | 2.61 | 2.37 | 2.35 | 2.66 | 3.34 | 3.14 | 3.20 | 1.96 | 1.96 |
| Liquids | 2.53 | 4.41 | 4.31 | 4.10 | 3.53 | 2.78 | 2.80 | 7.22 | 6.27 | 5.62 | 4.71 | 3.42 | 3.37 |
| Electricity | 0.83 | 1.63 | 1.69 | 1.72 | 1.45 | 1.22 | 1.21 | 2.88 | 3.14 | 3.03 | 2.34 | 1.79 | 1.72 |
| Other$^a$ | 1.16 | 2.72 | 2.62 | 2.80 | 2.49 | 2.18 | 2.17 | 4.25 | 4.72 | 5.37 | 3.93 | 2.88 | 2.84 |
| Primary energy, Gtoe | 8.98 | 15.38 | 15.37 | 15.36 | 13.55 | 11.43 | 11.43 | 24.83 | 24.84 | 24.66 | 19.83 | 14.25 | 14.25 |
| Coal | 2.18 | 3.71 | 4.31 | 2.91 | 3.39 | 2.29 | 2.28 | 3.79 | 7.83 | 2.24 | 4.14 | 1.50 | 1.47 |
| Oil | 3.06 | 4.66 | 4.50 | 4.26 | 3.78 | 3.02 | 3.02 | 7.90 | 4.78 | 4.33 | 4.04 | 2.67 | 2.62 |
| Gas | 1.68 | 3.62 | 3.41 | 3.84 | 3.18 | 3.06 | 2.96 | 4.70 | 5.46 | 7.91 | 4.50 | 3.92 | 3.34 |
| Nuclear | 0.45 | 0.91 | 0.58 | 1.03 | 0.90 | 0.67 | 0.85 | 2.90 | 1.09 | 2.82 | 2.74 | 0.52 | 1.77 |
| Renewables | 1.60 | 2.47 | 2.57 | 3.31 | 2.29 | 2.39 | 2.32 | 5.54 | 5.68 | 7.35 | 4.42 | 5.63 | 5.05 |
| Net carbon emissions, GtC$^b$ | 5.93 | 9.40 | 9.91 | 8.18 | 8.26 | 6.34 | 6.30 | 11.62 | 14.67 | 9.29 | 9.57 | 5.34 | 5.11 |
| Gross carbon emissions, GtC | 6.23 | 10.28 | 10.66 | 9.21 | 8.91 | 7.00 | 6.92 | 13.75 | 16.00 | 11.12 | 10.76 | 6.38 | 5.93 |
| Sulfur emissions, MtS | 59 | 53 | 61 | 51 | 59 | 34 | 34 | 54 | 64 | 45 | 55 | 22 | 22 |
| Cumulative energy investments | | | | | | | | | | | | | |
| 1990–2020, US(1990)\$, $10^{12}$ | | 14 | 15 | 16 | 12 | 10 | 9 | | | | | | |
| 2021–2050, US(1990)\$, $10^{12}$ | | | | | | | | 23 | 31 | 25 | 22 | 14 | 14 |

Note: Subtotals may not add due to independent rounding.
$^a$District heat, gas, and hydrogen.
$^b$Net carbon emissions do not include feedstocks and other non-energy emissions or carbon dioxide used for enhanced oil recovery.

# OECD

| | 1990 | 2020 | | | | | | 2050 | | | | | |
|---|---|---|---|---|---|---|---|---|---|---|---|---|---|
| | | A1 | A2 | A3 | B | C1 | C2 | A1 | A2 | A3 | B | C1 | C2 |
| Population, $10^9$ | 0.86 | 0.99 | 0.99 | 0.99 | 0.99 | 0.99 | 0.99 | 1.01 | 1.01 | 1.01 | 1.01 | 1.01 | 1.01 |
| GDP$_{ppp}$, US(1990)\$, $10^{12}$ | 14.1 | 27.6 | 27.6 | 27.6 | 25.0 | 23.2 | 23.2 | 44.7 | 44.7 | 44.7 | 36.6 | 32.1 | 32.1 |
| GDP$_{mer}$, US(1990)\$, $10^{12}$ | 16.4 | 31.6 | 31.6 | 31.6 | 28.6 | 26.6 | 26.6 | 51.1 | 51.1 | 51.1 | 41.7 | 36.7 | 36.7 |
| Agriculture | 0.6 | 0.8 | 0.8 | 0.8 | 0.8 | 0.8 | 0.8 | 0.8 | 0.8 | 0.8 | 0.8 | 0.8 | 0.8 |
| Industry | 5.5 | 8.7 | 8.7 | 8.7 | 8.3 | 8.0 | 8.0 | 11.1 | 11.1 | 11.1 | 10.2 | 9.7 | 9.7 |
| Services | 10.3 | 22.1 | 22.1 | 22.1 | 19.5 | 17.9 | 17.9 | 39.2 | 39.2 | 39.2 | 30.7 | 26.2 | 26.2 |
| Final energy, Gtoe | 2.88 | 3.84 | 3.83 | 3.83 | 3.49 | 2.45 | 2.44 | 3.90 | 4.15 | 4.03 | 3.35 | 1.83 | 1.69 |
| Solids | 0.31 | 0.10 | 0.15 | 0.08 | 0.16 | 0.10 | 0.09 | 0.01 | 0.02 | 0.01 | 0.02 | 0.01 | 0.01 |
| Liquids | 1.50 | 1.63 | 1.58 | 1.49 | 1.36 | 0.90 | 0.89 | 1.75 | 1.27 | 1.21 | 1.07 | 0.55 | 0.52 |
| Electricity | 0.51 | 0.91 | 0.95 | 0.95 | 0.85 | 0.63 | 0.63 | 1.21 | 1.39 | 1.27 | 1.10 | 0.68 | 0.63 |
| Other$^a$ | 0.55 | 1.19 | 1.16 | 1.31 | 1.13 | 0.82 | 0.83 | 0.93 | 1.47 | 1.54 | 1.16 | 0.59 | 0.54 |
| Primary energy, Gtoe | 4.18 | 5.71 | 5.71 | 5.72 | 5.25 | 3.73 | 3.73 | 6.67 | 6.69 | 6.68 | 5.59 | 3.02 | 3.03 |
| Coal | 0.91 | 1.20 | 1.50 | 0.76 | 1.24 | 0.54 | 0.49 | 1.06 | 2.32 | 0.12 | 0.95 | 0.08 | 0.05 |
| Oil | 1.72 | 1.72 | 1.64 | 1.55 | 1.41 | 0.97 | 0.96 | 1.82 | 0.96 | 0.77 | 0.93 | 0.44 | 0.42 |
| Gas | 0.79 | 1.56 | 1.48 | 1.81 | 1.41 | 1.14 | 1.14 | 1.32 | 1.71 | 2.50 | 1.39 | 1.07 | 0.60 |
| Nuclear | 0.36 | 0.68 | 0.50 | 0.72 | 0.66 | 0.55 | 0.62 | 1.55 | 0.50 | 1.58 | 1.45 | 0.38 | 0.98 |
| Renewables | 0.40 | 0.55 | 0.59 | 0.89 | 0.52 | 0.53 | 0.52 | 0.93 | 1.21 | 1.73 | 0.88 | 1.06 | 0.98 |
| Net carbon emissions, GtC$^b$ | 2.82 | 3.53 | 3.79 | 3.02 | 3.25 | 1.97 | 1.91 | 3.21 | 4.27 | 2.04 | 2.52 | 0.91 | 0.70 |
| Gross carbon emissions, GtC | 3.03 | 3.76 | 3.97 | 3.30 | 3.45 | 2.14 | 2.08 | 3.52 | 4.42 | 2.37 | 2.70 | 1.14 | 0.79 |
| Sulfur emissions, MtS | 21 | 4 | 10 | 4 | 11 | 3 | 3 | 3 | 2 | 1 | 2 | 1 | 1 |
| Cumulative energy investments | | | | | | | | | | | | | |
| 1990–2020, US(1990)\$, $10^{12}$ | | 6.76 | 7.60 | 7.23 | 6.54 | 4.12 | 4.00 | | | | | | |
| 2021–2050, US(1990)\$, $10^{12}$ | | | | | | | | 8.17 | 11.00 | 7.47 | 8.54 | 3.40 | 3.53 |

Note: Subtotals may not add due to independent rounding.

$^a$ District heat, gas, and hydrogen.

$^b$ Net carbon emissions do not include feedstocks and other non-energy emissions or carbon dioxide used for enhanced oil recovery.

## Reforming economies (REFs)

| | 1990 | 2020 | | | | | | 2050 | | | | | |
|---|---|---|---|---|---|---|---|---|---|---|---|---|---|
| | | A1 | A2 | A3 | B | C1 | C2 | A1 | A2 | A3 | B | C1 | C2 |
| Population, $10^9$ | 0.41 | 0.48 | 0.48 | 0.48 | 0.48 | 0.48 | 0.48 | 0.54 | 0.54 | 0.54 | 0.54 | 0.54 | 0.54 |
| GDP$_{ppp}$, US(1990)\$, $10^{12}$ | 2.6 | 2.2 | 2.2 | 2.2 | 2.0 | 2.3 | 2.3 | 8.1 | 8.1 | 8.1 | 5.1 | 5.1 | 5.1 |
| GDP$_{mer}$, US(1990)\$, $10^{12}$ | 1.1 | 1.7 | 1.7 | 1.7 | 1.4 | 1.5 | 1.5 | 7.9 | 7.9 | 7.9 | 4.1 | 3.9 | 3.9 |
| Agriculture | 0.1 | 0.1 | 0.1 | 0.1 | 0.1 | 0.1 | 0.1 | 0.4 | 0.4 | 0.4 | 0.3 | 0.3 | 0.3 |
| Industry | 0.5 | 0.7 | 0.7 | 0.7 | 0.6 | 0.7 | 0.7 | 3.1 | 3.1 | 3.1 | 1.8 | 1.8 | 1.8 |
| Services | 0.5 | 0.9 | 0.9 | 0.9 | 0.6 | 0.7 | 0.7 | 4.4 | 4.4 | 4.4 | 1.9 | 1.9 | 1.9 |
| Final energy, Gtoe | 1.20 | 1.80 | 1.81 | 1.79 | 1.29 | 1.31 | 1.31 | 2.84 | 2.90 | 2.83 | 1.80 | 1.28 | 1.27 |
| Solids | 0.22 | 0.22 | 0.24 | 0.20 | 0.17 | 0.15 | 0.15 | 0.05 | 0.05 | 0.05 | 0.07 | 0.05 | 0.05 |
| Liquids | 0.36 | 0.62 | 0.60 | 0.58 | 0.34 | 0.35 | 0.35 | 1.29 | 1.17 | 1.02 | 0.58 | 0.38 | 0.38 |
| Electricity | 0.14 | 0.22 | 0.23 | 0.23 | 0.17 | 0.16 | 0.16 | 0.46 | 0.46 | 0.45 | 0.27 | 0.19 | 0.19 |
| Other$^a$ | 0.48 | 0.74 | 0.74 | 0.78 | 0.62 | 0.65 | 0.64 | 1.03 | 1.22 | 1.31 | 0.89 | 0.66 | 0.66 |
| Primary energy, Gtoe | 1.74 | 2.28 | 2.28 | 2.29 | 1.66 | 1.68 | 1.69 | 3.72 | 3.73 | 3.73 | 2.38 | 1.67 | 1.67 |
| Coal | 0.44 | 0.47 | 0.49 | 0.30 | 0.37 | 0.23 | 0.27 | 0.61 | 0.65 | 0.10 | 0.37 | 0.05 | 0.05 |
| Oil | 0.49 | 0.63 | 0.62 | 0.62 | 0.38 | 0.40 | 0.40 | 1.33 | 1.17 | 0.98 | 0.57 | 0.38 | 0.38 |
| Gas | 0.64 | 0.93 | 0.93 | 0.98 | 0.70 | 0.87 | 0.79 | 0.79 | 1.36 | 1.86 | 0.97 | 0.93 | 0.89 |
| Nuclear | 0.06 | 0.09 | 0.05 | 0.10 | 0.08 | 0.05 | 0.10 | 0.47 | 0.12 | 0.23 | 0.24 | 0.02 | 0.09 |
| Renewables | 0.11 | 0.17 | 0.18 | 0.29 | 0.13 | 0.13 | 0.13 | 0.52 | 0.45 | 0.56 | 0.22 | 0.28 | 0.25 |
| Net carbon emissions, GtC$^b$ | 1.31 | 1.48 | 1.51 | 1.25 | 1.08 | 1.01 | 1.01 | 1.97 | 2.34 | 1.73 | 1.35 | 0.85 | 0.82 |
| Gross carbon emissions, GtC | 1.35 | 1.64 | 1.67 | 1.48 | 1.18 | 1.15 | 1.14 | 2.30 | 2.56 | 2.12 | 1.51 | 0.97 | 0.94 |
| Sulfur emissions, MtS | 15 | 7 | 8 | 7 | 7 | 5 | 5 | 12 | 7 | 3 | 5 | 2 | 2 |
| Cumulative energy investments | | | | | | | | | | | | | |
| 1990–2020, US(1990)\$, $10^{12}$ | | 2.58 | 2.61 | 2.72 | 1.91 | 1.66 | 1.68 | | | | | | |
| 2021–2050, US(1990)\$, $10^{12}$ | | | | | | | | 3.99 | 4.78 | 3.31 | 3.17 | 1.70 | 1.62 |

Note: Subtotals may not add due to independent rounding.
$^a$District heat, gas, and hydrogen.
$^b$Net carbon emissions do not include feedstocks and other non-energy emissions or carbon dioxide used for enhanced oil recovery.

# Developing countries (DCs)

| | 1990 | 2020 A1 | 2020 A2 | 2020 A3 | 2020 B | 2020 C1 | 2020 C2 | 2050 A1 | 2050 A2 | 2050 A3 | 2050 B | 2050 C1 | 2050 C2 |
|---|---|---|---|---|---|---|---|---|---|---|---|---|---|
| Population, $10^9$ | 3.99 | 6.45 | 6.45 | 6.45 | 6.45 | 6.45 | 6.45 | 8.51 | 8.51 | 8.51 | 8.51 | 8.51 | 8.51 |
| GDP$_{ppp}$, US(1990)\$, $10^{12}$ | 9.1 | 26.9 | 26.9 | 26.9 | 22.3 | 25.4 | 25.4 | 62.1 | 62.1 | 62.1 | 46.0 | 54.1 | 54.1 |
| GDP$_{mer}$, US(1990)\$, $10^{12}$ | 5.4 | 13.5 | 13.5 | 13.5 | 10.3 | 12.3 | 12.3 | 42.5 | 42.5 | 42.5 | 27.0 | 34.4 | 34.4 |
| Agriculture | 0.7 | 2.2 | 2.2 | 2.2 | 1.7 | 2.0 | 2.0 | 3.7 | 3.7 | 3.7 | 3.1 | 3.4 | 3.4 |
| Industry | 1.0 | 5.0 | 5.0 | 5.0 | 3.9 | 4.7 | 4.7 | 19.0 | 19.0 | 19.0 | 12.6 | 15.7 | 15.7 |
| Services | 1.7 | 6.3 | 6.3 | 6.3 | 4.6 | 5.6 | 5.6 | 19.8 | 19.8 | 19.8 | 11.4 | 15.3 | 15.3 |
| Final energy, Gtoe | 2.37 | 5.77 | 5.76 | 5.72 | 5.29 | 4.79 | 4.79 | 10.27 | 10.42 | 10.32 | 9.03 | 6.94 | 6.93 |
| Solids | 1.39 | 2.33 | 2.41 | 2.44 | 2.28 | 2.11 | 2.11 | 2.59 | 3.27 | 3.08 | 3.11 | 1.90 | 1.90 |
| Liquids | 0.67 | 2.16 | 2.13 | 2.03 | 1.83 | 1.54 | 1.55 | 4.18 | 3.84 | 3.39 | 3.07 | 2.49 | 2.48 |
| Electricity | 0.18 | 0.50 | 0.51 | 0.54 | 0.43 | 0.43 | 0.43 | 1.21 | 1.29 | 1.32 | 0.97 | 0.92 | 0.91 |
| Other[a] | 0.13 | 0.78 | 0.72 | 0.71 | 0.75 | 0.71 | 0.70 | 2.29 | 2.02 | 2.52 | 1.88 | 1.63 | 1.65 |
| Primary energy, Gtoe | 3.06 | 7.39 | 7.38 | 7.35 | 6.65 | 6.01 | 6.01 | 14.44 | 14.42 | 14.25 | 11.86 | 9.56 | 9.56 |
| Coal | 0.82 | 2.05 | 2.31 | 1.85 | 1.78 | 1.52 | 1.52 | 2.12 | 4.87 | 2.03 | 2.82 | 1.37 | 1.38 |
| Oil | 0.85 | 2.31 | 2.23 | 2.09 | 1.99 | 1.65 | 1.66 | 4.75 | 2.66 | 2.58 | 2.54 | 1.85 | 1.81 |
| Gas | 0.26 | 1.14 | 1.00 | 1.05 | 1.08 | 1.05 | 1.03 | 2.59 | 2.40 | 3.56 | 2.14 | 1.92 | 1.85 |
| Nuclear | 0.03 | 0.14 | 0.02 | 0.22 | 0.16 | 0.07 | 0.13 | 0.89 | 0.47 | 1.02 | 1.05 | 0.13 | 0.70 |
| Renewables | 1.09 | 1.76 | 1.81 | 2.13 | 1.64 | 1.72 | 1.67 | 4.09 | 4.02 | 5.07 | 3.32 | 4.30 | 3.82 |
| Net carbon emissions, GtC[b] | 1.80 | 4.39 | 4.61 | 3.92 | 3.92 | 3.36 | 3.38 | 6.45 | 8.06 | 5.52 | 5.70 | 3.59 | 3.59 |
| Gross carbon emissions, GtC | 1.85 | 4.88 | 5.02 | 4.43 | 4.28 | 3.70 | 3.70 | 7.93 | 9.02 | 6.63 | 6.55 | 4.26 | 4.20 |
| Sulfur emissions, MtS | 22 | 42 | 43 | 41 | 41 | 27 | 27 | 39 | 55 | 41 | 48 | 19 | 19 |
| Cumulative energy investments 1990–2020, US(1990)\$, $10^{12}$ | | 4.55 | 5.10 | 5.85 | 3.92 | 3.74 | 3.71 | | | | | | |
| 2021–2050, US(1990)\$, $10^{12}$ | | | | | | | | 11.28 | 15.53 | 13.89 | 10.55 | 9.02 | 8.96 |

Note: Subtotals may not add due to independent rounding.
[a] District heat, gas, and hydrogen.
[b] Net carbon emissions do not include feedstocks and other non-energy emissions or carbon dioxide used for enhanced oil recovery.

## North America (NAM)

| | 1990 | 2020 A1 | 2020 A2 | 2020 A3 | 2020 B | 2020 C1 | 2020 C2 | 2050 A1 | 2050 A2 | 2050 A3 | 2050 B | 2050 C1 | 2050 C2 |
|---|---|---|---|---|---|---|---|---|---|---|---|---|---|
| Population, $10^6$ | 280 | 340 | 340 | 340 | 340 | 340 | 340 | 360 | 360 | 360 | 360 | 360 | 360 |
| GDP$_{ppp}$, US(1990)\$, $10^9$ | 5913 | 12129 | 12129 | 12129 | 11109 | 10066 | 10066 | 19739 | 19739 | 19739 | 16612 | 14055 | 14055 |
| GDP$_{mer}$, US(1990)\$, $10^9$ | 6067 | 12146 | 12146 | 12146 | 11126 | 10084 | 10084 | 19741 | 19741 | 19741 | 16614 | 14057 | 14057 |
| Agriculture | 127 | 219 | 219 | 219 | 213 | 206 | 206 | 297 | 297 | 297 | 286 | 278 | 278 |
| Industry | 1809 | 3075 | 3075 | 3075 | 2947 | 2792 | 2792 | 4110 | 4110 | 4110 | 3847 | 3592 | 3592 |
| Services | 4131 | 8852 | 8852 | 8852 | 7967 | 7086 | 7086 | 15334 | 15334 | 15334 | 12481 | 10187 | 10187 |
| Final energy, Mtoe | 1496 | 2023 | 2033 | 2045 | 1853 | 1198 | 1199 | 2060 | 2193 | 2198 | 1806 | 792 | 766 |
| Solids | 132 | 57 | 103 | 35 | 97 | 59 | 53 | 4 | 13 | 3 | 15 | 6 | 4 |
| Liquids | 772 | 876 | 868 | 803 | 745 | 459 | 457 | 893 | 734 | 693 | 628 | 255 | 246 |
| Electricity | 262 | 480 | 491 | 489 | 442 | 303 | 300 | 633 | 707 | 670 | 565 | 296 | 281 |
| Other[a] | 331 | 610 | 571 | 718 | 569 | 377 | 389 | 530 | 739 | 833 | 598 | 235 | 234 |
| Primary energy, Mtoe | 2183 | 3026 | 3024 | 3029 | 2815 | 1827 | 1827 | 3535 | 3534 | 3533 | 3049 | 1348 | 1348 |
| Coal | 481 | 780 | 934 | 409 | 809 | 303 | 248 | 847 | 1635 | 68 | 730 | 25 | 11 |
| Oil | 834 | 899 | 905 | 854 | 781 | 488 | 483 | 879 | 665 | 566 | 623 | 266 | 261 |
| Gas | 495 | 812 | 738 | 1011 | 690 | 545 | 562 | 730 | 781 | 1698 | 609 | 400 | 313 |
| Nuclear | 153 | 258 | 163 | 268 | 258 | 212 | 258 | 705 | 118 | 420 | 737 | 205 | 366 |
| Renewables | 220 | 278 | 284 | 487 | 277 | 279 | 277 | 374 | 335 | 781 | 350 | 451 | 397 |
| Net carbon emissions, MtC[b] | 1490 | 2030 | 2170 | 1690 | 1880 | 1020 | 970 | 1930 | 2740 | 1400 | 1610 | 400 | 390 |
| Gross carbon emissions, MtC | 1600 | 2130 | 2260 | 1820 | 1990 | 1100 | 1040 | 2120 | 2830 | 1640 | 1700 | 510 | 430 |
| Sulfur emissions, MtS | 11 | 2 | 6 | 2 | 7 | 1 | 1 | 1 | 2 | 1 | 1 | 0 | 0 |
| Cumulative energy investments | | | | | | | | | | | | | |
| 1990–2020, US(1990)\$, $10^9$ | | 3670 | 3950 | 3800 | 3520 | 2000 | 1950 | | | | | | |
| 2021–2050, US(1990)\$, $10^9$ | | | | | | | | 4580 | 6050 | 3830 | 4890 | 1600 | 1540 |

Note: Subtotals may not add due to independent rounding.
[a] District heat, gas, and hydrogen.
[b] Net carbon emissions do not include feedstocks and other non-energy emissions or carbon dioxide used for enhanced oil recovery.

**Latin America and the Caribbean (LAM)**

| | 1990 | 2020 | | | | | | 2050 | | | | | |
|---|---|---|---|---|---|---|---|---|---|---|---|---|---|
| | | A1 | A2 | A3 | B | C1 | C2 | A1 | A2 | A3 | B | C1 | C2 |
| Population, $10^6$ | 430 | 670 | 670 | 670 | 670 | 670 | 670 | 840 | 840 | 840 | 840 | 840 | 840 |
| GDP$_{PPP}$, US(1990)$, $10^9$ | 2025 | 4310 | 4310 | 4310 | 4189 | 4262 | 4262 | 8648 | 8648 | 8648 | 7760 | 7990 | 7990 |
| GDP$_{mer}$, US(1990)$, $10^9$ | 1085 | 2718 | 2718 | 2718 | 2603 | 2672 | 2672 | 6989 | 6989 | 6989 | 5930 | 6198 | 6198 |
| Agriculture | 97 | 215 | 215 | 215 | 216 | 219 | 219 | 481 | 481 | 481 | 454 | 461 | 461 |
| Industry | 352 | 1049 | 1049 | 1049 | 1019 | 1042 | 1042 | 3128 | 3128 | 3128 | 2719 | 2825 | 2825 |
| Services | 636 | 1454 | 1454 | 1454 | 1368 | 1411 | 1411 | 3380 | 3380 | 3380 | 2757 | 2912 | 2912 |
| Final energy, Mtoe | 458 | 856 | 887 | 817 | 877 | 780 | 776 | 1291 | 1339 | 1276 | 1299 | 942 | 943 |
| Solids | 182 | 171 | 170 | 163 | 173 | 228 | 226 | 103 | 163 | 104 | 150 | 152 | 154 |
| Liquids | 191 | 469 | 465 | 377 | 461 | 314 | 322 | 583 | 557 | 540 | 605 | 366 | 367 |
| Electricity | 42 | 84 | 83 | 105 | 81 | 75 | 75 | 160 | 163 | 181 | 145 | 115 | 116 |
| Other$^a$ | 42 | 171 | 169 | 172 | 162 | 163 | 153 | 444 | 455 | 451 | 398 | 309 | 306 |
| Primary energy, Mtoe | 608 | 1170 | 1170 | 1147 | 1148 | 1012 | 1013 | 2020 | 2015 | 2021 | 1826 | 1386 | 1386 |
| Coal | 21 | 51 | 67 | 30 | 65 | 49 | 52 | 83 | 207 | 1 | 113 | 102 | 109 |
| Oil | 244 | 517 | 506 | 392 | 503 | 349 | 349 | 887 | 439 | 462 | 568 | 297 | 294 |
| Gas | 78 | 271 | 249 | 228 | 246 | 240 | 231 | 600 | 497 | 595 | 404 | 335 | 327 |
| Nuclear | 3 | 26 | 1 | 105 | 47 | 3 | 19 | 12 | 20 | 6 | 128 | 0 | 21 |
| Renewables | 262 | 305 | 348 | 392 | 287 | 371 | 362 | 439 | 851 | 956 | 614 | 651 | 635 |
| Net carbon emissions, MtC$^b$ | 280 | 580 | 590 | 420 | 590 | 430 | 440 | 1010 | 790 | 620 | 750 | 480 | 480 |
| Gross carbon emissions, MtC | 300 | 660 | 660 | 510 | 650 | 500 | 500 | 1220 | 910 | 770 | 860 | 570 | 570 |
| Sulfur emissions, MtS | 2 | 3 | 4 | 2 | 3 | 2 | 2 | 2 | 2 | 1 | 3 | 1 | 1 |
| Cumulative energy investments | | | | | | | | | | | | | |
| 1990–2020, US(1990)$, $10^9$ | 650 | 650 | 740 | 1000 | 630 | 630 | 620 | | | | | | |
| 2021–2050, US(1990)$, $10^9$ | | | | | | | | 1350 | 2500 | 2170 | 1690 | 1340 | 1350 |

Note: Subtotals may not add due to independent rounding.

$^a$ District heat, gas, and hydrogen.

$^b$ Net carbon emissions do not include feedstocks and other non-energy emissions or carbon dioxide used for enhanced oil recovery.

**Sub-Saharan Africa (AFR)**

| | 1990 | 2020 | | | | | | 2050 | | | | | |
|---|---|---|---|---|---|---|---|---|---|---|---|---|---|
| | | A1 | A2 | A3 | B | C1 | C2 | A1 | A2 | A3 | B | C1 | C2 |
| Population, $10^6$ | 490 | 1080 | 1080 | 1080 | 1080 | 1080 | 1080 | 1740 | 1740 | 1740 | 1740 | 1740 | 1740 |
| GDP$_{ppp}$, US(1990)\$, $10^9$ | 649 | 1631 | 1631 | 1631 | 1533 | 1572 | 1572 | 5031 | 5031 | 5031 | 3692 | 4112 | 4112 |
| GDP$_{mer}$, US(1990)\$, $10^9$ | 265 | 696 | 696 | 696 | 640 | 662 | 662 | 2727 | 2727 | 2727 | 1787 | 2071 | 2071 |
| Agriculture | 97 | 204 | 204 | 204 | 205 | 206 | 206 | 594 | 594 | 594 | 490 | 526 | 526 |
| Industry | 55 | 236 | 236 | 236 | 208 | 219 | 219 | 1285 | 1285 | 1285 | 788 | 938 | 938 |
| Services | 113 | 256 | 256 | 256 | 227 | 238 | 238 | 847 | 847 | 847 | 509 | 607 | 607 |
| Final energy, Mtoe | 222 | 518 | 511 | 525 | 518 | 534 | 534 | 1191 | 1155 | 1167 | 1026 | 917 | 938 |
| Solids | 165 | 273 | 276 | 289 | 280 | 262 | 262 | 475 | 486 | 477 | 458 | 342 | 343 |
| Liquids | 39 | 183 | 169 | 154 | 177 | 213 | 214 | 469 | 428 | 373 | 363 | 403 | 411 |
| Electricity | 16 | 41 | 44 | 56 | 42 | 39 | 39 | 123 | 125 | 143 | 108 | 93 | 88 |
| Other$^a$ | 1 | 22 | 22 | 26 | 19 | 19 | 19 | 124 | 115 | 174 | 97 | 80 | 96 |
| Primary energy, Mtoe | 287 | 693 | 694 | 693 | 663 | 663 | 663 | 1716 | 1719 | 1723 | 1394 | 1311 | 1308 |
| Coal | 85 | 240 | 216 | 103 | 180 | 169 | 170 | 440 | 656 | 240 | 425 | 340 | 356 |
| Oil | 41 | 177 | 116 | 127 | 170 | 216 | 216 | 435 | 100 | 223 | 216 | 191 | 188 |
| Gas | 4 | 31 | 20 | 68 | 17 | 27 | 26 | 162 | 146 | 264 | 113 | 106 | 114 |
| Nuclear | 2 | 3 | 0 | 3 | 3 | 0 | 0 | 92 | 0 | 156 | 78 | 0 | 26 |
| Renewables | 156 | 242 | 342 | 393 | 294 | 251 | 250 | 588 | 818 | 839 | 561 | 674 | 624 |
| Net carbon emissions, MtC$^b$ | 140 | 380 | 310 | 220 | 310 | 340 | 340 | 780 | 760 | 440 | 610 | 500 | 530 |
| Gross carbon emissions, MtC | 140 | 430 | 340 | 260 | 350 | 380 | 380 | 940 | 890 | 620 | 710 | 600 | 620 |
| Sulfur emissions, MtS | 2 | 3 | 4 | 2 | 4 | 2 | 2 | 4 | 9 | 4 | 7 | 3 | 3 |
| Cumulative energy investments | | | | | | | | | | | | | |
| 1990–2020, US(1990)\$, $10^9$ | | 520 | 630 | 630 | 390 | 320 | 320 | | | | | | |
| 2021–2050, US(1990)\$, $10^9$ | | | | | | | | 1420 | 2190 | 1610 | 1280 | 1330 | 1300 |

Note: Subtotals may not add due to independent rounding.
$^a$District heat, gas, and hydrogen.
$^b$Net carbon emissions do not include feedstocks and other non-energy emissions or carbon dioxide used for enhanced oil recovery.

## Middle East and North Africa (MEA)

| | 1990 | 2020 A1 | 2020 A2 | 2020 A3 | 2020 B | 2020 C1 | 2020 C2 | 2050 A1 | 2050 A2 | 2050 A3 | 2050 B | 2050 C1 | 2050 C2 |
|---|---|---|---|---|---|---|---|---|---|---|---|---|---|
| Population, $10^6$ | 270 | 580 | 580 | 580 | 580 | 580 | 580 | 920 | 920 | 920 | 920 | 920 | 920 |
| GDP$_{ppp}$, US(1990)\$, $10^9$ | 1141 | 2956 | 2956 | 2956 | 2825 | 2825 | 2825 | 7406 | 7406 | 7406 | 5939 | 5997 | 5997 |
| GDP$_{mer}$, US(1990)\$, $10^9$ | 573 | 1646 | 1646 | 1646 | 1536 | 1536 | 1536 | 5217 | 5217 | 5217 | 3723 | 3779 | 3779 |
| Agriculture | 120 | 245 | 245 | 245 | 244 | 243 | 243 | 459 | 459 | 459 | 405 | 407 | 407 |
| Industry | 122 | 553 | 563 | 563 | 536 | 536 | 536 | 2458 | 2458 | 2458 | 1805 | 1830 | 1830 |
| Services | 331 | 838 | 838 | 838 | 756 | 757 | 757 | 2300 | 2300 | 2300 | 1514 | 1542 | 1542 |
| Final energy, Mtoe | 238 | 644 | 638 | 661 | 611 | 584 | 585 | 1503 | 1544 | 1519 | 1294 | 1075 | 1073 |
| Solids | 25 | 28 | 26 | 28 | 32 | 29 | 28 | 20 | 26 | 21 | 23 | 24 | 24 |
| Liquids | 151 | 374 | 400 | 456 | 301 | 310 | 310 | 893 | 885 | 993 | 653 | 562 | 563 |
| Electricity | 23 | 48 | 51 | 49 | 49 | 46 | 46 | 139 | 148 | 138 | 116 | 111 | 105 |
| Other$^a$ | 39 | 194 | 161 | 128 | 229 | 199 | 200 | 451 | 486 | 368 | 502 | 377 | 381 |
| Primary energy, Mtoe | 548 | 870 | 870 | 876 | 838 | 780 | 780 | 2004 | 1999 | 2002 | 1676 | 1391 | 1394 |
| Coal | 6 | 2 | 2 | 1 | 7 | 2 | 2 | 1 | 1 | 0 | 1 | 0 | 0 |
| Oil | 204 | 460 | 475 | 553 | 335 | 350 | 350 | 1081 | 930 | 1095 | 692 | 575 | 576 |
| Gas | 111 | 364 | 332 | 268 | 441 | 384 | 384 | 617 | 855 | 769 | 853 | 695 | 706 |
| Nuclear | 0 | 0 | 0 | 0 | 0 | 0 | 0 | 43 | 19 | 18 | 42 | 0 | 6 |
| Renewables | 27 | 44 | 60 | 54 | 55 | 44 | 44 | 262 | 194 | 120 | 89 | 121 | 105 |
| Net carbon emissions, MtC$^b$ | 290 | 560 | 560 | 560 | 530 | 490 | 490 | 1050 | 1170 | 1210 | 1020 | 760 | 820 |
| Gross carbon emissions, MtC | 300 | 620 | 610 | 640 | 570 | 540 | 540 | 1300 | 1330 | 1410 | 1130 | 930 | 940 |
| Sulfur emissions, MtS | 2 | 2 | 2 | 3 | 1 | 1 | 1 | 2 | 2 | 2 | 1 | 1 | 1 |
| Cumulative energy investments | | | | | | | | | | | | | |
| 1990–2020, US(1990)\$, $10^9$ | | 600 | 630 | 710 | 580 | 600 | 610 | | | | | | |
| 2021–2050, US(1990)\$, $10^9$ | | | | | | | | 1310 | 1510 | 1840 | 1200 | 1070 | 1100 |

Note: Subtotals may not add due to independent rounding.

$^a$ District heat, gas, and hydrogen.

$^b$ Net carbon emissions do not include feedstocks and other non-energy emissions or carbon dioxide used for enhanced oil recovery.

## Western Europe (WEU)

| | 1990 | 2020 A1 | 2020 A2 | 2020 A3 | 2020 B | 2020 C1 | 2020 C2 | 2050 A1 | 2050 A2 | 2050 A3 | 2050 B | 2050 C1 | 2050 C2 |
|---|---|---|---|---|---|---|---|---|---|---|---|---|---|
| Population, $10^6$ | 430 | 490 | 490 | 490 | 490 | 490 | 490 | 490 | 490 | 490 | 490 | 490 | 490 |
| GDP$_{ppp}$, US(1990)\$, $10^9$ | 5740 | 11188 | 11188 | 11188 | 10096 | 9477 | 9477 | 18585 | 18585 | 18585 | 15011 | 13346 | 13346 |
| GDP$_{mer}$, US(1990)\$, $10^9$ | 7009 | 13660 | 13660 | 13660 | 12327 | 11570 | 11570 | 22692 | 22692 | 22692 | 18328 | 16295 | 16295 |
| Agriculture | 397 | 504 | 504 | 504 | 476 | 460 | 460 | 391 | 391 | 391 | 377 | 372 | 372 |
| Industry | 2311 | 3833 | 3833 | 3833 | 3649 | 3528 | 3528 | 5251 | 5251 | 5251 | 4807 | 4563 | 4563 |
| Services | 4300 | 9322 | 9322 | 9322 | 8202 | 7583 | 7583 | 17049 | 17049 | 17049 | 13143 | 11360 | 11360 |
| Final energy, Mtoe | 1017 | 1343 | 1330 | 1316 | 1213 | 932 | 927 | 1399 | 1484 | 1378 | 1187 | 819 | 714 |
| Solids | 130 | 28 | 28 | 28 | 43 | 28 | 28 | 3 | 4 | 3 | 3 | 3 | 3 |
| Liquids | 518 | 515 | 486 | 468 | 412 | 292 | 291 | 628 | 327 | 313 | 287 | 210 | 187 |
| Electricity | 174 | 315 | 334 | 324 | 295 | 243 | 242 | 436 | 519 | 445 | 410 | 299 | 264 |
| Other$^a$ | 195 | 486 | 482 | 496 | 463 | 369 | 367 | 332 | 634 | 617 | 488 | 307 | 260 |
| Primary energy, Mtoe | 1455 | 1982 | 1981 | 1983 | 1800 | 1424 | 1425 | 2366 | 2386 | 2375 | 1930 | 1294 | 1296 |
| Coal | 316 | 326 | 438 | 274 | 346 | 179 | 183 | 132 | 580 | 42 | 148 | 55 | 34 |
| Oil | 604 | 561 | 502 | 465 | 424 | 322 | 320 | 690 | 141 | 103 | 188 | 105 | 91 |
| Gas | 229 | 576 | 553 | 610 | 543 | 468 | 454 | 403 | 730 | 613 | 667 | 589 | 237 |
| Nuclear | 166 | 332 | 276 | 337 | 321 | 278 | 290 | 768 | 300 | 975 | 591 | 123 | 527 |
| Renewables | 140 | 186 | 211 | 297 | 167 | 178 | 178 | 372 | 635 | 642 | 336 | 423 | 406 |
| Net carbon emissions, MtC$^b$ | 960 | 1110 | 1190 | 980 | 1020 | 690 | 690 | 910 | 1180 | 460 | 700 | 410 | 230 |
| Gross carbon emissions, MtC | 1030 | 1200 | 1260 | 1080 | 1080 | 770 | 760 | 990 | 1220 | 520 | 750 | 520 | 270 |
| Sulfur emissions, MtS | 9 | 2 | 3 | 2 | 3 | 1 | 1 | 1 | 1 | 1 | 1 | 1 | 0 |
| Cumulative energy investments | | | | | | | | | | | | | |
| 1990–2020, US(1990)\$, $10^9$ | | 2360 | 2720 | 2610 | 2270 | 1660 | 1610 | 2770 | 3570 | 2750 | 2690 | 1380 | 1550 |
| 2021–2050, US(1990)\$, $10^9$ | | | | | | | | | | | | | |

Note: Subtotals may not add due to independent rounding.

$^a$ District heat, gas, and hydrogen.

$^b$ Net carbon emissions do not include feedstocks and other non-energy emissions or carbon dioxide used for enhanced oil recovery.

# Central and Eastern Europe (EEU)

| | 1990 | 2020 A1 | 2020 A2 | 2020 A3 | 2020 B | 2020 C1 | 2020 C2 | 2050 A1 | 2050 A2 | 2050 A3 | 2050 B | 2050 C1 | 2050 C2 |
|---|---|---|---|---|---|---|---|---|---|---|---|---|---|
| Population, $10^6$ | 120 | 130 | 130 | 130 | 130 | 130 | 130 | 140 | 140 | 140 | 140 | 140 | 140 |
| $GDP_{ppp}$, US(1990)\$, $10^9$ | 712 | 740 | 740 | 740 | 600 | 670 | 670 | 2295 | 2295 | 2295 | 1524 | 1538 | 1538 |
| $GDP_{mer}$, US(1990)\$, $10^9$ | 296 | 589 | 589 | 589 | 388 | 438 | 438 | 2295 | 2295 | 2295 | 1105 | 1124 | 1124 |
| Agriculture | 25 | 38 | 38 | 38 | 29 | 33 | 33 | 101 | 101 | 101 | 73 | 73 | 73 |
| Industry | 160 | 274 | 274 | 274 | 191 | 215 | 215 | 892 | 892 | 892 | 489 | 496 | 496 |
| Services | 111 | 278 | 278 | 278 | 168 | 190 | 190 | 1302 | 1302 | 1302 | 544 | 555 | 555 |
| Final energy, Mtoe | 227 | 334 | 339 | 332 | 261 | 243 | 243 | 453 | 445 | 435 | 348 | 246 | 228 |
| Solids | 51 | 19 | 33 | 19 | 36 | 15 | 15 | 6 | 6 | 6 | 26 | 6 | 6 |
| Liquids | 55 | 89 | 80 | 78 | 66 | 59 | 59 | 185 | 138 | 141 | 83 | 63 | 58 |
| Electricity | 30 | 52 | 55 | 55 | 39 | 37 | 37 | 90 | 95 | 97 | 61 | 48 | 45 |
| Other[a] | 90 | 174 | 171 | 180 | 120 | 132 | 131 | 172 | 205 | 191 | 178 | 129 | 120 |
| Primary energy, Mtoe | 337 | 415 | 415 | 421 | 323 | 308 | 308 | 591 | 593 | 594 | 449 | 326 | 326 |
| Coal | 159 | 183 | 181 | 142 | 130 | 106 | 110 | 187 | 188 | 64 | 135 | 48 | 40 |
| Oil | 79 | 83 | 79 | 76 | 70 | 66 | 66 | 174 | 81 | 83 | 65 | 25 | 23 |
| Gas | 66 | 110 | 102 | 143 | 85 | 99 | 95 | 88 | 161 | 228 | 120 | 151 | 103 |
| Nuclear | 13 | 21 | 21 | 21 | 21 | 21 | 21 | 93 | 37 | 83 | 80 | 18 | 80 |
| Renewables | 19 | 18 | 33 | 40 | 17 | 17 | 17 | 47 | 126 | 135 | 49 | 84 | 79 |
| Net carbon emissions, MtC[b] | 280 | 310 | 300 | 250 | 240 | 210 | 210 | 370 | 360 | 240 | 250 | 150 | 110 |
| Gross carbon emissions, MtC | 290 | 350 | 330 | 320 | 260 | 240 | 240 | 410 | 380 | 290 | 280 | 170 | 130 |
| Sulfur emissions, MtS | 5 | 2 | 2 | 2 | 2 | 2 | 2 | 2 | 1 | 1 | 1 | 1 | 1 |
| Cumulative energy investments | | | | | | | | | | | | | |
| 1990–2020, US(1990)\$, $10^9$ | | 580 | 580 | 620 | 410 | 340 | 350 | | | | | | |
| 2021–2050, US(1990)\$, $10^9$ | | | | | | | | 810 | 920 | 670 | 710 | 340 | 370 |

Note: Subtotals may not add due to independent rounding.

[a] District heat, gas, and hydrogen.

[b] Net carbon emissions do not include feedstocks and other non-energy emissions or carbon dioxide used for enhanced oil recovery.

# Newly independent states of the former Soviet Union (FSU)

| | 1990 | 2020 | | | | | | 2050 | | | | | |
|---|---|---|---|---|---|---|---|---|---|---|---|---|---|
| | | A1 | A2 | A3 | B | C1 | C2 | A1 | A2 | A3 | B | C1 | C2 |
| Population, $10^6$ | 290 | 350 | 350 | 350 | 350 | 350 | 350 | 390 | 390 | 390 | 390 | 390 | 390 |
| GDP$_{ppp}$, US(1990)\$, $10^9$ | 1840 | 1500 | 1500 | 1500 | 1400 | 1600 | 1600 | 5850 | 5850 | 5850 | 3600 | 3600 | 3600 |
| GDP$_{mer}$, US(1990)\$, $10^9$ | 785 | 1134 | 1134 | 1134 | 971 | 1075 | 1075 | 5561 | 5561 | 5561 | 2951 | 2816 | 2816 |
| Agriculture | 93 | 94 | 94 | 94 | 95 | 105 | 105 | 269 | 269 | 269 | 229 | 219 | 219 |
| Industry | 314 | 455 | 455 | 455 | 417 | 462 | 462 | 2235 | 2235 | 2235 | 1357 | 1295 | 1295 |
| Services | 378 | 585 | 585 | 585 | 458 | 508 | 508 | 3057 | 3057 | 3057 | 1365 | 1302 | 1302 |
| Final energy, Mtoe | 973 | 1469 | 1469 | 1453 | 1029 | 1065 | 1065 | 2384 | 2457 | 2391 | 1452 | 1031 | 1045 |
| Solids | 171 | 200 | 205 | 179 | 133 | 138 | 137 | 46 | 42 | 44 | 44 | 42 | 42 |
| Liquids | 307 | 527 | 517 | 500 | 272 | 287 | 294 | 1104 | 1031 | 879 | 493 | 314 | 319 |
| Electricity | 107 | 171 | 176 | 176 | 127 | 124 | 123 | 373 | 364 | 349 | 206 | 143 | 144 |
| Other$^a$ | 388 | 570 | 570 | 598 | 497 | 517 | 511 | 862 | 1020 | 1119 | 710 | 532 | 540 |
| Primary energy, Mtoe | 1402 | 1867 | 1864 | 1868 | 1339 | 1377 | 1377 | 3133 | 3140 | 3135 | 1929 | 1339 | 1339 |
| Coal | 285 | 288 | 313 | 161 | 241 | 126 | 159 | 418 | 457 | 31 | 236 | 5 | 5 |
| Oil | 409 | 544 | 544 | 540 | 311 | 334 | 335 | 1158 | 1085 | 902 | 506 | 358 | 361 |
| Gas | 572 | 817 | 831 | 838 | 613 | 767 | 693 | 706 | 1197 | 1629 | 849 | 780 | 791 |
| Nuclear | 47 | 70 | 33 | 78 | 63 | 33 | 78 | 373 | 81 | 148 | 165 | 0 | 9 |
| Renewables | 89 | 147 | 142 | 252 | 111 | 117 | 113 | 477 | 320 | 425 | 174 | 197 | 174 |
| Net carbon emissions, MtC$^b$ | 1030 | 1160 | 1210 | 1000 | 840 | 800 | 800 | 1600 | 1990 | 1500 | 1100 | 700 | 710 |
| Gross carbon emissions, MtC | 1060 | 1300 | 1330 | 1160 | 920 | 910 | 900 | 1880 | 2180 | 1830 | 1230 | 800 | 810 |
| Sulfur emissions, MtS | 10 | 5 | 6 | 5 | 5 | 3 | 3 | 11 | 6 | 2 | 4 | 1 | 1 |
| Cumulative energy investments | | | | | | | | | | | | | |
| 1990–2020, US(1990)\$, $10^9$ | | 2000 | 2030 | 2100 | 1500 | 1320 | 1330 | | | | | | |
| 2021–2050, US(1990)\$, $10^9$ | | | | | | | | 3180 | 3860 | 2640 | 2460 | 1360 | 1250 |

Note: Subtotals may not add due to independent rounding.
$^a$ District heat, gas, and hydrogen.
$^b$ Net carbon emissions do not include feedstocks and other non-energy emissions or carbon dioxide used for enhanced oil recovery.

**Centrally planned Asia and China (CPA)**

| | 1990 | 2020 A1 | 2020 A2 | 2020 A3 | 2020 B | 2020 C1 | 2020 C2 | 2050 A1 | 2050 A2 | 2050 A3 | 2050 B | 2050 C1 | 2050 C2 |
|---|---|---|---|---|---|---|---|---|---|---|---|---|---|
| Population, $10^6$ | 1240 | 1710 | 1710 | 1710 | 1710 | 1710 | 1710 | 1980 | 1980 | 1980 | 1980 | 1980 | 1980 |
| $GDP_{ppp}$, US(1990)\$, $10^9$ | 2438 | 9652 | 9652 | 9652 | 6611 | 8771 | 8771 | 21930 | 21930 | 21930 | 14181 | 18807 | 18807 |
| $GDP_{mer}$, US(1990)\$, $10^9$ | 474 | 3852 | 3852 | 3852 | 2038 | 3279 | 3279 | 13871 | 13871 | 13871 | 6660 | 10711 | 10711 |
| Agriculture | 179 | 852 | 852 | 852 | 485 | 742 | 742 | 893 | 893 | 893 | 654 | 801 | 801 |
| Industry | 115 | 1315 | 1315 | 1315 | 736 | 1146 | 1146 | 6099 | 6099 | 6099 | 3190 | 4886 | 4886 |
| Services | 180 | 1686 | 1686 | 1586 | 817 | 1391 | 1391 | 6879 | 6879 | 6879 | 2816 | 5023 | 5023 |
| Final energy, Mtoe | 751 | 2044 | 2051 | 2055 | 1758 | 1514 | 1512 | 3250 | 3149 | 3289 | 2750 | 1858 | 1839 |
| Solids | 581 | 1275 | 1302 | 1320 | 1156 | 985 | 980 | 1234 | 1495 | 1393 | 1420 | 756 | 747 |
| Liquids | 93 | 411 | 398 | 366 | 323 | 263 | 266 | 940 | 793 | 722 | 630 | 425 | 422 |
| Electricity | 51 | 162 | 163 | 167 | 127 | 131 | 131 | 391 | 384 | 419 | 265 | 285 | 282 |
| Other[a] | 26 | 196 | 187 | 202 | 152 | 135 | 135 | 684 | 477 | 755 | 435 | 392 | 388 |
| Primary energy, Mtoe | 945 | 2533 | 2533 | 2510 | 2102 | 1829 | 1829 | 4496 | 4487 | 4318 | 3549 | 2568 | 2566 |
| Coal | 557 | 1477 | 1649 | 1426 | 1198 | 1017 | 1007 | 1297 | 3238 | 1230 | 1671 | 749 | 740 |
| Oil | 127 | 371 | 363 | 327 | 354 | 283 | 289 | 951 | 126 | 252 | 277 | 176 | 160 |
| Gas | 13 | 165 | 137 | 214 | 111 | 136 | 124 | 760 | 363 | 1098 | 341 | 339 | 287 |
| Nuclear | 0 | 51 | 0 | 51 | 51 | 19 | 51 | 433 | 4 | 480 | 358 | 64 | 370 |
| Renewables | 247 | 469 | 385 | 492 | 388 | 375 | 358 | 1055 | 754 | 1258 | 902 | 1240 | 1009 |
| Net carbon emissions, MtC[b] | 690 | 1870 | 2060 | 1800 | 1560 | 1330 | 1320 | 2290 | 3630 | 2030 | 2050 | 1020 | 970 |
| Gross carbon emissions, MtC | 700 | 2020 | 2180 | 1950 | 1670 | 1430 | 1410 | 2690 | 3840 | 2250 | 2260 | 1180 | 1120 |
| Sulfur emissions, MtS | 11 | 27 | 25 | 28 | 25 | 17 | 17 | 23 | 27 | 25 | 27 | 11 | 11 |
| Cumulative energy investments | | | | | | | | | | | | | |
| 1990–2020, US(1990)\$, $10^9$ | | 1820 | 1740 | 1800 | 1170 | 1070 | 1070 | | | | | | |
| 2021–2050, US(1990)\$, $10^9$ | | | | | | | | 3460 | 4980 | 3970 | 3420 | 2650 | 2640 |

Note: Subtotals may not add due to independent rounding.

[a] District heat, gas, and hydrogen.

[b] Net carbon emissions do not include feedstocks and other non-energy emissions or carbon dioxide used for enhanced oil recovery.

## South Asia (SAS)

| | 1990 | 2020 | | | | | | 2050 | | | | | |
|---|---|---|---|---|---|---|---|---|---|---|---|---|---|
| | | A1 | A2 | A3 | B | C1 | C2 | A1 | A2 | A3 | B | C1 | C2 |
| Population, $10^6$ | 1130 | 1780 | 1780 | 1780 | 1780 | 1780 | 1780 | 2280 | 2280 | 2280 | 2280 | 2280 | 2280 |
| GDP$_{ppp}$, US(1990)\$, $10^9$ | 1363 | 3385 | 3385 | 3385 | 3184 | 3301 | 3301 | 9078 | 9078 | 9078 | 6901 | 8309 | 8309 |
| GDP$_{mer}$, US(1990)\$, $10^9$ | 377 | 1175 | 1175 | 1175 | 1071 | 1131 | 1131 | 4570 | 4570 | 4570 | 3026 | 4001 | 4001 |
| Agriculture | 120 | 285 | 285 | 285 | 283 | 282 | 282 | 767 | 767 | 767 | 637 | 724 | 724 |
| Industry | 100 | 443 | 443 | 443 | 397 | 425 | 425 | 2234 | 2234 | 2234 | 1436 | 1942 | 1942 |
| Services | 157 | 446 | 446 | 446 | 391 | 424 | 424 | 1570 | 1570 | 1570 | 952 | 1335 | 1335 |
| Final energy, Mtoe | 356 | 809 | 826 | 819 | 789 | 731 | 731 | 1648 | 1785 | 1702 | 1446 | 1276 | 1286 |
| Solids | 256 | 414 | 427 | 439 | 419 | 381 | 383 | 659 | 755 | 756 | 721 | 460 | 470 |
| Liquids | 64 | 180 | 198 | 184 | 168 | 141 | 140 | 425 | 500 | 205 | 263 | 337 | 329 |
| Electricity | 22 | 62 | 64 | 63 | 61 | 55 | 55 | 179 | 210 | 191 | 162 | 142 | 143 |
| Other$^a$ | 15 | 152 | 137 | 133 | 140 | 154 | 153 | 385 | 320 | 550 | 301 | 337 | 344 |
| Primary energy, Mtoe | 444 | 984 | 985 | 985 | 941 | 885 | 885 | 2195 | 2192 | 2197 | 1787 | 1662 | 1664 |
| Coal | 108 | 194 | 194 | 187 | 194 | 182 | 178 | 285 | 433 | 370 | 433 | 157 | 166 |
| Oil | 75 | 187 | 228 | 187 | 185 | 137 | 135 | 399 | 436 | 80 | 253 | 245 | 236 |
| Gas | 25 | 165 | 165 | 165 | 165 | 165 | 165 | 172 | 358 | 450 | 290 | 260 | 265 |
| Nuclear | 1 | 18 | 18 | 18 | 18 | 8 | 18 | 155 | 301 | 202 | 233 | 40 | 135 |
| Renewables | 234 | 421 | 381 | 428 | 380 | 393 | 389 | 1185 | 664 | 1095 | 578 | 960 | 863 |
| Net carbon emissions, MtC$^b$ | 190 | 410 | 450 | 390 | 420 | 380 | 380 | 530 | 890 | 620 | 700 | 470 | 470 |
| Gross carbon emissions, MtC | 190 | 470 | 510 | 470 | 470 | 420 | 410 | 750 | 1060 | 760 | 870 | 540 | 550 |
| Sulfur emissions, MtS | 2 | 3 | 4 | 3 | 4 | 3 | 3 | 5 | 8 | 5 | 7 | 3 | 3 |
| Cumulative energy investments | | | | | | | | | | | | | |
| 1990–2020, US(1990)\$, $10^9$ | | 600 | 590 | 710 | 550 | 540 | 540 | | | | | | |
| 2021–2050, US(1990)\$, $10^9$ | | | | | | | | 2250 | 2400 | 2540 | 1480 | 1600 | 1510 |

Note: Subtotals may not add due to independent rounding.
$^a$District heat, gas, and hydrogen.
$^b$Net carbon emissions do not include feedstocks and other non-energy emissions or carbon dioxide used for enhanced oil recovery.

# Other Pacific Asia (PAS)

| | 1990 | 2020 | | | | | | 2050 | | | | | |
|---|---|---|---|---|---|---|---|---|---|---|---|---|---|
| | | A1 | A2 | A3 | B | C1 | C2 | A1 | A2 | A3 | B | C1 | C2 |
| Population, $10^6$ | 430 | 620 | 620 | 620 | 620 | 620 | 620 | 750 | 750 | 750 | 750 | 750 | 750 |
| GDP$_{ppp}$, US(1990)\$, $10^9$ | 1511 | 4972 | 4972 | 4972 | 3934 | 4627 | 4627 | 9998 | 9998 | 9998 | 7544 | 8913 | 8913 |
| GDP$_{mer}$, US(1990)\$, $10^9$ | 656 | 3421 | 3421 | 3421 | 2373 | 3058 | 3058 | 9164 | 9164 | 9164 | 5902 | 7659 | 7659 |
| Agriculture | 103 | 360 | 360 | 360 | 270 | 332 | 332 | 491 | 491 | 491 | 417 | 460 | 460 |
| Industry | 274 | 1428 | 1428 | 1428 | 1030 | 1298 | 1298 | 3817 | 3817 | 3817 | 2657 | 3302 | 3302 |
| Services | 279 | 1634 | 1634 | 1634 | 1073 | 1428 | 1428 | 4856 | 4856 | 4856 | 2828 | 3897 | 3897 |
| Final energy, Mtoe | 343 | 856 | 850 | 840 | 742 | 649 | 649 | 1391 | 1449 | 1363 | 1212 | 868 | 855 |
| Solids | 183 | 172 | 204 | 205 | 221 | 228 | 228 | 101 | 343 | 328 | 336 | 164 | 160 |
| Liquids | 127 | 539 | 500 | 489 | 402 | 299 | 300 | 873 | 675 | 558 | 551 | 395 | 389 |
| Electricity | 25 | 101 | 101 | 101 | 76 | 80 | 80 | 218 | 260 | 251 | 177 | 176 | 174 |
| Other[a] | 9 | 44 | 45 | 47 | 43 | 42 | 42 | 198 | 172 | 226 | 147 | 133 | 132 |
| Primary energy, Mtoe | 424 | 1136 | 1133 | 1135 | 954 | 843 | 843 | 2008 | 2006 | 1990 | 1630 | 1239 | 1241 |
| Coal | 46 | 82 | 187 | 102 | 134 | 99 | 113 | 13 | 330 | 187 | 171 | 19 | 7 |
| Oil | 163 | 592 | 548 | 505 | 446 | 318 | 319 | 998 | 626 | 464 | 537 | 362 | 357 |
| Gas | 30 | 144 | 100 | 109 | 97 | 103 | 104 | 277 | 177 | 384 | 137 | 184 | 152 |
| Nuclear | 19 | 43 | 4 | 43 | 43 | 35 | 43 | 153 | 129 | 156 | 209 | 24 | 141 |
| Renewables | 166 | 275 | 295 | 376 | 234 | 288 | 264 | 566 | 744 | 800 | 576 | 650 | 583 |
| Net carbon emissions, MtC[b] | 210 | 580 | 640 | 520 | 510 | 390 | 410 | 790 | 830 | 600 | 570 | 360 | 330 |
| Gross carbon emissions, MtC | 220 | 680 | 720 | 600 | 580 | 440 | 460 | 1030 | 1000 | 840 | 720 | 440 | 400 |
| Sulfur emissions, MtS | 3 | 3 | 5 | 3 | 4 | 2 | 2 | 2 | 6 | 4 | 4 | 1 | 1 |
| Cumulative energy investments | | | | | | | | | | | | | |
| 1990–2020, US(1990)\$, $10^9$ | | 760 | 770 | 860 | 600 | 580 | 550 | | | | | | |
| 2021–2050, US(1990)\$, $10^9$ | | | | | | | | 1490 | 1950 | 1760 | 1480 | 1030 | 1060 |

Note: Subtotals may not add due to independent rounding.

[a] District heat, gas, and hydrogen.

[b] Net carbon emissions do not include feedstocks and other non-energy emissions or carbon dioxide used for enhanced oil recovery.

**Pacific OECD (PAO)**

| | 1990 | 2020 A1 | 2020 A2 | 2020 A3 | 2020 B | 2020 C1 | 2020 C2 | 2050 A1 | 2050 A2 | 2050 A3 | 2050 B | 2050 C1 | 2050 C2 |
|---|---|---|---|---|---|---|---|---|---|---|---|---|---|
| Population, $10^6$ | 140 | 160 | 160 | 160 | 160 | 160 | 160 | 150 | 150 | 150 | 150 | 150 | 150 |
| GDP$_{ppp}$, US(1990)\$, $10^9$ | 2413 | 4293 | 4293 | 4293 | 3784 | 3669 | 3669 | 6395 | 6395 | 6395 | 4991 | 4664 | 4664 |
| GDP$_{mer}$, US(1990)\$, $10^9$ | 3280 | 5835 | 5835 | 5835 | 5143 | 4988 | 4988 | 8692 | 8692 | 8692 | 6784 | 6339 | 6339 |
| Agriculture | 82 | 115 | 115 | 115 | 109 | 108 | 108 | 124 | 124 | 124 | 117 | 116 | 116 |
| Industry | 1370 | 1799 | 1799 | 1799 | 1671 | 1644 | 1644 | 1727 | 1727 | 1727 | 1572 | 1530 | 1530 |
| Services | 1827 | 3921 | 3921 | 3921 | 3364 | 3236 | 3236 | 6841 | 6841 | 6841 | 5095 | 4694 | 4694 |
| Final energy, Mtoe | 368 | 470 | 470 | 467 | 420 | 317 | 316 | 443 | 469 | 453 | 354 | 223 | 213 |
| Solids | 51 | 14 | 14 | 14 | 16 | 14 | 14 | 5 | 6 | 5 | 5 | 5 | 5 |
| Liquids | 213 | 242 | 226 | 219 | 199 | 146 | 146 | 228 | 204 | 209 | 153 | 88 | 82 |
| Electricity | 79 | 115 | 126 | 133 | 109 | 85 | 85 | 142 | 161 | 151 | 124 | 83 | 82 |
| Other$^a$ | 25 | 99 | 103 | 101 | 96 | 73 | 71 | 69 | 98 | 88 | 71 | 47 | 43 |
| Primary energy, Mtoe | 543 | 702 | 704 | 710 | 630 | 478 | 478 | 771 | 769 | 775 | 612 | 381 | 381 |
| Coal | 111 | 92 | 130 | 75 | 90 | 60 | 59 | 82 | 102 | 9 | 71 | 3 | 3 |
| Oil | 285 | 265 | 237 | 235 | 206 | 158 | 158 | 248 | 150 | 99 | 115 | 68 | 68 |
| Gas | 62 | 169 | 184 | 187 | 173 | 125 | 123 | 184 | 194 | 184 | 116 | 81 | 47 |
| Nuclear | 45 | 87 | 61 | 110 | 81 | 61 | 71 | 78 | 82 | 179 | 118 | 47 | 89 |
| Renewables | 40 | 88 | 91 | 103 | 81 | 74 | 67 | 179 | 240 | 303 | 191 | 183 | 173 |
| Net carbon emissions, MtC$^b$ | 370 | 390 | 420 | 350 | 350 | 250 | 250 | 370 | 340 | 170 | 220 | 100 | 80 |
| Gross carbon emissions, MtC | 410 | 430 | 460 | 400 | 380 | 280 | 280 | 420 | 360 | 210 | 250 | 110 | 90 |
| Sulfur emissions, MtS | 1 | 0 | 0 | 0 | 1 | 0 | 0 | 0 | 0 | 0 | 0 | 0 | 0 |
| Cumulative energy investments | | | | | | | | | | | | | |
| 1990–2020, US(1990)\$, $10^9$ | | 730 | 930 | 820 | 750 | 460 | 440 | | | | | | |
| 2021–2050, US(1990)\$, $10^9$ | | | | | | | | 820 | 1380 | 890 | 960 | 420 | 440 |

Note: Subtotals may not add due to independent rounding.
$^a$ District heat, gas, and hydrogen.
$^b$ Net carbon emissions do not include feedstocks and other non-energy emissions or carbon dioxide used for enhanced oil recovery.

## Appendix D: Regional Reviewers and Experts

### Regional Review Coordinators

| | |
|---|---|
| North America (NAM) | Richard E. Balzhiser |
| Latin America and the Caribbean (LAM) | José Luiz Alquéres |
| Sub-Saharan Africa (AFR) | Steve J. Lennon |
| Middle East and North Africa (MEA) | Hisham Khatib |
| Western Europe (WEU) | Claude Destival |
| Central and Eastern Europe (EEU) | Klaus Brendow |
| Newly independent states of the former Soviet Union (FSU) | Klaus Brendow |
| Centrally planned Asia and China (CPA) | Yuan Shuxun |
| South Asia (SAS) | Rajendra K. Pachauri |
| Other Pacific Asia (PAS) | Yoichi Kaya |
| Pacific OECD (PAO) | Yoichi Kaya |

## Reviewers and Experts

Keigo Akimoto
José Luiz Alquéres
Martin Altstätter
Richard E. Balzhiser
J.A. Basson
Simon Batov
Mariano Bauer
Natan Bernot
Anatoly A. Beschinsky
Kankar Bhattacharya
Georges Bouchard
Klaus Brendow
Angus A. Bruneau
Vitaly V. Bushuev
Margaret Carson
Pradeep Chaturvedi
Zhu Chengzhang
Alessandro Clerici
E. Philip Cockshutt
Jean Constantinescu
C.J. Cooper
Mike R.V. Corrigall
Thérèse de Mazancourt
Adilson de Oliveira
Jörg Debelius
Louis D. DeMouy
Vittorio D'Ermo
Claude Destival
R.J. Dinning
R.K. Dutkiewicz
A.R. Du Toit
A.R. Dykes
James A. Edmonds
Stefan Fecko
Zhou Fengqi
Luiz A.M. Fonseca
John S. Foster
Jean-Romain Frisch
Yasumasa Fujii

János Gács
Lin Gan
Anatoly A. Gordukalov
Malcolm Grimston
Sujata Gupta
John G. Hollins
Marek Jaczewski
Michael Jefferson
Anne Johnson
Poong Eil Juhn
Yoichi Kaya
Arshad M. Khan
Hisham Khatib
Tony Kimpton
Mico Klepo
I.A. Kotze
Ostoj Kristan
Thomas Kuhn
Rainer Kurz
Phil C. Lachambre
Guida Lami
Nicolae Liciu
Vladimir L. Likhachev
Jack E. Little
Alexey A. Makarov
Alexej M. Mastepanov
Edgar McCarthy
Lorne G. McConnell
Elena A. Medvedeva
Victor V. Nechaev
Alexandr S. Nekrasov
Gurgen G. Olkhovsky
Rajendra K. Pachauri
Jyoti K. Parikh
Damir Pesut
Carlos O. Pierro
Max Pilegaard
Paul R. Portney

Robert Priddle
Wang Qingyi
Keywan Riahi
Richard Richels
Alexander Röhrl
Edward S. Rubin
Angelo Saullo
Andreas Schäfer
Michael Schultz
David S. Scott
Chen Shutong
Yuan Shuxun
Oskar T. Sigvaldson
Kari Sinivuori
Yuri Sinyak
Charles Slagorsky
A.A. Sonalov
Chauncey Starr
V.A. Stukalov
Carlos Suarez
Miroslav Sumpik
Kae Takase
Gregory Tosen
C. van Horen
Petr Veselsky
Thierry Villaron
Dimitry B. Volfberg
Miroslav Vrba
John H. Walsh
John P. Weyant
Zhu Yuezhong
Yao Yufang
C. Pierre Zaleski
H.F.M. Zewald
Wei Zhihong
Huang Zhijie
Wu Zhonghu
Nada Zupanč

## Appendix E: Accessibility of Study Results via the Internet

The most important assumptions and results of the IIASA–WEC study can also be accessed electronically for those interested in greater detail.

- Access the IIASA home page at http://www.iiasa.ac.at.
- Select "Research" from the selection list to the left of the screen.
- Select "ECS" from the "Research Projects" list.
- Click on "IIASA–WEC Database" to enter the database.

The database contains

- Definitions with short description of the regions, scenarios, and energy carriers considered in the study.
- Parameters such as GDP (both $GDP_{mer}$ and $GDP_{ppp}$) and population — the main driving variables in the scenarios.
- Results, showing

  - the development of primary and final energy, and electricity production;

  - carbon dioxide and sulfur dioxide emissions;

  - use of natural resources;

  - required cumulative investments in the energy sector.

- Derived indicators that allow comparisons among, for example, in energy intensities, emission intensities, and GDP per capita trends for different world regions and scenarios.

Standard queries are offered to generate tables and graphics for energy consumption, emissions, and other variables either as world totals or for selected regions, scenarios, or energy carriers. In addition individual customized queries can be defined by clicking on "User-defined query" on the left selection menu. User-generated tables can be imported directly into EXCEL spreadsheets and graphics can be downloaded as postscript files by clicking the "Download data" button. The database contains over 100 time series for each region and scenario, totaling nearly 10,000 time series including aggregated totals for the world and the three macroregions, OECD, REFs, DCs. In total, the database contains approximately 100,000 data entries. The selection menu also includes options for downloading graphics included in this book, as well as answers to frequently asked questions.

The following table gives an overview of the data variables contained in the database and presents a short, explanatory description of the entries.

Data availability matrix – Version 1.0

| | BIO | CUM | DPE | ELE | EMI | EPE | EXP | GDP | GRF | IMP | LPS | NEN | NET | NTR | POP | PPP | PRO | RSC | SFT | TFC | TPE |
|---|---|---|---|---|---|---|---|---|---|---|---|---|---|---|---|---|---|---|---|---|---|
| AGR | | | | | | | | y | | | | | | | | | | | | | |
| BCO | | | | | | | | | | | | | | y | | | y | | | y | y |
| BNC | | | | | | | | | | | | | | y | | | y | | | y | y |
| CH4 | | | | | y | | | | | | | | | | | | | | | | |
| CO2 | y | | | | y | | | | y | | | y | y | | | | | y | y | | y |
| COA | | | y | | | | y | | | y | | | | y | | | y | | | y | y |
| ELE | | | | | | | y | | | y | | | | | | | | | | y | |
| GAF | | | | | | | | | | | | | | | | | | | | | |
| GAS | | | y | | | | y | | | y | | | | y | | | y | | | y | y |
| GRD | | | | | | | | | | | | | | | | | | | | y | |
| H2_ | | | | | | | y | | | y | | | | | | | | | | y | |
| HEA | | | | | | | | | | | | | | | | | | | | y | |
| HYD | | | y | | | | | | | | | | | y | | | y | | | y | y |
| IND | | | | | | | | y | | | | | | | | | | | | y | |
| LIF | | | | | | | | | | | | | | | | | | | | y | |
| MET | | | | | | | y | | | y | | | | | | | | | | y | |
| NUC | | | y | | | | | | | | | | | y | | | y | | | | y |
| OIL | | | y | | | | y | | | y | | | | y | | | y | | | y | y |
| OTF | | | | | | | | | | | | | | | | | | | | y | y |
| OTH | | | y | | | | | | | | | | | y | | | y | | | y | y |
| OTR | | | | | | | | | | | | | | y | | | y | | | y | y |
| RAC | | | | | | | | | | | | | | | | | | | | y | |
| SER | | | | | | | | y | | | | | | | | | | | | | |
| SO2 | | | | | y | | | | | | y | | | | | | | | | | |
| SOF | | | | | | | | | | | | | | | | | | | | y | |
| SOL | | | | | | | | | | | | | | y | | | y | | | | y |
| SOT | | | | | | | | | | | | | | y | | | y | | | y | y |
| TOT | y | | y | | | y | y | y | | y | | | | y | y | y | y | | | y | y |
| TRP | | | | | | | | | | | | | | | | | | | | y | |
| | BIO | CUM | DPE | ELE | EMI | EPE | EXP | GDP | GRF | IMP | LPS | NEN | NET | NTR | POP | PPP | PRO | RSC | SFT | TFC | TPE |

# Description:

## Products

| | |
|---|---|
| AGR | Agriculture |
| BCO | Biomass commercial |
| BNC | Biomass noncommercial |
| CH4 | CH4 emissions |
| CO2 | CO2 emissions (as C) |
| COA | Coal |
| ELE | Electricity |
| GAF | Gaseous fuels |
| GAS | Natural gas |
| GRD | Grids |
| H2_ | Hydrogen |
| HEA | District heat |
| HYD | Hydropower |
| IND | Industry |
| LIF | Liquid fuels |
| MET | Methanol |
| NUC | Nuclear energy |
| OIL | Oil products |
| OTF | Other final consumption (all TFC excluding noncommercial, coal, liquids, gas, and electricity) |
| OTH | Other |
| | in TPE: renewables and trade of synfuels |
| | in PRO: renewables |
| | in ELE: all excluding coal, oil, gas, nuclear, hydro |
| | in TFC: all TFC excluding solids, liquids, electricity |
| OTR | Other renewables (wind, geothermal, wastes) |
| RAC | Residential and commercial |
| SER | Services |
| SO2 | SO2 emissions (as S) |
| SOF | Solid fuels |
| SOL | Solar energy |
| SOT | Solar thermal energy |
| TOT | Total |
| TRP | Transport |

## Flows

| | | |
|---|---|---|
| BIO | MtC | CO2 in biomass used for energy purposes |
| CUM | Mtoe | Cumulative extraction |
| DPE | Mtoe | Total primary energy consumption (direct equivalence method, see *Box 5.3*) |
| ELE | Mtoe | Electricity generation |
| EMI | Mt | Emissions |
| EPE | Mtoe | Total primary energy consumption (engineering method, see *Box 5.3*) |
| EXP | Mtoe | Exports |
| GDP | bill | US$ GDP at market exchange rates |
| GRF | MtC | Gross fossil emissions: CO2 in all fossil primary energy consumption without clearing synfuels trade, i.e., exported synfuels have emissions in the exporting region, and without considering CO2 sequestration (tertiary oil recovery, feedstock uses) |
| IMP | Mtoe | Imports |
| LPS | Mt | SO2 from "large point sources" |
| NEN | MtC | CO2 in fuels used for non-energy uses |
| NET | MtC | CO2 net emissions: GRF minus reinjection/scrubbing minus non-energy uses; in NET = EMI |
| NTR | Mtoe | Net trade (TPE - PRO) |
| POP | mln | Population |
| PPP | bill | US$ GDP at purchasing power parities |
| PRO | Mtoe | Domestic production |
| RSC | MtC | CO2 reduced by reinjection and scrubbing |
| SFT | MtC | CO2 in trade of synthetic fuels |
| TFC | Mtoe | Total final energy consumption |
| TPE | Mtoe | Total primary energy consumption (substituion equivalence method, see *Box 5.3*) |

# References

Adelman, M.A., 1992, The International Energy Outlook, paper presented at the International Energy Workshop, 22–24 June 1992, Boston, MA, USA.

Aghion, Ph., and Howitt, P., 1998, *Endogenous Growth Theory,* The MIT Press, Cambridge, MA, USA.

Alcamo, J., Shaw, R., and Hordijk, L., eds., 1990, *The RAINS Model of Acidification, Science and Strategies in Europe,* Kluwer Academic Publishers, Dordrecht, Netherlands.

Alcamo, J., Bouwman, A., Edmonds, J., Grübler, A., Morita, T., and Sugandhy, A., 1995, An evaluation of the IPCC IS92 emission scenarios, in *Climate Change 1994,* Intergovernmental Panel on Climate Change Special Report, Cambridge University Press, Cambridge, UK, pp. 247–304.

Alexandratos, N., 1995, *World Agriculture towards 2010, an FAO Study,* John Wiley & Sons Ltd., Chichester, UK.

Amann, M., Cofala, J., Dörfner, P., Gyarfas, F., and Schöpp, W., 1995, *Impacts of Energy Scenarios on Regional Acidification,* report to the World Energy Council Project 4 on Environment, International Institute for Applied Systems Analysis, Laxenburg, Austria.

Argote, L., and Epple, D., 1990, Learning curves in manufacturing, *Science,* **247**(23 February):920–924.

Arthur, W.B., 1983, On Competing Technologies and Historical Small Events: The Dynamics of Choice under Increasing Returns, WP-83-90, International Institute for Systems Analysis, Laxenburg, Austria.

Arthur, W.B., 1989, Competing technologies, increasing returns, and lock-in by historical events, *The Economic Journal,* **99**:116–131.

Aslanyan, G.S., and Volfberg, D.B., 1996, The State and Outlooks for the Development of Power Generation in the Countries within the Commonwealth of Independent States, EU/CIS – Energy policy for the XXI century, 13–14 July, Helsinki, Finland.

Ayres, R.U., 1989, *Energy Efficiency in the US Economy: A New Case for Conservation,* RR-89-12, International Institute for Systems Analysis, Laxenburg, Austria.

Barro, R.J., 1997, *Determinants of Economic Growth,* The MIT Press, Cambridge, MA, USA.

Bashmakov, I., 1990, *Energy and Europe: The Global Dimension,* USSR Academy of Sciences, Moscow, Russia.

Beck, P., 1994, *Prospects and Strategies for Nuclear Power, Global Boon or Dangerous Diversion?* Royal Institute of International Affairs and Earthscan, London, UK.

Beck, R.J., 1995, *Oil Industry Outlook*, PennWell Publishing, Tulsa, OK, USA.

Berry, B.J.L., 1990, Urbanization, in B. Turner, II, W.C. Clark, R.W. Kates, J.F. Richards, J.T. Mathews, and W.B. Meyers, eds., *The Earth as Transformed by Human Action: Global and Regional Changes in the Biosphere over the Past 300 Years,* Cambridge University Press, Cambridge, UK, pp. 103–119.

BGR (Bundesanstalt für Geowissenschaften und Rohstoffe), 1989, *Reserven, Ressourcen und Verfügbarkeit von Energierohstoffen*, BGR, Hannover, Germany.

Bhattacharya, K., 1997, Analysis of Energy Supply Scenarios for South Asia, Indira Gandhi Institute of Development Research, Mumbai, India (mimeo).

Bolin, B., 1998, The Kyoto negotiations on climate change: A science perspective, *Science*, Vol. 279, 16 January, pp. 330–331.

Bongaards, J., 1996, Global trends in AIDS mortality, in W. Lutz, ed., *The Future Population of the World: What Can We Assume Today?* Revised and updated edition, Earthscan, London, UK, pp. 170–195.

Bos, E., and Vu, M.T., 1994, *World Population Projections: Estimates and Projections with Related Demographic Statistics 1994–1995 Edition*, The World Bank, Washington, DC, USA.

Bos, E., Vu, M.T., Leven, A., and Bulatao, R.A., 1992, *World Population Projections 1992–1993*, Johns Hopkins University Press, Baltimore, MD, USA.

Bowman, M., 1995, New York State Energy Office, personal communication.

BP (British Petroleum), 1995 and earlier volumes, *BP Statistical Review of World Energy*, BP, London, UK.

BP (British Petroleum), 1997, *BP Statistical Review of World Energy*, BP, London, UK.

China's Agenda 21, 1994, White Paper on China's Population, Environment and Development in the 21st Century, adopted at the 16th Executive Meeting of the State Council of the People's Republic of China, Beijing, China.

Christiansson, L., 1995, Diffusion and Learning Curves of Renewable Energy Technologies, WP-95-126, International Institute for Applied Systems Analysis, Laxenburg, Austria.

Cleland, C., 1996, A regional review of fertility trends in developing countries: 1960 to 1995, in W. Lutz, ed., *The Future Population of the World: What Can We Assume Today?* Revised and updated edition, Earthscan, London, UK, pp. 47–72.

Commission for the Energy Strategy of Russia, 1995, Russian Energy Strategy: Main Statement, INEN Institute, Russian Academy of Sciences, Moscow, Russia (in Russian).

Committee on Science, Engineering, and Public Policy (COSEPUP), 1991, *Policy Implications of Greenhouse Warming*, National Academy Press, Washington, DC, USA.

Criqui, P., 1998, Institut d'Economie et de Politique de l'Energie (IEPE), Grenoble, France, personal communication based on input data of the POLES model.

Delahaye, C., and Grenon, M., 1983, *Conventional and Unconventional World Natural Gas Resources*, CP-83-S4, International Institute for Applied Systems Analysis, Laxenburg, Austria.

Desai, V.V., 1996, Mechanisms of financing Asian development, in *Europe-Asia Science and Technology for their Future*, Forum Engelberg VdF Verlag, Zürich, Switzerland.

Dienes, L., Dobozi, I., and Radetzki, M., 1994, *Energy and Economic Reform in the Former Soviet Union: Implications for Production, Consumption and Exports, and for the International Energy Markets*, Macmillan Press, London, UK.

Dyakov A.V., 1996, United Power System of Russia and its Importance for Energy Security of the CIS, in Energy Security of the Commonwealth of Independent States, Papers of International Consultative Meeting "Energy Security of the Commonwealth of Independent States," Moscow, Russia, 13 May, pp. 125–154.

EC (European Commission), 1996, Directorate General for Energy (DG XVII), European energy to 2020: A scenario approach, *Energy in Europe*, Special Issue, Spring.

EC (European Commission), 1997, *Energy in Europe: 1997 – Annual Energy Review*, Special Issue, European Commission, DG XVII, September.

EIA (Energy Information Administration), 1994, *Annual Energy Outlook 1994 with Projections to 2010*, DOE/EIA-0383(94) US Department of Energy, Washington, DC, USA.

EIA (Energy Information Administration), 1995, Japan: Energy and the Environment, found on www.iea.doe.gov/emeu/env/japan.html

EIA (Energy Information Administration), 1997, International Energy Database, found on www.iea.doe.gov/pub/international/tablee1.xls

EIA (Energy Information Administration), 1998a, *International Energy Annual*, EIA, Washington, DC, USA.

EIA (Energy Information Administration), 1998b, *International Energy Outlook 1998*, EIA, DOE/EIA-048(98), Washington, DC, USA.

EPA (Environmental Protection Agency), 1996, *National Air Pollutant Emission Trends 1900–1995*, EPA-454/R-96-007, EPA, Washington, DC, USA.

ESCIS (Energy Security of the Commonwealth of Independent States), 1996, Papers of International Consultative Meeting "Energy Security of the Commonwealth of Independent States," Moscow, Russia, 13 May.

Fengqi, Z., 1998, Energy Research Institute, State Department Planning Commission P.R. China, Beijing, China, personal communication.

Fettweis, G.B., 1979, *World Coal Resources: Methods of Assessment and Results,* Elsevier, Amsterdam, Netherlands.

Fischer, G., and Rosenzweig, C., 1996, The Impacts of Climate Change, $CO_2$, and $SO_2$ on Agricultural Supply and Trade: An Integrated Assessment, WP-96-05, International Institute for Applied Systems Analysis, Laxenburg, Austria.

Fischer G., Frohberg, K., Keyzer, M.A., and Parikh, K.S., 1988, *Linked National Models: A Tool for International Policy Analysis*, Kluwer Academic Publishers, Dordrecht, Netherlands.

Fischer, G., Frohberg, K., Parry, M.L., and Rosenzweig, C., 1994, Climate change and world food supply, demand and trade: Who benefits, who loses? *Global Environmental Change*, **4/1**:7–23.

Fisher, J.C., 1974, *Energy Crisis In Perspective,* John Wiley & Sons, New York, NY, USA.

Foell, W., Amann, M., Carmichael, G., Chadwick, M., Hettelingh, J.-P., Hordijk, L., Dianwo, Z., eds., 1995, *RAINS Asia: An Assessment Model for Air Pollution in Asia*, Report on the World Bank sponsored project "Acid Rain and Emission Reductions in Asia," World Bank, Washington, DC, USA.

Freeman, C., ed., 1983, *Long Waves in the World Economy*, Buttersworth, London, UK.

Gaffin, S.R., 1998, *Population Growth and Greenhouse Gas Emissions Scenarios: Background Paper for the IPCC Special Report on Emissions Scenarios*, IPCC Special Report: Submitted Manuscript, January 15, Environmental Defense Fund, New York, NY, USA.

Gan, L., 1998, *Implementing China's Agenda 21: From National Strategy to Local Actions*, Working Paper 1998:4, CICERO, Oslo, Norway.

Garenne, M., 1996, Mortality in sub-Saharan Africa: Trends and prospects, in W. Lutz, ed., *The Future Population of the World: What Can We Assume Today?* Revised and updated edition, Earthscan, London, UK, pp. 149–169.

General Agreement on Tariffs and Trade (GATT), 1994, The Final Act of the Uruguay Round, Press Summary, GATT Secretariat, Geneva, Switzerland.

Gilli, P.V., Nakićenović, N., Grübler, A., Bodda, F.L., 1990, *Technologischer Fortschritt, Strukturwandel und Effizienz der Energieanwendung: Trends weltweit und in Österreich*, Band 6, Schriftenreihe der Forschungsinitiative des Verbundkonzerns, Vienna, Austria.

Goldemberg, J., 1991. "Leap-frogging": A new energy policy for developing countries, *WEC Journal*, December:27–30.

Goldemberg, J., 1994, Universidade de São Paulo, Instituto de Electrotécnica Energia, personal communication.

Goldemberg, J., 1996, The evolution of ethanol costs in Brazil, *Energy Policy*, **24**(12):1127–1128.

Goldemberg, J., Monaco, L.C., and Macedo, C., 1993, The Brazilian fuel-alcohol program, in T.B. Johansson, H. Kelly, A.K.N. Reddy, and R.H. Williams, eds., *Renewable Energy: Sources for Fuels and Electricity*, Island Press, Washington, DC, USA, pp. 841–863.

Goldemberg, J., Johansson, T.B., Reddy, A.K., and Williams, R.H., 1988, *Energy for a Sustainable World*, Wiley Eastern, New Dehli, India.

Government of the Russian Federation, 1995, *Russian Economic Trends 1995*, Vol. 4, Nos. 1 and 2, Whurr Publishers, London, UK.

Green, C., Legler, J., Ashok, S., and Foell, W.K., 1995, Energy module, in W. Foell, M. Amann, G. Carmichael, M. Chadwick, J.-P. Hettelingh, L. Hordijk, Z. Dianwo, eds., *RAINS Asia: An Assessment Model for Air Pollution in Asia*, Report on the World Bank sponsored project "Acid Rain and Emission Reductions in Asia," World Bank, Washington, DC, USA.

Grossman, G., and Helpman, E., 1991, *Innovation and Growth in the Global Economy*, The MIT Press, Cambridge, MA, USA.

Grubb, M., Koch, M., Thomson, K., Munson, A., and Sullivan, F., 1993, *The Earth Summit Agreements: A Guide and Assessment*, The Royal Institute of International Affairs and Earthscan, London, UK.

Grübler, A., 1990, Technology diffusion in a long-wave context: The case of the steel and coal industries, in T. Vasko, R.U. Ayres, and L. Fontvieille, eds., *Life Cycles and Long Waves*, Springer-Verlag, Berlin, Germany, pp. 117–146.

Grübler, A., 1991, Energy in the 21st century: From resources to environmental and lifestyle constraints, *Entropie*, **164/165**:29–34.

Grübler, A., 1996, Time for a change: On the patterns of diffusion and innovation, *Dædalus*, **125**(3):19–42.

Grübler, A., 1998, *Technology and Global Change*, Cambridge University Press, Cambridge, UK.

Grübler, A., and Messner, S., 1996, Technological uncertainty, in N. Nakićenović, W.D. Nordhaus, R. Richels, and F.L. Toth, eds., *Climate Change: Integrating Science, Economics, and Policy*, CP-96-1, International Institute for Applied Systems Analysis, Laxenburg, Austria, pp. 295–314.

Grübler, A., and Nakićenović, N., 1991, Long waves, technology diffusion, and substitution, *Review XIV*, (2):313 342.

Grübler, A., and Nakićenović, N., 1994, *International Burden Sharing in Greenhouse Gas Reduction*, RR-94-9, International Institute for Applied Systems Analysis, Laxenburg, Austria.

Grübler, A., and Nakićenović, N., 1996, Decarbonizing the global energy system, *Technological Forecasting and Social Change*, **53**(1):97–110.

Gürer, N., and Ban, J., 1997, Factors affecting energy-related $CO_2$ emissions: Past levels and present trends, *OPEC Review*, December:309–350.

Häfele, W., program leader, 1981, *Energy in a Finite World: A Global Systems Analysis*, Ballinger, Cambridge, MA, USA.

Häfele, W., 1990, Energy from nuclear power, *Scientific American*, Special Issue, September, **263**(3):137–144.

Hall, D.O., and Rosillo-Calle, F., 1991, Why biomass matters: Energy and the environment, *Biomass Users Network News*, **5**(4):4–14.

Hanke, T., 1995, Die Märkte spielen verrückt, *Die Zeit*, Nr. 18, p. 33.

Hettelingh, J.-P., Chadwick, M., Sverdrup, H., and Zhao, D., 1995, Assessment of environmental effects of acidic deposition, in W. Foell, M. Amann, G. Carmichael, M. Chadwick, J.-P. Hettelingh, L. Hordijk, Z. Dianwo, eds., *RAINS Asia: An Assessment Model for Air Pollution in Asia*, Report on the World Bank sponsored project "Acid Rain and Emission Reductions in Asia," World Bank, Washington, DC, USA.

Hyman, L.S., 1994, Financing electricity expansion, *World Energy Council Journal*, July:15–20.

IAEA (International Atomic Energy Agency), 1998, Reference Data Series No. 2, April.

IEA (International Energy Agency), 1993, *Energy Statistics and Balances of OECD and Non-OECD Countries 1971–1991*, OECD, Paris, France.

IEA (International Energy Agency), 1994, *Energy in Developing Countries: A Sectorial Analysis*, OECD, Paris, France.

IEA (International Energy Agency), 1996, *World Energy Outlook*, 1996 edition, IEA/OECD, Paris, France.

IEA (International Energy Agency), 1997a, *Australia 1997 Review*.

IEA (International Energy Agency), 1997b, *Indicators of Energy Use and Efficiency: Understanding the Link between Energy and Human Activity.* IEA/ OECD, Paris, France.

IEA (International Energy Agency) 1998, Energy Investment, Joint paper by the Energy Charter Secretariat and the International Energy Agency presented to the G8 Energy Ministerial in Moscow, Russia, April 1.

IGBP TCWG (International Geosphere Biosphere Program Terrestrial Carbon Working Group), 1998, The terrestrial carbon cycle: Implications for the Kyoto Protocol, *Science*, **280**(29 May):1393–1394.

IIASA–WEC (International Institute for Applied Systems Analysis–World Energy Council) 1995, *Global Energy Perspectives to 2050 and Beyond*, WEC, London, UK.

IMF (International Monetary Fund), 1995, *International Financial Statistics,* IMF, Washington, DC, USA.

IMF (International Monetary Fund), 1998, *World Economic Outlook*, IMF, Washington, DC, USA, May.

INFRAS, 1996, *Structural Transformation: Process Towards Sustainable Development in India and Switzerland*, INFRAS, Zurich, Switzerland.

IPCC (Intergovernmental Panel on Climate Change), Working Group I, 1995, Radiative forcing of climate change, in *Climate Change 1994*, Intergovernmental Panel on Climate Change Special Report, Cambridge University Press, Cambridge, UK, pp. 1–247.

IPCC (Intergovernmental Panel on Climate Change), 1996a, *Climate Change 1995. The Science of Climate Change,* Contribution of Working Group I to the Second Assessment Report of the Intergovernmental Panel on Climate Change, J.T. Houghton, L.G. Meira Filho, B.A. Callander, N. Harris, A. Kattenberg, and K. Maskell, eds., Cambridge University Press, Cambridge, UK.

IPCC (Intergovernmental Panel on Climate Change), 1996b, *Climate Change 1995. Impacts, Adaptations and Mitigation of Climate Change: Scientific-Technical Analyses*, Contribution of Working Group II to the Second Assessment Report of the Intergovernmental Panel on Climate Change, R.T. Watson, M.C. Zinyowera, and R.H. Moss, eds., Cambridge University Press, Cambridge, UK.

IPCC (Intergovernmental Panel on Climate Change), 1996c, *Climate Change 1995. Economic and Social Dimensions of Climate Change*, Contribution of Working Group III to the Second Assessment Report of the Intergovernmental Panel on Climate Change, J.P. Bruce, H. Lee, and E.F. Haites, eds., Cambridge University Press, Cambridge, UK.

IPCC (Intergovernmental Panel on Climate Change), 1996d, Second Assessment Synthesis of Scientific-Technical Information Relevant to Interpreting Article 2 of the UN Framework Convention on Climate Change, IPCC Second Assessment, Climate Change 1995, IPCC, Geneva, Switzerland.

Ito, T., 1997, Japan's Economy Needs Structural Change, Finance&Development, IMF, Washington, DC, USA, June.

Kvenvolden, K.A., 1988, Methane hydrate: A major reservoir of carbon in the shallow geosphere? *Chemical Geology*, **71**:41–51.

Kvenvolden, K.A., 1993, A primer on gas hydrates, in D.G. Howell, ed., *The Future of Energy Gases*, US Geological Survey (USGS) Professional Paper 1570, USGS, Denver, USA, pp. 661–675.

Leggett, J., Pepper, W.J., Swart, R.J., Edmonds, J., Meira Filho, L.G., Mintzer, I., Wang, M.X., and Watson, J., 1992, Emissions Scenarios for the IPCC: An Update, in *Climate Change 1992: The Supplementary Report to the IPCC Scientific Assessment*, Cambridge University Press, Cambridge, UK.

Lucas, R., 1988, On the mechanics of economic development, *Journal of Monetary Economics*, 22(1):3–42.

Lutz, W., ed., 1994, *The Future Population of the World: What Can We Assume Today?* Earthscan, London, UK.

Lutz, W., ed., 1996, *The Future Population of the World: What Can We Assume Today?* Revised and updated edition, Earthscan, London, UK.

Lutz, W., Sanderson, W., and Scherbov, S., 1997, Doubling of world population unlikely, *Nature*, **387**(6635):803–805.

MacDonald, G.J., 1990, The future of methane as an energy resource, *Annual Review of Energy*, **15**:53–83.

MacGregor, P.R., Maslak, C.E., and Stoll, H.G., 1991, *The Market Outlook for Integrated Gasification Combined Cycle Techology*, General Electric Company, New York, NY, USA.

MacKellar, L., Goujon, A., Lutz, W., and Prinz, C., 1995, The "Population versus Consumption" Debate in Household Terms, International Institute for Applied Systems Analysis, Laxenburg, Austria (internal paper).

MacKellar, L., Lutz, W., Munn, R.E., Wexler, L., O'Neill, B.C., McMichael, A., and Suhrke, A., 1998, *Population and Global Warming*, Book Manuscript, International Institute for Applied Systems Analysis, Laxenburg, Austria.

Maddison, A., 1989, *The World Economy in the 20th Century*, Development Centre Studies, Organisation for Economic Co-Operation and Development, OECD, Paris, France.

Maddison, A., 1995, *Explaining the Economic Performance of Nations,* Edward Elgar, Aldershot, UK.

Makarov, A., 1996, Long-Term Aspects of Assuring Energy Security of the CIS Countries, in Energy Security of the Commonwealth of Independent States, Papers of International Consultative Meeting "Energy Security of the Commonwealth of Independent States," Moscow, Russia, 13 May, pp. 113–124.

Makarov, A., Mastepanov, A., and Volfberg, D., 1995, *Energy Strategy of Russia,* paper prepared for the 16th WEC Congress, Institute of Energy Research, Moscow, Russia.

Malhotra, A.K., 1997, Private Participation in Infrastructure: Lessons from Asia's Power Sector. Finance & Development, The World Bank, Washington, DC, USA, December (http://www.worldbank.org/html/extpb/annrep97).

Manne, A., and Richels, R., 1992, *Buying Greenhouse Insurance: The Economic Costs of $CO_2$ Emission Limits*, The MIT Press, Cambridge, MA, USA.

Manne, A.S., and Schrattenholzer, L., 1997, International Energy Workshop: Summary of Poll Responses, International Institute for Applied Systems Analysis, Laxenburg, Austria and Stanford University, Stanford, CA, USA.

Marchetti, C., 1978, *Energy Systems: The Broader Context,* RM-78-18, International Institute for Applied Systems Analysis, Laxenburg, Austria.

Marchetti, C., and Nakićenović, N., 1979, *The Dynamics of Energy Systems and the Logistic Substitution Model*, RR-79-13, International Institute for Applied Systems Analysis, Laxenburg, Austria.

Martin, J-M., 1988, L'intensité énergétique de l'activité economique dans les pays industrialisés: Les evolutions de très longue periode liverent-elles des enseignements utiles? *Economies et Societés,* **4**:9–27.

Martin, J-M., 1994, Politiques énergétiques et développement durable, *Futuribles,* **189**:29–46.

Masters, C.D., Root, D.H., and Attanasi, E.D., 1991, World resources of crude oil and natural gas, *Preprints of the 13th World Petroleum Congress*, **25**:1–14, John Wiley & Sons Ltd., Chichester, UK.

Masters, C.D., Root, D.H., and Attanasi, E.D., 1994, World petroleum assessment and analysis, *Proceedings of the 14th World Petroleum Congress*, John Wiley & Sons Ltd., Chichester, UK.

McDevitt, T.M., 1996, World Population Profile: 1996, US Census Bureau, Washington, DC, USA (http://www.census.gov/ipc/www/wp96.html)

McKelvey, V.E., 1972, Mineral resource estimates and public policy, *American Scientist*, **60**:32–40.

Messner, S., 1997, Endogenized technological learning in an energy systems model, *Journal of Evolutionary Economics*, **7**:291–313.

Messner, S., and Nakićenović, N., 1992, A Comparative Assessment of Different Options to Reduce $CO_2$ Emissions, WP-92-27, International Institute for Applied Systems Analysis, Laxenburg, Austria.

Messner, S., and Strubegger, M., 1991, User's Guide to CO2DB: The IIASA $CO_2$ Technology Database Version 1.0, WP-91-31a, International Institute for Applied Systems Analysis, Laxenburg, Austria.

Messner, S., and Strubegger, M., 1994, The energy model MESSAGE III, in J-F. Hake *et al.* eds., *Advances in Systems Analysis: Modelling Energy-Related Emissions on a National and Global Level*, Forschungszentrum Jülich Gmbh, Jülich, Germany, pp. 29–47.

Messner, S., and Strubegger, M., 1995, User's Guide for MESSAGE III, WP-95-69, International Institute for Applied Systems Analysis, Laxenburg, Austria.

Messner, S., Golodnikov, A., and Gritsevskyi, A., 1996, A stochastic version of the dynamic linear programming model MESSAGE III, *Energy*, **21**(9):775–785.

Moreira, J.R., and Poole, A.D., 1993, Hydropower and its constraints, in Johansson, T.B., H. Kelly, A.K.N. Reddy, and R.H. Williams, eds., *Renewable Energy Sources for Fuels and Electricity*, Island Press, Washington, DC, USA, pp. 73–120.

Nakićenović, N., 1984, Growth to Limits: Long Waves and the Dynamics of Technology, International Institute for Applied Systems Analysis, Laxenburg, Austria.

Nakićenović, N., 1987, Technological Substitution and Long Waves in the USA, in T. Vasko, ed., *The Long-Wave Debate*, Springer Verlag, Berlin, Germany, pp. 76–104.

Nakićenović, N., 1996, Freeing energy from carbon, *Dædalus*, **125**(3):95–112.

Nakićenović, N., Amann, M., and Fischer, G., 1998, Global energy supply and demand and their environmental effects, mimeo, International Institute for Applied Systems Analysis, Laxenburg, Austria.

Nakićenović, N., and Grübler, A., eds., 1991, *Diffusion of Technologies and Social Behaviors*, Springer-Verlag, Berlin, Germany.

Nakićenović, N., A. Grübler, A. Inaba, S. Messner, S. Nilsson, Y. Nishimura, H.-H. Rogner, A. Schäfer, L. Schrattenholzer, M. Strubegger, J. Swisher, D. Victor, and D. Wilson, 1993. Long-term strategies for mitigating global warming, *Energy*, **18**(5):401–609.

Nakićenović, N., Grübler, A., Ishitani, H., Johansson, T., Marland, G., Moreira, J.R., and Rogner, H.-H., 1996, Energy primer, in *Climate Change 1995: Impacts, Adaptations and Mitigation of Climate Change: Scientific-Technical Analysis*, Contribution of Working Group II to the Second Assessment Report of the IPCC. Cambridge University Press, Cambridge, UK, pp. 75–92.

NRC (National Research Council), 1986, *Population Growth and Economic Development: Policy Questions,* National Academy Press, Washington, DC, USA.

OECD (Organisation for Economic Co-operation and Development), 1991, The resistance to agricultural reform, *OECD Observer*, **171**(August/September): 4–8.

OECD (Organisation for Economic Co-operation and Development), 1996, *Energy Policies of OECD Countries: 1996 Review,* OECD, Paris, France.

OECD (Organisation for Economic Co-operation and Development), 1997, *Short-Term Economic Indicators Transition Economies*, 4/1997, Organization for economic co-operation and development, The center for co-operation with the economies in transition, Paris, France.

OECD (Organisation for Economic Co-operation and Development), 1998a, *National Accounts 1960–1996,* Vol. I, Main Aggregates, Statistics Directorate, OECD, Paris, France.

OECD (Organization for Economic Cooperation and Development), 1998b, IEA Energy Prices and Taxes, Fourth Quarter 1997, OECD, Paris, France.

OECD/NEA and IAEA (Organisation for Economic Cooperation and Development/ Nuclear Energy Agency and International Atomic Energy Agency), 1995 and earlier volumes, *Uranium Resources, Production and Demand*, OECD, Paris, France.

OLADE-SIEE, 1997, *Energy Economic Information System*, Latin American Energy Organization, Quito, Ecuador, October–December.

Pachauri, R.K., 1997, Notes on South Asia Component Report, TERI, New Dehli, India (mimeo).

Parikh, Y., 1998, The emperor needs new clothes: Long-range energy use scenarios by IIASA–WEC and IPCC, *Energy*, **23**(1):69–70.

PCAST (President's Committee of Advisors on Science and Technology), Panel on Federal Energy R&D, 1997, "Federal Energy Research and Development for the Challenges of the Twenty-First Century," Executive Office of the President of the United States, Washington, DC, USA. September.

Pepper, W., Leggett, J., Swart, R., Wasson, J., Edmonds, J., and Mintzer, I., 1992, Emission Scenarios for the IPCC. An Update: Assumptions, Methodology, and Results, paper prepared for IPCC Working Group I, Geneva, Switzerland.

Progress on China's Agenda 21, 1996, The Administrative Centre for China's Agenda 21, Beijing, China.

Reddy, A.K.N., and Goldemberg, J., 1990, Energy for the developing world, *Scientific American*, Special Issue, September, **263**(3):110–118.

Richards, J.F., 1990, Land transformations, in B.L. Turner II, *et al.*, eds., *The Earth As Transformed by Human Action*, Cambridge University Press, Cambridge, UK, pp. 163–178.

Rogner, H.-H., 1990, *Analyse der Förderpotentiale und langfristige Verfügbarkeit von Kohle, Erdgas und Erdöl*, Studie A.3.1., Enquête-Kommission "Vorsorge zum Schutz der Erdatmosphäre", Deutscher Bundestag, Economica-Verlag, Bonn, Germany.

Rogner, H.-H., 1997, An assessment of world hydrocarbon resources, *Annual Review of Energy and Environment*, **22**:217–262.

Romer, P.M., 1986, Increasing returns and long-run growth, *Journal of Political Economy*, **94**(5):1002–1037.

Rosa, L.P., and La Rovere, E.L., 1997, Overview of Latin American technology development for avoiding greenhouse gases emissions and for mitigating climate change, paper presented at the Latin American Regional Workshop on Technological Choices and New Opportunities for Sustainable Development within the Implementation of the United Nations Framework on Climate Change, 3–4 October 1997, Rio de Janeiro, Brazil.

Rostow, W.W., 1980, *Why the Poor Get Richer and the Rich Slow Down*, The MacMillan Press Ltd, London, UK.

Russian Academy of Sciences, 1995, Energeticheskaya Strategia Rossii (Osnovnye Polozhenia), Moscow, Russia.

Russian Energy Pictures, 1997, Statistical Bulletin, Center of Energy Efficiency, Moscow, Russia.

Safranek, Yu.K., 1996, The energy security of Russia and the CIS, in Energy Security of the Commonwealth of Independent States, Papers of International Consultative Meeting "Energy Security of the Commonwealth of Independent States," Moscow, Russia, 13 May, pp. 27–42.

Sathaye, J., and Meyers, S., 1990, Urban Energy Use in Developing Countries, in A.V. Desai and M.A. Kettani, eds., *Patterns of Energy Use in Developing Countries*, International Development Research Centre UN University, Wiley Eastern Limited, New Delhi, India, pp. 177–204.

Sathaye, J., Ketoff, A., Schipper, L., and Lele, S., 1989, *An End-use Approach to Development of Long-term Energy Demand Scenarios for Developing Countries*. LBNL-25611, Lawrence Berkeley National Laboratory, Berkeley, CA, USA.

Schäfer, A., 1995, Trends in Global Motorized Mobility: The Past 30 Years and Implications for the Next Century, WP-95-49, International Institute for Applied Systems Analysis, Laxenburg, Austria.

Schäfer, A., and Victor, D., 1997, The past and future of global mobility, *Scientific American*, (October):36–39.

Schipper, L., and Cooper, C., 1991, Energy use and conservation in the USSR: Patterns, Prospects, and Problems, Lawrence Berkeley Laboratory, Berkeley, CA, USA.

Schipper, L., and Meyers, S., 1992, *Energy Efficiency and Human Activity: Past Trends, Future Prospects*, Cambridge University Press, Cambridge, UK.

Schrattenholzer, L., and Marchant, K., 1996, The 1995 International Energy Workshop: The poll results and a review of papers, *OPEC Review*, Vol. XX, No. 1, 25–45.

Schrattenholzer, L., and Schäfer, A., 1996, World Regional Scenarios Described With the 11R Model of Energy-Economy-Environment Interactions, WP-96-108, International Institute for Applied Systems Analysis, Laxenburg, Austria.

Schumpeter, J.A., 1939, *Business Cycles: A Theoretical, Historical and Statistical Analysis of the Capitalist Process*, 2 volumes, McGraw Hill, New York, NY, USA.

Siddiqi, T., ed., 1994, *National Response Strategy for Global Climate Change: People's Republic of China*, Asian Development Bank and State Science and Technology Commission of China, Bejing, China.

Sinton, J.E., Fridley, D.G., Levine, M.D., Yang, F., Zhenping, J., Xing, Z., Kejun, J., and Xiaofeng, L., 1992, China Energy Databook, LBL-32822 Ref. 3, UC-900, Ernest Orlando Lawrence Berkeley National Laboratory, University of California, Berkeley, CA, USA.

Sinyak, Y., and Yamaji, K., eds., 1990, *Energy Efficiency and Prospects for the USSR and Eastern Europe*, Central Research Institute of Electric Power Industry (CRIEPI), Tokyo, Japan.

Smil, V., 1994, *Energy in World History*, Westview Press, Boulder, CO, USA.

Smith, K.R., 1993, Fuel combustion, air pollution exposure, and health: The situation in developing countries, *Annual Review of Energy and the Environment*, **18**:529–66.

Sørensen, B., 1979, *Renewable Energy,* Academic Press, London, UK.

Strubegger, M., and Reitgruber, I., 1995, Statistical Analysis of Investment Costs for Power Generating Technologies, WP-95-109, International Institute for Applied Systems Analysis, Laxenburg, Austria.

TERI (Tata Energy Research Institute), 1994, *TERI Energy Data Directory Yearbook*, Pauls Press, New Delhi, India.

UN (United Nations), 1992, *Long-Range World Population Projections: Two Centuries of Population Growth 1950–2150,* Department of International Economic and Social Affairs, UN, New York, NY, USA.

UN (United Nations), 1993a, *Human Development Report 1993*, UN, New York, NY, USA.

UN (United Nations), 1993b, *Macroeconomic Data Systems, MSPA Data Bank of World Development Statistics, MSPA Handbook of World Development Statistics*, MEDS/DTA/1 and 2 June, UN, New York, NY, USA.

UN (United Nations), 1993c, *UN MEDS Macroeconomic Data System, MSPA Data Bank of World Development Statistics,* MEDS/DTA/1 MSPA-BK.93, Long-Term Socio-Economic Perspectives Branch, Department of Economic and Social Information & Policy Analysis, UN, New York, NY, USA.

UN (United Nations), 1994, *World Urbanization Prospects: The 1992 Revision*, Population Division, Division of Economic Development, UN, New York, NY, USA.

UN (United Nations), 1996, Department for Economic and Social Information and Policy Analysis, Population Division, World Population Prospects: The 1996 Revision – Annex I: Demographic Indicators, UN, New York, NY, USA.

UN (United Nations), 1997, *World Economic and Social Survey 1997: Trends and Policies in the World Economy*, UN, New York, NY, USA.

UN (United Nations), 1998, *World Population Projections to 2150*, UN Department of Economic and Social Affairs, Population Division, UN, New York, NY, USA.

UNDP (United Nations Development Programme), 1993. *Human Development Report 1993*, Oxford University Press, Oxford, UK.

UN/ECE (United Nations Economic Commission for Europe), 1994, *Protocol to the 1979 Convention on Long-range Transboundary Air Pollution on Future Reduction of Sulfur Emissions*, ECE/EB.AIR/40, Geneva, Switzerland.

UN/ECE (United Nations Economic Commission for Europe), 1996, Economic and Social Council, Economic Commission for Europe, Committee on Energy, Sixth session, 6–8 November 1996, Present Situation and Prospects for the fuel and Energy Complexes in the Countries of the Commonwealth of independent States, Submitted by an ad hoc group of experts from CIS countries on energy policy and strategy, ENERGY/R.131/Add.1, 26 September 1996, UNDP, New York, NY, USA.

UNEP (United Nations Environmental Programme), 1997, *Global Environment Outlook*, Oxford University Press, New York, NY, USA.

UN/FCCC (United Nations Framework Convention on Climate Change), 1992, Convention Text, IUCC, Geneva, Switzerland.

UN/FCCC (United Nations Framework Convention on Climate Change), 1995, Conclusion of Outstanding Issues and Adoption of Decisions, Appendix 1, First session 28 March–7th April, Berlin, Germany.

UN/FCCC (United Nations Framework Convention on Climate Change), 1997, Kyoto Protocol to the United Nations Framework Convention on Climate Change, FCCC/CP/L7/Add.1, 10 December 1997, UN, New York, NY, USA.

USDOC (US Department of Commerce), 1975 and consecutive volumes, *Historical Statistics of the United States, Colonial Times to 1970, Bicentennial Edition, Parts 1 and 2*, US Government Printing Office, Washington, DC, USA.

USDOE (US Department of Energy), 1998, *Annual Energy Outlook*, DOE, Washington, DC, USA.

Vasko, T., Ayres, R., and Fontvielle, L., eds., 1990, *Life Cycles and Long Waves*, Springer-Verlag, Berlin, Germany.

Vyakhirev, R.I., 1996, Problems of reliable operation of interconnected gas supply system of Russia, in Energy Security of the Commonwealth of Independent States, Papers of International Consultative Meeting "Energy Security of the Commonwealth of Independent States," Moscow, Russia, 13 May, pp. 155-171.

WEC (World Energy Council), 1992, *Survey of Energy Resources 1992*, WEC, London, UK.

WEC (World Energy Council), 1993, *Energy for Tomorrow's World: The Realities, the Real Options and the Agenda for Achievements*, Kogan Page, London, UK.

WEC (World Energy Council), 1994, *New Renewable Energy Resources: A Guide to the Future*, Kogan Page, London, UK.

WEC (World Energy Council), 1995a, Post-Rio '92: Developments Relating to Climate Change, Report on the Proceedings of the First Conference of the Parties to the UN Framework Convention on Climate Change, 28 March–7 April, Berlin, Germany, WEC, London, UK.

WEC (World Energy Council), 1995b, Climate Change: Scientific, Technical, and Institutional Developments Since 1992, WEC, London, UK.

WEC (World Energy Council), 1995c, Local and Regional Energy-Related Environmental Issues, WEC, London, UK.

WEC (World Energy Council), 1995d, The Energy Economy in Central & Eastern Europe in Transition, WEC, London, UK.

WEC (World Energy Council), 1995e, Energy Transition in Central and Eastern Europe: Investment Needs & Financing Possibilities, WEC, London, UK.

WEC (World Energy Council), 1998, Survey of Energy Resources 1998, WEC, London, UK.

Wene, C.-O., 1995, Energy–Economy Analysis: Linking the Macroeconomic and Systems-Engineering Approaches, WP-92-42, International Institute for Applied Systems Analysis, Laxenburg, Austria.

Wigley, T.M.L., and Raper, S.C.B., 1993, Future changes in global mean temperature and sea level, in R.A. Warwick, E.M. Barrow, and T.M.L. Wigley, eds., *Climate and Sea Level Change: Observations, Projections and Implementation*, Cambridge University Press, Cambridge, UK, pp. 111–133.

Wigley, T.M.L., Salmon, M., and Raper, S.C.B., 1994, *Model for the Assessment of Greenhouse-gas Induced Climate Change*, Version 1.2, Climate Research Unit, University of East Anglia, UK.

Williams, R., 1995, A Low Emissions Energy Supply System (LESS) for the World (Low-Nuclear Variant), review draft, prepared for the IPCC Working Group IIa, Energy Supply Mitigation Options, April, Princeton University, Princeton, NJ, USA.

World Bank, 1992, *World Development Report 1992: Development and the Environment*. Oxford University Press, London, UK.

World Bank, 1993, *World Development Report 1993*, Oxford University Press, New York, NY, USA.

World Bank, 1994, *World Development Report 1994*, Johns Hopkins University Press, Baltimore, MD, USA.

World Bank, 1995, *World Tables 1995*, Johns Hopkins University Press, Baltimore, MD, USA.

World Bank, 1996, *World Development Report 1996: From Plan to Market*, Oxford University Press, New York, NY, USA.

World Bank, 1997a, *Global Economic Prospects and the Developing Countries,* World Bank, Washington, DC, USA.

World Bank, 1997b, World Development Report 1997, Oxford University Press, New York, NY, USA.

World Bank, 1998a, Regional Brief: Latin America and the Caribbean, http://www.worldbank.org/html/extdr/offrep/lac/lacrb.htm

World Bank, 1998b, Regional Brief: South Asia, http://www.worldbank.org /html/extdr/offrep/sas/southas.htm.

World Health Organization (WHO), 1992, *Urban Air Pollution in Megacities of the World*, reprint 1993, Blackwell, Oxford, UK.

Yeager, K., 1998, EPRI and the Success of Collaborative R&D, Testimony to the Subcommittee on Energy and the Environment of the Committee on Science of the US House of Representatives, March 31.

## Further Reading

Arrow, K., Bolin, B., Constanza, R., Dasgupta, P., Folke, C., Holling, C.S., Jansson, B.O., Levin, S., Maler, K.G., Perrings, C., and Pimentel, D., 1995, Economic growth, carrying capacity, and the environment, *Science*, **268**(April 28):520–521.

Ausubel, J.H., and Sladovich, H.E., eds., 1989, *Technology and Environment*, National Academy Press, Washington, DC, USA.

Bolin, B., 1995, Politics of climate change, *Nature*, **374**(March 16):208.

Clark, W.C., 1990, Energy and environment: Strategic perspectives on policy design, in J.W. Tester *et al.* eds., *Energy and the Environment in the Twentyfirst Century*, The MIT Press, Cambridge, MA, USA, pp. 63–78.

Cline, W.R., 1992, *The Economics of Global Warming*, Institute for International Economics, Washington, DC, USA.

Colombo, U., 1992, Sustainable energy development, in *Science and Sustainability*, International Institute for Applied Systems Analysis, Laxenburg, Austria, pp. 95–119.

Davis, G.R., 1990, Energy for planet earth, *Scientific American*, Special Issue, September, **263**(3):55–62.

Desai, A.V., and Kettani, M.A., eds., 1990, *Patterns of Energy Use in Developing Countries*, International Development Research Centre UN University, Wiley Eastern Limited, New Delhi, India.

Dosi, G., Freeman, C., Nelson, R., Silverberg, G., and Soete, L., eds., 1988, *Technical Change and Economic Theory*, Pinter Publishers, London, UK.

Dupont-Roc, G., 1995, A new century plenty of (alternative) energy, *Shell Venster*, March/April:3–7 (in Dutch).

Edmonds, J.E., Wise, M.A., and MacCracken, C.N., 1994, Advanced Energy Technologies and Climate Change: An Analysis Using the Global Change Assessment Model (GCAM), WP-62, Fondazione ENI Enrico Matttei, Milano, Italy.

Flavin, C., and Lenssen, N., 1994, *Power Surge: Guide to the Coming Energy Revolution*, W.W. Norton & Co., New York, NY, USA.

Frisch, J.R., Bredow, K., and Saunders, R., 1989, *World Energy Horizons 2000–2020*, WEC, London, Editions Technip, Paris, France.

Grubb, M., Walker, J., Herring, H., and Rouse, K., 1992, *Emerging Energy Technologies: Impacts and Policy Implications*, The Royal Institute of International Affairs, London, UK.

Heal, G., and Chichilnisky, G., 1991, *Oil and the International Economy*, Clarendon Press, Oxford, UK.

Helm, J.L., ed., 1990, *Energy Production, Consumption and Consequences*, National Academy Press, Washington, DC, USA.

Holdren, J.P., and Pachauri, R.K., 1992, Energy, in *An Agenda of Science for Environment and Development into the 21st Century*, International Council of Scientific Unions, Cambridge University Press, Cambridge, UK, pp. 103–118.

Hordijk, L., 1991, Persistent pollutants: Integrated environmental economic systems, in J.B. Opschoor and D.W. Pearce, eds., *Persistent Pollutants: Economics and Policy*, Kluwer Academic Publishers, Dordrecht, Netherlands.

EA (International Energy Agency), 1991, *Energy Efficiency and the Environment*, IEA/OECD, Paris, France.

Johansson, T.B., Kelly, H., Reddy, A.K.N., and Williams, R.H., eds., 1993, *Renewable Energy Sources for Fuels and Electricity*, Island Press, Washington, DC, USA.

Kaya, Y., Nakićenović, N., Nordhaus, W.D., and Toth, F.L., eds., 1993, *Costs, Impacts, and Benefits of $CO_2$ Mitigation*, CP-93-2, International Institute for Applied Systems Analysis, Laxenburg, Austria.

Lovins, A.B., 1991, Energy, people and industrialization, in P.E. Trudeau (ed), *Energy for a Habitable World: A Call for Action*, Crane Russak & Co., New York, NY, USA, pp. 117–150.

Lazarus, M.L., Greber, L., Hall, J., Bartels, C., Bernow, S., Hansen, E., Raskin, P., and von Hippel, D., 1993, *Towards a Fossil Free Energy Future: The Next Energy Transition, A Technical Analysis for Greenpeace International*, Stockholm Environmnetal Institute Boston Center, Boston, USA.

Lee, T.H., Ball, B.C., and Tabors, R.D., 1990, *Energy Aftermath: How We Can Learn from the Blunders of the Past to Create a Hopeful Energy Future*, Harvard Business School Press, Boston, MA, USA.

Munasinghe, M., and Meier, P., 1993, *Energy Policy Analysis and Modeling*, Cambridge University Press, Cambridge, UK.

Nordhaus, W.D., 1994, *Managing the Global Commons: The Economics of Climate Change*, The MIT Press, Cambridge, MA, USA.

Office for Technology Assessment US Congress (OTA), 1992, *Fueling Development: Energy Technologies for Developing Countries*, OTA, Washington, DC, USA.

Pachauri, R.K., and Srivaslana, L., eds., 1990, *Energy, Environment, Development,* Vikas Publishers, New Delhi, India.

Skea, J., 1995, Energy, in M. Parry and R. Duncan, eds., *Economic Implications of Climate Change in Britain*, Earthscan, London, UK.

Starr, C., 1993, Global energy and electricity futures, *Energy*, **18**(3):225–237.

Tyler, S., Sathaye, J., and Goldman, N., eds., 1994, Energy use in Asian cities, *Energy*, Special Issue, **19**(5):503–600.